Advances in Biosensor Technology for Mycobacterium Tuberculosis Detection

Advances in Biosensor Technology for Mycobacterium Tuberculosis Detection

Edited by

MANSI CHATURVEDI
Academic of Scientific and Innovative Research (AcSIR),
CSIR-Advanced Materials and Processes Research Institute,
Bhopal, Madhya Pradesh, India

CHETNA DHAND
Academic of Scientific and Innovative Research (AcSIR),
CSIR-Advanced Materials and Processes Research Institute,
Bhopal, Madhya Pradesh, India

NEERAJ DWIVEDI
Academic of Scientific and Innovative Research (AcSIR),
CSIR-Advanced Materials and Processes Research Institute,
Bhopal, Madhya Pradesh, India

JAMANA P. CHAURASIA
Academic of Scientific and Innovative Research (AcSIR),
CSIR-Advanced Materials and Processes Research Institute,
Bhopal, Madhya Pradesh, India

ELSEVIER

ACADEMIC PRESS
An imprint of Elsevier

Academic Press is an imprint of Elsevier
125 London Wall, London EC2Y 5AS, United Kingdom
50 Hampshire Street, 5th Floor, Cambridge, MA 02139, United States

ISBN: 978-0-443-44589-7

For Information on all Academic Press publications
visit our website at https://www.elsevier.com/books-and-journals

Publisher: Andre Wolff
Acquisitions Editor: Kattie Washington
Editorial Project Manager: Shreya Jain
Production Project Manager: Rajeswari Rajkumar
Cover Designer: Arumugam Kothandan

Typeset by MPS Limited, Chennai, India

Contents

List of contributors

Archana
Nanobio Laboratory, Special Centre for Nanoscience, Jawaharlal Nehru University, New Delhi, Delhi, India

Mohd. Arshan
Centre for Scientific Research and Development, People's University, Bhopal, Madhya Pradesh, India

Kanishka Bhatia
Amity Institute of Applied Sciences, Amity University, Uttar Pradesh, India

Neha Bisht
Academic of Scientific and Innovative Research (AcSIR), CSIR-Advanced Materials and Processes Research Institute, Bhopal, Madhya Pradesh, India

Mansi Chaturvedi
Academic of Scientific and Innovative Research (AcSIR), CSIR-Advanced Materials and Processes Research Institute, Bhopal, Madhya Pradesh, India

Chetna Dhand
Academic of Scientific and Innovative Research (AcSIR), CSIR-Advanced Materials and Processes Research Institute, Bhopal, Madhya Pradesh, India

Neeraj Dwivedi
Academic of Scientific and Innovative Research (AcSIR), CSIR-Advanced Materials and Processes Research Institute, Bhopal, Madhya Pradesh, India

Rajni Gautam
Physics Department, School of Basic and Applied Sciences, K. R. Mangalam University, Gurugram, Haryana, India

Priyanshi Gupta
Amity Institute of Applied Sciences, Amity University, Uttar Pradesh, India

Richa Jain
Centre for Scientific Research and Development, People's University, Bhopal, Madhya Pradesh, India

Sapna Kaithwas
Innovative Materials and Processes Division, CSIR-Advanced Materials and Processes Research Institute (AMPRI), Bhopal, Madhya Pradesh, India

Sahil Kohli
Research and Development Cell, Department of Chemistry, Manav Rachna International Institute of Research and Studies, Faridabad, Haryana, India

Neeraj Kumari
Chemistry Department, School of Basic and Applied Sciences, K. R. Mangalam University, Gurugram, Haryana, India

Vanshika Kumari
Amity Institute of Applied Sciences, Amity University, Uttar Pradesh, India

Priti Kushwaha
Centre for Scientific Research and Development, People's University, Bhopal, Madhya Pradesh, India

Akansha Lodhi
Centre for Scientific Research and Development, People's University, Bhopal, Madhya Pradesh, India

Maumita Das Mukherjee
Amity Institute of Applied Sciences, Amity University, Uttar Pradesh, India

Sneh Nema
Academic of Scientific and Innovative Research (AcSIR), CSIR-Advanced Materials and Processes Research Institute, Bhopal, Madhya Pradesh, India

Anju Nigam
Department of Physics, Govt PG College, Kunidhar, Manila, Almora, Uttarakhand, India

Antonio Puertas-Segura
Department of Chemical Engineering, Universitat Politècnica de Catalunya, Barcelona, Spain

Garima Rathee
Department of Chemical Engineering, Universitat Politècnica de Catalunya, Barcelona, Spain

Jyotsna Rathee
Computer Science & Engineering Department, Deenbandhu Chhotu Ram University of Science and Technology, Sonipat, Haryana, India

Himani Shukla
Chemistry Department, School of Basic and Applied Sciences, K. R. Mangalam University, Gurugram, Haryana, India

Pratima R. Solanki
Nanobio Laboratory, Special Centre for Nanoscience, Jawaharlal Nehru University, New Delhi, Delhi, India

Shikha Teotia
Amity Institute of Applied Sciences, Amity University, Uttar Pradesh, India

Kajal Tiwari
CSIR-AMPRI, Bhopal, Madhya Pradesh, India

About the authors

Ms. Mansi Chaturvedi is a research scholar at the CSIR–AMPRI, Bhopal, India. Her research mainly focused on the development of biosensing nanomaterials (2D materials such as graphene and MXenes) for the detection of various infectious diseases like tuberculosis, dengue, COVID-19, and Parkinson's disease.

Dr. Chetna Dhand is a Scientist-E at the CSIR–Advanced Materials and Processes Research Institute (CSIR–AMPRI), Bhopal, India, and associate professor at the Academy of Scientific and Innovative Research (AcSIR), Ghaziabad, India. She obtained her PhD in chemistry from the University of Delhi and CSIR–National Physical Laboratory, New Delhi, India. She has previously worked as a postdoctoral research fellow at the Singapore Eye Research Institute (SERI) and as an adjunct assistant professor at Duke–NUS Medical School, Singapore. Her research interests include biomaterials, nanobiomaterials, bio-inspired materials, 2D layered nanomaterials, antimicrobial coatings, electrochemical sensors and biosensors, 3D bioprinting, drug delivery, and tissue engineering.

Dr. Neeraj Dwivedi is presently a Scientist F and associate professor at the CSIR–AMPRI, Bhopal, India, and AcSIR, Ghaziabad, India, respectively. His research interests include carbon materials, interface engineering, carbon nanocoatings, and 2D materials such as graphene-based materials, MXenes, metal oxides, and nitrides for various applications, including energy, sensors, tribology, corrosion, and antimicrobial.

Dr. Jamana P. Chaurasia is presently working as a Chief scientist and Head PPD at CSIR-AMPRI, Bhopal, India. His research interests include management of science and technology: R&D support activities.

Acknowledgment

We would like to express our sincere gratitude to all those who contributed, directly or indirectly, to the completion of this book. This work would not have been possible without the encouragement and support of numerous individuals and institutions.

First and foremost, we extend our heartfelt appreciation to the Director of CSIR–Advanced Materials and Processes Research Institute, Dr. Thallada Bhaskar, for his visionary leadership and continued encouragement. We are equally grateful to the heads of departments, Dr. Asokan Papu, Dr. N. Sathish, and Dr. Hemant Kumar Shukla, for their unwavering support, valuable guidance, and institutional facilitation, all of which were instrumental in bringing this work to fruition.

We sincerely thank our colleagues and collaborators for their insightful suggestions, stimulating discussions, and generous sharing of knowledge. Their intellectual contributions and professional camaraderie have significantly enriched the quality and depth of this manuscript.

Our appreciation also extends to the publishers, editors, and reviewers for their professionalism, patience, and constructive feedback, which greatly strengthened and refined this work.

I, Mansi Chaturvedi, would like to express my profound gratitude to my mentors and teachers for their invaluable guidance, thoughtful discussions, and constant encouragement throughout this journey. Their academic insight, constructive criticism, and sustained support have substantially shaped the direction and quality of this book.

Finally, and most importantly, I express my deepest gratitude to my parents, Mr. Mankar and Mrs. Sangeeta Chaturvedi, whose unconditional love, patience, and unwavering support have been my greatest source of strength and inspiration. I feel truly blessed to have the unwavering support and encouragement of my siblings, Rudraksha and Chetna Chaturvedi. Their faith in me and constant motivation have been instrumental in the successful completion of this work.

CHAPTER 1

Origin, spread, pathogenesis and demography of *Mycobacterium tuberculosis*: a most dangerous pathogenic bacterium

Richa Jain
Centre for Scientific Research and Development, People's University, Bhopal, Madhya Pradesh, India

1.1 *Mycobacterium tuberculosis*

Mycobacterium tuberculosis (MTB), the etiological agent of tuberculosis (TB), is an obligately aerobic, acid-fast bacillus belonging to the family Mycobacteriaceae. The organism exhibits a distinctive cell structure (Fig. 1.1) and is characterized by rod-shaped, catalase-positive cells that often form straight or slightly branched chains. MTB is a slow-growing pathogen, requiring approximately 2 to 40 days to produce visible colonies on enriched solid media. A defining feature of MTB is its lipid-rich cell wall, which contains long-chain fatty acids known as mycolic acids (60 to 90 carbon atoms). These complex fatty acids possess an α-carbon linked to an aliphatic chain and a β-carbon bearing a hydroxyl group. The abundance of mycolic acids and associated lipids external to the peptidoglycan layer confers remarkable resistance to host immune defenses, desiccation, and many antimicrobial agents, thereby contributing to the pathogen's persistence and pathogenicity.

TB is a complex airborne infectious disease caused by members of the *Mycobacterium tuberculosis* complex (MTBC). While MTB predominantly infects the lungs, leading to pulmonary TB, it can disseminate through lymphatic and hematogenous routes to extrapulmonary sites, resulting in diverse clinical manifestations (Pai et al., 2016). TB pathogenesis exists along a dynamic continuum, ranging from latent infection where bacilli persist within granulomatous lesions under immune control, to active disease, characterized by uncontrolled bacterial replication, tissue damage, and transmissibility. This spectrum reflects the heterogeneity of TB outcomes and presents substantial challenges for diagnosis, treatment, and disease control (Pai et al., 2016). Despite the availability of preventive and therapeutic strategies, TB remains a major global public health concern. The World Health Organization (WHO) estimates that approximately one-quarter of the global population nearly two billion individuals harbors latent TB infection, of whom 5%–10% will progress to active disease during their lifetime .

Advances in Biosensor Technology for Mycobacterium Tuberculosis Detection
DOI: https://doi.org/10.1016/B978-0-443-44589-7.00010-8

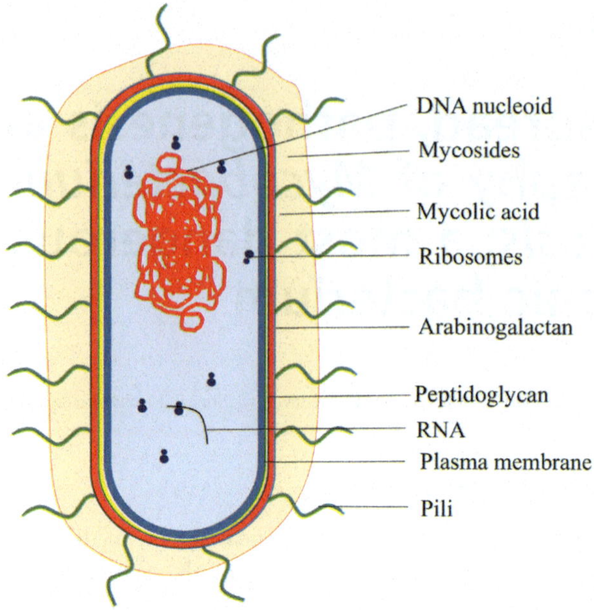

Figure 1.1 *Mycobacterium tuberculosis* cell structure. A diagram of Mycobacterium tuberculosis cell structure with labeled components.

Recent epidemiological data indicate a resurgence in TB burden, with an estimated 10.8 million new cases reported globally in 2023 and an incidence rate of 134 cases per 100,000 population, underscoring the ongoing gap in achieving End TB targets (World Health Organization, 2024a, 2024b). TB continues to cause approximately 1.0–1.1 million deaths annually, disproportionately affecting low- and middle-income countries, where access to timely diagnosis and effective treatment remains limited (Pai et al., 2016).

At the immunological level, recent advances have expanded the understanding of host–pathogen interactions in TB. Innate immune mechanisms—including tumor necrosis factor-α (TNF-α) signaling, neutrophil-mediated inflammation, reactive oxygen and nitrogen species, autophagy, inflammasome activation, and antimicrobial peptide regulation play pivotal roles in determining host resistance or susceptibility to MTB infection (Andrade et al., 2024). Additionally, emerging evidence suggests that, alongside $CD4^+$ T cells, TH17 cells contribute significantly to protective immunity against MTB, offering promising directions for next-generation TB vaccine development (Diedrich et al., 2024).

Concurrently, translational research efforts are driving innovation in TB diagnostics. Artificial intelligence (AI)-based frameworks that integrate chest radiography, cough sound analysis, and multimodal vision–language models have demonstrated high diagnostic accuracy in distinguishing active TB from other respiratory diseases. These approaches hold particular promise for scalable, cost-effective deployment in resource-limited settings (Batra et al., 2025). Collectively, these advances build upon the foundational insights outlined by Pai et al. (2016), emphasizing the necessity of a multilayered strategy that integrates epidemiological surveillance, immunological discovery, and technological innovation to effectively address the global TB burden.

1.2 Origin of *Mycobacterium tuberculosis*

Genomic and phylogeographic evidence suggests that MTB evolved from an ancestral pool of Mycobacterium canettii–like organisms in East Africa and subsequently coevolved with human populations during their migration out of Africa approximately 70,000 years ago (Comas et al., 2013). This longstanding association between MTB and humans is further supported by ancient DNA recovered from Egyptian mummies, indicating the pathogen's presence in early human civilizations. A comprehensive review by Brites and Gagneux (2020a) highlights the evolutionary history and global diversity of the MTBC, emphasizing its East African origin, structured lineage diversification, and adaptive evolution in parallel with human dispersal. In 1882, Robert Koch identified MTB as the causative agent ofTB (Kaufmann and Schaible, 2005). Since then, TB remains a major global health problem, continuing to spread in diverse populations.

Today, seven major human-adapted lineages and several animal-adapted strains are recognized, each showing distinct geographic distributions, virulence traits, and transmission patterns. Unlike many bacteria, MTBC evolves clonally without horizontal gene transfer, relying instead on genome decay, regulatory adaptations, and antigenic variation to persist and spread. Certain lineages, such as the Beijing family (Lineage 2), are strongly associated with drug resistance and epidemic expansion, whereas others remain geographically restricted and are less transmissible. The paper also highlights the role of host–pathogen coevolution, noting that MTB lineages show associations with specific human populations, reflecting deep evolutionary interactions. Importantly, the authors stress that lineage diversity has significant implications for public health, because it influences disease severity, immune evasion, diagnostic accuracy, drug response, and vaccine effectiveness. They conclude that integrating evolutionary biology, genomics, and epidemiology is crucial for refining global TB control strategies and tailoring interventions to the lineage-specific dynamics of this pathogen.

1.3 Pathogenesis of *Mycobacterium tuberculosis*

Transmission occurs via the inhalation of aerosolized droplets. Once inside the lungs, MTB is phagocytosed by alveolar macrophages, where it survives intracellularly by inhibiting phagosome–lysosome fusion and resisting reactive oxygen/nitrogen intermediates (Cambier et al., 2014). Although the lungs are the primary site of infection in pulmonary TB, the pathogen can disseminate to other organs, leading to extrapulmonary disease (WHO, 2022). Host immune responses wall off bacilli in granulomas, leading to latent TB infection (LTBI) (Fig. 1.2). In this condition, bacilli remain viable in host tissues without producing symptoms, making individuals silent carriers of latent TB infection (Horsburgh et al., 2011). Reactivation occurs under immunosuppression, progressing to active disease characterized by chronic cough, fever, and tissue necrosis (Flynn and Chan, 2001).

The immunopathogenesis of (MTB) represents a dynamic interplay in which the pathogen subverts host immunity to ensure persistence. Following inhalation, bacilli are engulfed by alveolar macrophages but evade destruction by inhibiting phagosome–-lysosome fusion, neutralizing reactive intermediates, and altering host signaling (Flynn and Chan, 2001). These adaptations promote granuloma formation, a hallmark of TB pathology, which simultaneously restricts bacterial spread and provides a niche for long-term latency (Cambier et al., 2014; Ernst, 2020). MTB also manipulates host cell death, favoring necrosis over apoptosis, thereby enhancing tissue destruction and facilitating transmission (Ramakrishnan, 2020). Table 1.1 summarizes molecular aspects of MTB pathogenesis. Importantly, strain-specific variation in immune modulation underscores the evolutionary flexibility of the pathogen (Coscolla and Gagneux, 2014; Brites and Gagneux, 2020b). Current therapy relies on prolonged multidrug regimens, usually

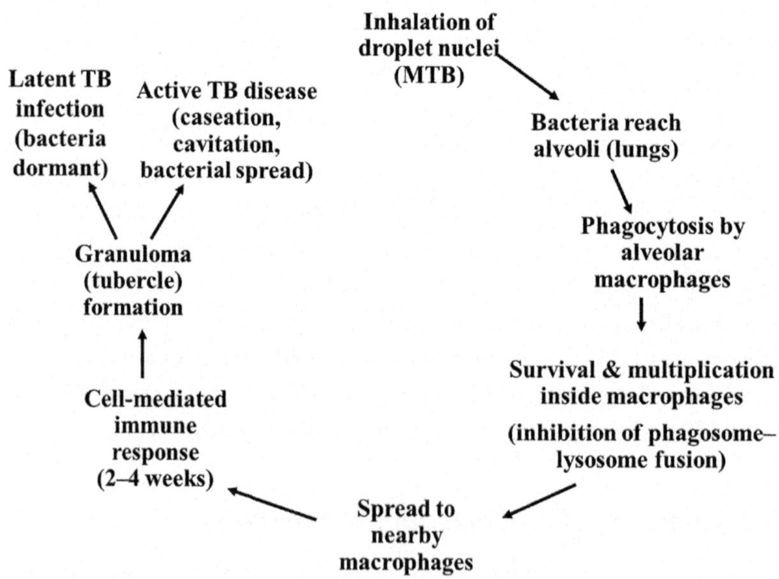

Figure 1.2 *Mycobacterium tuberculosis* (MTB) infection cycle in human.

Table 1.1 Molecular Pathogenesis of ***Mycobacterium tuberculosis*** (MTB).

Stage	Cellular & Molecular Events	Host Immunological Response	MTB Evasion & Persistence Mechanisms	Resultant Pathophysiology
1. Aerosol Deposition and Initial Phagocytosis	Inhalation of aerosolized droplet nuclei (1–5 μm) containing bacilli, which reach the alveolar spaces. Deposition on alveolar epithelial surfaces.	Resident alveolar macrophages (AMs) perform surveillance, recognizing pathogen-associated molecular patterns (PAMPs) via Toll-like receptors (TLRs) and C-type lectin receptors. Complement opsonization facilitates phagocytosis via CR3 andCR4.	The highly lipidic, mycolic acid-rich cell envelope provides intrinsic resistance to desiccation, osmotic lysis, and many disinfectants.	Establishment of a primary infection focus within the pulmonary parenchyma.
2. Intracellular Survival and Replication	MTB is internalized into a phagosome. The bacilli actively arrest phagosomal maturation, preventing its progression to a phagolysosome.	Macrophage activation attempts to generate reactive oxygen intermediates (ROIs) and reactive nitrogen intermediates (RNIs) to kill the pathogen. Proinflammatory cytokines (TNF-α, IL-1, IL-6) are secreted.	• Phagosome Maturation Arrest: Secretion of effectors (e.g., via the ESX-1 secretion system) that manipulate host Rab GTPases and phosphatidylinositol 3-phosphate (PI3P) signaling. • Neutralization of ROIs/RNIs: Expression of bacterial enzymes such as KatG and AhpC.	Successful intracellular replication, transforming the macrophage into a vehicle for dissemination. Formation of an early, unstructured lesion.

(*Continued*)

Table 1.1 (Continued)

Stage	Cellular & Molecular Events	Host Immunological Response	MTB Evasion & Persistence Mechanisms	Resultant Pathophysiology
			• Inhibition of Apoptosis: Expression of the nuoG gene to suppress host cell apoptosis, promoting a necrotic death that facilitates dissemination.	
3. Early Dissemination and Adaptive Immune Priming	Infected macrophages and dendritic cells (DCs) migrate via lymphatics to regional (hilar and mediastinal) lymph nodes. DCs process MTB antigens and present them.	Antigen Presentation: DCs present MTB peptide antigens on major histocompatibility complex (MHC) class II molecules to naïve $CD4^+$ T cells in lymph nodes. Th1 Polarization: IL-12 production by DCs and macrophages drives the differentiation of naïve $CD4^+$ T cells into Th1 effector cells.	The bacilli continue to replicate within the newly recruited monocytes and macrophages that are arriving at the initial infection site.	Formation of the Ghon focus (pulmonary lesion) and the Ghon complex (Ghon focus plus involved lymph nodes). This marks the incubation period before adaptive immunity is fully engaged.
4. Granuloma Formation and Latency	A structured, organized immune cell aggregate (granuloma) forms, isolating the bacilli. The core consists of infected macrophages, surrounded	• Cell-Mediated Immunity (CMI): Th1 cells release key cytokines: • Interferon-gamma (IFN-γ): Potently	Dormancy program: In response to the hypoxic, acidic, and nutrient-deprived granulomatous core, MTB induces the DosR regulon, shifting	Latent tuberculosis Infection (LTBI): A dynamic equilibrium is established. The host contains the

	by epithelioid histiocytes and multinucleated giant (Langhans) cells, all encircled by a lymphocyte cuff.	activates macrophages to enhance their mycobactericidal activity. • Tumor Necrosis Factor-alpha (TNF-α): Essential for recruiting immune cells and maintaining the structural integrity of the granuloma.	metabolism into a nonreplicating persistent (NRP) state to ensure long-term survival.	infection but does not eradicate it. The individual is asymptomatic and noninfectious. The granuloma may eventually calcify.
5. Reactivation and Transmission (Active TB)	In approximately 5%–10% of LTBI cases, immunological control fails (e.g., due to immunosuppression, aging, TNF-α blockade). The granuloma's caseous center undergoes liquefaction.	Loss of granuloma integrity. Host-derived matrix metalloproteinases (MMPs) degrade the extracellular matrix, contributing to tissue destruction and liquefaction.	Extracellular replication: The liquefied caseum provides a nutrient-rich, aerobic environment that allows for rapid extracellular bacillary proliferation.	Active pulmonary TB: Clinical symptoms manifest (e.g., cough, fever, and weight loss). The liquefied necrotic material erodes into a bronchus, forming a cavity. This allows massive numbers of bacilli to be expelled in aerosol droplets, completing the transmission cycle.

combining isoniazid, rifampin, pyrazinamide, and ethambutol for six months or longer. However, the emergence of MDR and extensively drug-resistant (XDR) strains poses significant challenges (WHO, 2023). Novel approaches, including host-directed therapies, next-generation antimicrobials, and advanced vaccine candidates, are under investigation to improve compliance, shorten treatment, and overcome resistance (Zumla et al., 2023; Ghosh et al., 2024).

1.4 Demographic load and prevalence of *Mycobacterium tuberculosis*

In world highest burden of MTB lies in Southeast Asia (45%), Africa (24%), and the Western Pacific (17%), with India contributing approximately 26% of global cases (World Health Organization, 2024a). Adults account for approximately 88% of cases, men bear a higher burden than women, and ~12% of cases occur in children and adolescents (World Health Organization, 2024b). Drug-resistant TB (MDR and XDR) remains a major threat, complicating control strategies.

Dheda et al. (2017) provide a detailed synthesis of the epidemiology, pathogenesis, transmission dynamics, and clinical management of multidrug-resistant (MDR), extensively drug-resistant (XDR), and effectively "incurable" tuberculosis (TB). MDR-TB, defined by resistance to at least isoniazid and rifampin, and XDR-TB, additionally resistant to fluoroquinolones and second-line injectable agents, emerge predominantly from inadequate therapeutic regimens, suboptimal adherence, and the ongoing transmission of resistant strains. These forms of TB are particularly concerning because resistant bacilli display minimal fitness costs, enabling efficient person-to-person transmission comparable to drug-susceptible strains. Diagnosis remains a critical barrier because rapid molecular assays such as GeneXpert and line probe assays are not uniformly accessible in high-burden regions, delaying the initiation of effective therapy. Therapeutic options are limited to prolonged, toxic, and costly regimens with variable efficacy, and the emergence of strains resistant to all available drugs highlights the specter of truly untreatable TB. The socioeconomic consequences are profound, disproportionately affecting populations with fragile health systems and limited access to comprehensive care. According to the WHO Global TB Report 2024, approximately 400,000 incident cases of MDR/RR-TB (rifampin-resistant and multidrug-resistant TB) occurred in 2023, resulting in an estimated 150,000 deaths. MDR/RR-TB accounted for 3.2% of new TB cases and 16% of previously treated cases, with India, China, and the Russian Federation contributing the highest burdens. These data underscore the persistence of drug-resistant TB as a major obstacle to global TB elimination targets. Addressing this crisis necessitates a multipronged strategy involving universal access to rapid drug-susceptibility testing, the deployment of shorter all-oral regimens (e.g., BPaLM), strengthened infection control, and sustained investment in novel drugs, host-directed therapies, and vaccines, coupled with socioeconomic interventions to mitigate the structural determinants of transmission and treatment failure.

1.5 Strategies and policies to control *Mycobacterium tuberculosis*

The WHO End TB Strategy aims to reduce TB incidence by 80% and deaths by 90% by 2030. Management of MTB involves strategic planning at the community level, primarily, sensitization of people about the causes, symptoms, diagnosis, prevention, and control of MTB is of prime importance. Nutritional status and immunity directly affects the process of infection. Prevention of transmission along with early diagnosis and proper treatment regimens can prevent MTB infection up to much extent and also help in combating drug resistance. Early diagnosis followed by treatment with standard drug regimens prevents development of drug resistance. Now a days personalized medicine concept has been gained importance as compared to directly observed treatment strategies. Other key measures include rapid diagnostics (GeneXpert MTB/RIF), universal access to rifapentine-based preventive therapy, improved case detection, and vaccine development (World Health Organization, 2024a). Renewed optimism surrounds the goal of tuberculosis (TB) elimination, fueled by significant advances in vaccine candidates, host-directed therapies, and drug development. The success of these innovations depends on robust global partnerships and integrated health strategies to achieve mid-century elimination goals. Strengthening health systems, social protection, and global collaboration are also emphasized (Tiberi et al., 2018). Recognizing that scientific breakthroughs must be met with accessible implementation, the World Health Organization has convened a Global Finance and Access Working Group. This initiative is designed to ensure that forthcoming TB vaccines are distributed equitably and sustainably across high-burden regions, a critical step toward achieving the population-level impact necessary to meet the WHO End TB targets (WHO, 2025). This new era is largely defined by breakthroughs in vaccine research, which are poised to address the limitations of the century-old Bacillus Calmette–Guérin (BCG) vaccine.

1.6 Limitations in controlling *Mycobacterium tuberculosis*

Major challenges include the emergence of MDR and XDR-TB, underdiagnosis in children, stigma, weak health infrastructure in high-burden countries, and limited access to novel drugs (bedaquiline, pretomanid) (Dheda et al., 2017). The COVID-19 pandemic disrupted TB control efforts, leading to a resurgence in mortality (World Health Organization, 2024b).

1.7 Conclusion and future prospects

MTB remains the world's deadliest bacterial pathogen, sustained by its evolutionary success, airborne transmission, and ability to persist in latent form. Despite global efforts, TB continues to impose a heavy health and socioeconomic burden. Effective elimination requires scientific innovation and equitable healthcare access. Mitigation of the global tuberculosis burden is poised for a paradigm shift, moving beyond conventional microbiological methods and protracted chemotherapy toward a framework of precision medicine. This transformation is driven by the convergence of advanced molecular diagnostics, rational drug design, and a deeper understanding of host–pathogen interactions, offering the potential to disrupt transmission cycles and accelerate progress toward eradication milestones. Future diagnostic modalities are transitioning from pathogen-centric, sputum-dependent methods to decentralized, host-centric platforms. The development of nonsputum-based assays, including those detecting host transcriptomic signatures in peripheral blood or pathogen-specific volatile organic compounds (VOCs) in breath, promises to enhance case detection in pediatric and extrapulmonary TB. Complementing this, the application of next-generation sequencing (NGS) is revolutionizing the field; metagenomic NGS (mNGS) enables culture-independent identification of MTB complex directly from clinical specimens, while whole-genome sequencing (WGS) provides high-resolution prediction of antimicrobial resistance (AMR) by identifying mutations across the entire mycobacterial genome. These genomic data are being integrated with artificial intelligence (AI) and deep learning-algorithms, which can perform sophisticated analyses of digital radiographs to augment radiological screening and predictive modeling. In parallel, therapeutic strategies are undergoing a radical evolution aimed at regimen shortening, safety, and efficacy against drug-resistant strains. The objective is to supplant the standard six-month regimen with ultra-short courses, potentially as brief as 1 to 2 months, through the optimization of novel drug combinations. This is facilitated by the discovery of new chemical entities targeting unique bacterial pathways, such as the inhibition of ATP synthase (bedaquiline), the decaprenylphosphoryl-β-D-ribofuranose 2′-epimerase (DprE1) enzyme, and the caseinolytic protease (ClpP1P1) complex. A complementary approach involves the use of host-directed therapies (HDTs), which employ immunomodulatory agents to enhance pathogen containment and mitigate immunopathology, thereby acting as potent adjuncts to antimicrobial chemotherapy. The ultimate convergence of these advances is the advent of personalized treatment, in which where WGS-derived AMR profiles guide the immediate deployment of tailored therapeutic regimens. This precision will be further supported by innovations in drug delivery, such as long-acting injectable formulations and nanoparticle-based carriers designed for targeted pulmonary delivery and optimized pharmacokinetics, thereby ensuring maximal adherence and therapeutic efficacy.

References

Andrade BB, Amaral EP, Mayer-Barber KD, Sher A. Innate immunity in tuberculosis: lessons from animal models and implications for human disease. Frontiers in Immunology 2024;15:1467016. Available from https://doi.org/10.3389/fimmu.2024.1467016.

Batra S, Kumar A, Zhang X, Lee J. Multimodal AI for tuberculosis screening: vision–language and cough-audio models in clinical translation [Manuscript submitted for publication]. arXiv 2025. arXiv https://arxiv.org/abs/2503.14538.

Brites D, Gagneux S. The nature and evolution of genomic diversity in *Mycobacterium tuberculosis*. Cold Spring Harbor Perspectives in Medicine 2020;10(4):a037069. Available from https://doi.org/10.1101/cshperspect.a037069.

Brites D, Gagneux S. The origin and evolution of *Mycobacterium tuberculosis* complex. Pathogens 2020;9(10):1–14. Available from https://doi.org/10.3390/pathogens9100829.

Cambier CJ, Falkow S, Ramakrishnan L. Host evasion and exploitation schemes of *Mycobacterium tuberculosis*. Cell 2014;159(7):1497–509. Available from https://doi.org/10.1016/j.cell.2014.11.024.

Comas I, Coscolla M, Luo T, et al. Out-of-africa migration and neolithic coexpansion of *Mycobacterium tuberculosis* with modern humans. Nature Genetics 2013;45(10):1176–82. Available from https://doi.org/10.1038/ng.2744.

Coscolla M, Gagneux S. Consequences of genomic diversity in *Mycobacterium tuberculosis*. Seminars in immunology 2014;26(6):431–44. Available from https://doi.org/10.1016/j.smim.2014.09.012.

Dheda K, Gumbo T, Maartens G, Dooley KE, McNerney R, Murray M, et al. The epidemiology, pathogenesis, transmission, diagnosis, and management of multidrug-resistant, extensively drug-resistant, and incurable tuberculosis. The Lancet Respiratory Medicine 2017;5(4):291–360. Available from https://doi.org/10.1016/S2213-2600(17)30079-6.

Diedrich CR, O'Garra A, Kaufmann SHE. TH17-like CD4$^+$ T cell programs in tuberculosis resistance. Nature Immunology 2024;25(6):876–89. Available from https://doi.org/10.1038/s41590-024-01897-8.

Ernst JD. Mechanisms of *M. tuberculosis* immune evasion as challenges to TB vaccine design. Cell Host & Microbe 2020;27(1):34–48. Available from https://doi.org/10.1016/j.chom.2019.12.001.

Flynn JL, Chan J. Immunology of tuberculosis. Annual Review of Immunology 2001;19:93–129. Available from https://doi.org/10.1146/annurev.immunol.19.1.93.

Ghosh S, Joshi M, Mitra D. Host-directed therapies for tuberculosis: current status and future directions. Frontiers in Immunology 2024;15:1392487. Available from https://doi.org/10.3389/fimmu.2024.1392487.

Kaufmann SH, Schaible UE. 100th anniversary of Robert Koch's nobel prize for the discovery of the tubercle bacillus. Trends in Microbiology 2005;13:469–75.

Pai M, Behr MA, Dowdy D, Dheda K, Divangahi M, Boehme CC, et al. Tuberculosis. Nature Reviews Disease Primers 2016;2:16076. Available from https://doi.org/10.1038/nrdp.2016.76.

Ramakrishnan L. Revisiting the role of the granuloma in tuberculosis. Nature Reviews. Immunology 2020;20(11):612–22. Available from https://doi.org/10.1038/s41577-020-0282-9.

Tiberi S, du Plessis N, Walzl G, Vjecha MJ, Rao M, Ntoumi F, et al. Tuberculosis: progress and advances in development of new drugs, treatment regimens, and host-directed therapies. The Lancet Infectious Diseases 2018;18(7):e183–98. Available from https://doi.org/10.1016/S1473-3099(18)30110-5.

World Health Organization. (2024a). Global tuberculosis report 2024: TB incidence. https://www.who.int/teams/global-tuberculosis-programme/tb-reports/global-tuberculosis-report-2024/tb-disease-burden/1-1-tb-incidence.

World Health Organization. (2024b). Global tuberculosis report 2024: TB mortality. https://www.who.int/teams/global-tuberculosis-programme/tb-reports/global-tuberculosis-report-2024/tb-disease-burden/1-2-tb-mortality.

Zumla A, Marais BJ, McHugh TD, Maeurer M, Goletti D. Advancing tuberculosis treatment and prevention in the 21st century. Nature Reviews. Microbiology 2023;21(5):307–25. Available from https://doi.org/10.1038/s41579-022-00797-2.

CHAPTER 2

Immune response towards *Mycobacterium tuberculosis* infection

Richa Jain, Akansha Lodhi, Mohd. Arshan, and Priti Kushwaha
Centre for Scientific Research and Development, People's University, Bhopal, Madhya Pradesh, India

2.1 Introduction

In 2022, more than 1.3 million deaths were recorded due to tuberculosis (TB), which remains the biggest global health challenge. Individuals with weakened immunity are vulnerable when exposed to droplets containing *Mycobacterium tuberculosis* (MTB) (1–10 bacilli), which can cause TB infection. Upon the entry of MTB into the nasal cavity of host, the first line of defense (epithelial cells and phagocytes i.e., neutrophils, monocytes, and dendritic cells) becomes active. Usually, first line of defense eliminates the MTB thus preventing infection. The fate of MTB inside the host body is governed by variety of factors like immunity, nutritional status, cellular environment, and so forth. In condition where MTB bypasses the first line of defense and enters phagocyte, it uses host cell machinery for its survival and multiplication. Various factors like host's genetic makeup, immunity, etc. influence the virulence of MTB, and thus clinical outcome of exposure cannot be accurately determined. If MTB remains a silent guest inside host body, it results in a latent TB infection, on the other hand, if MTB is aggressive, it causes development of active TB infection. The duration for which MTB remains in the latent stage depends on the complex relation between bacterial virulence and host factors, such as immunity, nutritional status, and environmental conditions. As per surveys conducted in various countries, MTB infections causes high rates of morbidity and mortality, particularly in low- and middle-income nations (World Health Organization, 2023). The rapid increase in TB cases around the world demands rapid diagnosis of MTB, effective treatment, and control strategies. A critical understanding of these interacting determinants is paramount for developing the next-generation vaccines, diagnostics, and therapeutics to combat this global health issue. This chapter highlights the basic of immune response upon exposure to MTB.

2.2 Host and bacterial determinants in *Mycobacterium tuberculosis*

MTB has evolved multiple strategies to escape the host immune system (Coscolla and Gagneux, 2014; Stanley et al., 2003). Infection with MTB is governed by a dynamic balance between host determinant and bacterial determinant. ESAT-6, CFP-10, and

Advances in Biosensor Technology for Mycobacterium Tuberculosis Detection
DOI: https://doi.org/10.1016/B978-0-443-44589-7.00001-7

cell wall lipids (mycolic acids) of MTB help the bacteria invade the immune system of host and survive inside macrophage. Host determinants, including genetic polymorphisms, age, gender, and comorbidities like diabetes, malnutrition, and anemia and immune competence, play major roles in determining susceptibility or resistance to infection. TNF-α, IL-2, and Th1 response regulate the extent of infection caused by the bacteria (Brennan and Nikaido, 1995). MTB alters host cell signaling (SLAM), thereby making a niche in granulomas, while a weakened immune system facilitates bacteria to grow inside the host. Studies have reported that polymorphisms in chromosomal regions 11p13 and 18q11.2 moderately increase the risk of infection in individuals (Thye et al., 2010). It has also been reported that germline mutations in Mendelian Susceptibility to Mycobacterial Diseases (MSMD) involve a minimum of 20 genes (including *IFNGR1*, *IL12B*, *IL12RB1*, *STAT1*, *ISG15*, and *IRF8*) that impair the IL-12/IFN-γ axis. This pathway is crucial for immunity against weakly virulent MTB (Thye et al., 2010). Mutations in genes like IFNGR1 and IL12B hampers the Th1 response, leading to disseminated MTB infections even after BCG vaccination (Sia and Rengarajan, 2019). Deletion of FadB4, a gene involved in cell envelope biosynthesis, has been observed in MTB strains associated with outbreaks. In macrophage infection models, cells harboring these deletions induce the rapid accumulation of lipid droplets and IL-1β secretion, both hallmarks of the immune response (Farzand et al., 2022). On the other hand, the expression of osteopontin (OPN) protein prevents the dissemination of MTB infection (Nau et al., 2000). SerB2, a protein of MTB, inhibits IL-8 production, while CysA2 increases the expression of the co-stimulatory molecule CD86 on antigen-presenting cells (Yelamanchi and Surolia, 2021). Lipoarabinomannan (LAM) present in cell wall acts like an immunomodulator and interacts with the DC-SIGN receptor on dendritic cells, thereby modulating cytokine production and impairing T-cell priming. Another modulin secreted by MTB is EsxW, which possesses strong immunogenic properties (Abdallah et al., 2021). The ESX-1 system secreted by MTB has an ESAT-6/CFP-10 complex that helps the bacteria to perforate the phagosome membrane, allowing them to escape into the nutrient-rich cytosol (Stanley et al., 2003). While PhoP helps MTB to tolerate extreme conditions (acidic pH, signaling biomolecules) prevailing inside macrophages (Gupta et al., 2006; Rustad et al., 2009; Walters et al., 2006), MTB in the granuloma stage shifts its metabolism to host-derived lipids like cholesterol (Pandey and Sassetti, 2008). In response to this hostile environment, the DosR regulon induces a non-replicating, dormant state that renders the bacteria highly tolerant to many frontline antibiotics (Mokrousov et al., 2022). MTB is not a monolithic pathogen; different phylogenetic lineages exhibit variations in virulence, with the "Beijing" lineage, for example, being associated with disease outbreaks and a potentially dysregulated host immune response (Gopalaswamy et al., 2020).

2.3 Immune response against *Mycobacterium tuberculosis*

Upon the entry of MTB inside the host, various immune cells get activated. The activation of CD4+ Th1 cells triggers the release of IFN-γ and TNF-α, which in turn activate macrophages, leading to the formation of granulomas (Weiss and Schaible, 2015). A granuloma has dual roles: primarily, it controls the MTB infection, but simultaneously, it allows MTB to be in a dormant stage for years. On the other hand, CD8+ T cells serve as a secondary defense mechanism and directly kill infected cells. However, the response is delicately balanced; because neutrophils cause damage to healthy tissues, and regulatory T cells can reduce protective immunity, which may aid in the survival of the pathogen (Fig. 2.1). In addition, dendritic cells are necessary for a specific T-cell response. It is essential to understand the specific roles of these cells to develop effective vaccines and host-directed treatments (Table 2.1). Understanding of underlying mechanisms of cellular immunity against MTB provides the basis for modern TB prevention strategies. The BCG vaccine works by creating memory Th1 cells, while next-generation vaccines aim to improve this response, often by stimulating CD8+ T cells. Host-directed therapies are new approaches that aim to adjust the body's immune response—by modulating macrophage killing or reducing pathological inflammation—to work in combination with antibiotics and improve treatment outcomes.

2.4 Factors affecting immune response against *Mycobacterium tuberculosis*

In addition to static genetics, the functional role of the immune system is the most important host factor. HIV co-infection is the strongest risk factor for the development

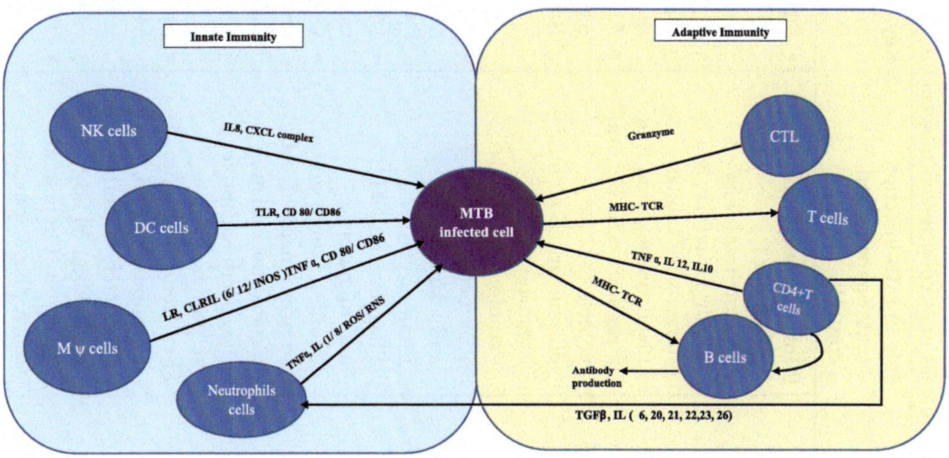

Figure 2.1 Immune response against *Mycobacterium tuberculosis* (MTB) infection, the cells involved are natural killer (NK) cells, dendritic cells (DC), macrophage (M), C-type lectin receptors (CLRs), toll-like receptors(TLR), reactive oxygen species (ROS), reactive nitrogen species (RNS).

Table 2.1 Roles of immune cells in *Mycobacterium tuberculosis* (MTB) infection and prevention.

Immune cell	Role in MTB infection (pathogenesis and control)	Key mechanisms and molecules	Role in prevention (vaccines & host-directed therapy)	References
Alveolar macrophage	Primary host cell and first responder. Phagocytoses inhaled bacilli. However, MTB subverts its killing mechanisms, using it as a replicative niche and vehicle for dissemination. Central organizer of the granuloma.	• Phagosome maturation arrest: MTB blocks phagosome-lysosome fusion • Cytokine production: TNF-α, IL-1β, IL-12 (initiates immunity). • Antimicrobial effectors: reactive oxygen/nitrogen species (ROS/RNS), autophagy.	• Vaccination (BCG): BCG is taken up by macrophages, initiating the immune cascade that leads to T-cell priming. • Host-directed therapy (HDT): agents like Vitamin D aim to enhance macrophage antimicrobial pathways (e.g., cathelicidin production).	Wolf et al. (2008)
Dendritic Cell (DC)	Bridge between innate and adaptive immunity. Phagocytoses MTB and migrates to draining lymph nodes to prime naïve T cells. Crucial for initiating the specific adaptive response.	• Antigen presentation: presents MTB antigens on MHC-II to CD4+ T cells • Migration: CCR7-mediated migration to lymph nodes • Cytokine production: IL-12, critical for driving Th1 differentiation.	• Vaccination: effective vaccines, including BCG and novel candidates, rely on DCs to present antigens and generate robust T-cell memory. Some vaccine strategies are designed to specifically target antigens to DCs.	Cooper (2009)
CD4+ Th1 Cell	The "General" of the immune response. Absolutely essential for controlling MTB infection. Orchestrates macrophage activation and granuloma formation. Depletion (e.g., in HIV) leads to rapid TB reactivation.	• Interferon-gamma (IFN-γ): Master cytokine that "super-activates" macrophages to enhance their killing capacity • Tumor Necrosis Factor-alpha (TNF-α): critical for the formation and structural maintenance of the granuloma.	• Vaccination (BCG): The primary goal of BCG is to induce a population of memory Th1 cells that can rapidly produce IFN-γ upon MTB exposure • Correlates of Protection: IFN-γ production by Th1 cells is a key immunological readout in vaccine trials.	Behar et al. (2010)

CD8+ Cytotoxic T Lymphocyte (CTL)	The "Assassin." Recognizes and kills MTB-infected cells, particularly those presenting antigens on MHC-I (e.g., from cytosolic MTB). Provides a secondary source of IFN-γ and helps control bacterial load, especially in chronic infection.	• Cytotoxicity: releases perforin and granzymes to induce apoptosis of infected cells • IFN-γ Production: contributes to macrophage activation • Recognition: binds to MTB peptide-MHC-I complexes on infected cells.	• Vaccination: next-generation vaccine candidates aim to induce strong CD8+ T-cell responses alongside CD4+ responses, as they may target infected cells that are less accessible to CD4+ cells.	Egan et al. (2021)
Neutrophil	The "Controversial Cavalry." Rapidly recruited to the site of infection. Role is dual and context-dependent. Can contribute to early bacterial control but is also associated with severe tissue damage and disease progression.	• Phagocytosis & ROS: can engulf and kill MTB • Neutrophil Extracellular Traps: Can trap bacilli • Pathology: Release of proteases and inflammatory mediators can cause lung tissue destruction. Can act as a "Trojan horse" for disseminating bacteria.	• Vaccination: generally not a direct target of vaccine-induced immunity • HDT: modulating neutrophil responses (e.g., reducing their pathological inflammation) is a potential therapeutic strategy to limit lung damage in active TB.	Chen et al. (2020)
Regulatory T Cell (Treg)	The "Brakes" of the immune system. Suppresses the anti-TB immune response to prevent excessive inflammation and tissue damage. While this protects the host, it also benefits MTB by facilitating its persistence and the establishment of latency.	• Anti-inflammatory cytokines: IL-10 and TGF-β, which dampen macrophage and T-cell function • Cell-Cell Contact: expression of inhibitory molecules like CTLA-4 and PD-1.	• Vaccination: an ideal vaccine would induce strong protective immunity while minimizing excessive Treg induction • HDT: transiently blocking Treg function (e.g., anti-IL-10 antibodies) during antibiotic treatment is a strategy being explored to boost pathogen clearance.	Kumar et al. (2022)

of active tuberculosis, since the loss of CD4+ Th1 cells leads to the destruction of the structural and functional integrity of the granuloma, with a consequent loss of bacterial control and rapid progression of the disease (Hamada et al., 2021). On similar lines, the presence of co-existing comorbidities has been shown to significantly alter the risk profile. Type 2 Diabetes Mellitus (T2DM), for instance, impairs neutrophil chemotaxis, macrophage activation, and cytokine production, thus creating a permissive environment for the replication of MTB (Restrepo, 2016). Malnutrition and Vitamin D deficiency are linked to weakend cell-mediated immunity. The susceptibility differs with gender and age; the chances of males getting infected by MTB are comparatively higher. The behavioral factors, along with biological factors, contribute to MTB infection (Hanekom et al., 2011). MTB is reported to alter the host epigenetic factors, thereby affecting transcription. Both the upregulation and downregulation of genes related to immune system, inflammation, and cell survival may decide the fate of MTB inside the host (Gauba et al., 2021; Piras et al., 2019). Macrophages infected with MTB display significant modification in the methylation status of inflammation-related genes, with the promoter region of IL-17 displaying a remarkably high increase in methylation (Zheng et al., 2016). Additionally, this methylation status is influenced by the host genetic background and the MTB strain, thereby emphasizing the intricate relationship between the two (Choudhary et al., 2020).

Immune system dysregulation plays a significant role in the course of tuberculosis. The following conditions weaken the immune system and make it more susceptible to severe or extrapulmonary tuberculosis: human immunodeficiency virus infection, diabetes mellitus, and immunosuppressive therapy (Loddenkemper et al., 2016; Getahun et al., 2021). Weakened immune responses lead to uncontrolled bacterial replication and spread. Neutrophil extracellular traps (NETs) are essential components of the host response; however, their dysfunction is linked to inflammatory and autoimmune diseases. In tuberculosis, NETs have immunomodulatory properties that determine the level of pro-inflammatory and anti-inflammatory reactions, thus affecting the immune microenvironment and the course of the disease (Papayannopoulos, 2018; Brinkmann and Zychlinsky, 2012; Oehlers et al., 2020).

2.5 Treatment of *Mycobacterium tuberculosis* infections

2.5.1 Diagnosis: the prerequisite for treatment

Early diagnosis of TB infection is challenging due to the fact that only 5–10% of infected people develop clinical symptoms such as cough, hemoptysis, fever, night sweats, weight loss, and fatigue. *In vitro* diagnosis of MTB involves sputum smear microscopy using Ziehl-Neelsen stain. It is comparatively inexpensive and easy technique, but its sensitivity is very low. Isolation of bacteria on Löwenstein-Jensen agar is a highly specific and sensitive test and remains the gold standard test till date. The only limitation is the slow growth of the bacteria (2 days to few weeks). Staining of bacteria by acid fast stain is generally used to detect infection microscopically. Further confirmation of pulmonary MTB infection is done by Chest X-ray for the detection of caseous lesions and Ghoncomplexes. Molecular diagnosis of MTB employs highly specific nucleic acid amplification tests (NAATs)—primarily GeneXpert MTB/RIF and Truenat—to detect DNA within 2hours. These tests simultaneously detect MTB and rifampicin resistance from sputum or tissue samples. Other diagnostic methods include BACTEC NAP (p-nitro-α-acetylamino-β-hydroxypropiophenone) test to differentiate the MTB complex from other *Mycobacteria*. Immunological tests such as QuantiFERON-TB Gold and Mantoux (tuberculin skin test) are commonly used for diagnosing active pulmonary TB but not latent TB (Barik et al., 2025).

It is essential to have an early and accurate diagnosis of tuberculosis (TB) in order to control its transmission and the emergence of drug-resistant strains.

2.5.2 Drug regimens and the challenge of resistance

The treatment strategy for TB depends basically on the type of MTB strain responsible for the infection and disease. For drug-susceptible TB, initially antibiotics such as isoniazid (INH), rifampicin (RIF), pyrazinamide (PZA), and ethambutol (EMB) are given to patient for six months, thereafter INH and RIF treatment is extended for four months (Cavalcante-Silva et al., 2023). The success of the treatment depends mainly on patient adherence to the given treatment and the maintenance of proper hygiene; treatment adherence achieves 85% curability in most of the cases. Success in treatment clearly reflects the high potency of these drugs but also how well they work alongside the body's immune system to eliminate the infection. These drugs target different biomolecules of the MTB bacteria, INH weakens mycolic acid synthesis and hence making the cell vulnerable to immune attack. RIF blocks RNA transcription while EMB destabilizes the cell wall thus preventing the development of drug resistance. PZA remains active inside microenvironment of immune cells and granulomas, where MTB exists in a semidormant state (Zhang and Yew, 2015). Since these drugs target different cellular components of MTB thereby reducing probability of development of drug resistance in bacteria.

In case of latent tuberculosis infection (LTBI), the MTB remains silent (without showing any symptoms) for a long time but may reactivate in weakened immunity. Prolonged isoniazid monotherapy has been shown to significantly reduce this risk by clearing dormant bacteria before they can cause active disease (Menzies et al., 2011).

Recently, the emergence of Multidrug-resistant TB (MDR-TB) has rendered strains resistant to both INH and RIF, while fluoroquinolones (second-line drug) are ineffective against extensively drug-resistant TB (XDR-TB) (World Health

Organization, 2023). Hence, elimination of drug-resistant strains remains the biggest challenge in TB treatment. In MDR-TB treatment may take 9 to 20 months, requiring higher dosages of more powerful drugs (Kanabalan et al., 2021). The emergence of totally drug-resistant TB strains further highlights how fragile current treatment options can be (Bailo et al., 2015).

Several new drugs are available in the market today that target MDR-TB, among which Bedaquiline have been reported to improve survival of patients having MDR-TB. Bedaquiline disrupts energy production in MTB (Diacon et al., 2014). Newer agents such as Felamanid and Pretomanid have further extended treatment options. A combination of Bedaquiline, Pretomanid, and Linezolid has been found to be active against XDR-TB (Conradie et al., 2020). These newer regimens not only improve treatment outcomes but also shorten therapy and support better immune recovery. The World Health Organization has recently recommended treatment using Bedaquiline, Pretomanid, and Linezolid for 9–12 months against MDR-TB, thereby reducing the physical and psychological burden of long-term treatment while maintaining effectiveness (World Health Organization, 2023).

2.5.3 Adjunctive and host-directed therapies

Adjunctive and host-directed therapies (HDTs) have gained attention recently due to their specificity and effectivity. Adjunctive corticosteroid is generally prescribed in TB meningitis and Pericardial TB to avoid prolonged complications (Thwaites et al., 2004; Mahomed et al., 2023). HDTs regulate immunological pathways to eliminate MTB infection while minimizing damage to underlying tissues. Other approaches under HDTs include statins, which promote autophagy and phagolysosomal maturation; vitamin D supplementation, which enhances macrophage antimicrobial activity; and cytokine-based therapies designed to augment T-cell–mediated immunity (Barik et al., 2025). Although many HDTs remain in experimental or early clinical stages, they represent a promising adjunct to conventional antitubercular chemotherapy.

2.5.4 Treatment in special populations

Treatment of TB in immunocompromised HIV patients is highly challenging and complicated since drug-drug interactions can occur between rifampicin and antiretroviral therapy (ART), as well as the risk of Immune Reconstitution Inflammatory Syndrome (IRIS) (Menzies et al., 2011). Combinations of two or more drugs are recommended for Spinal TB and neurological deficits or destroyed lung syndrome.

In children, the management of the disease requires careful dose adjustments and is aided by preparations with acceptable taste, which helps in improving patient compliance, apart from the need for prolonged treatment regimens and corticosteroids in severe cases like meningitis (Mahomed et al., 2023). Lastly, surgery remains a very important, although rare, adjunctive therapy in complicated cases of tuberculosis, such as spinal tuberculosis with neurological impairment or destroyed lung syndrome, and is always used in combination with effective antitubercular therapy.

2.5.5 Future prospects

Looking forward, the future of TB treatment and control hinges on creating shorter, safer, and more efficient regimens, especially for drug-resistant TB. The wider implementation of new drugs like bedaquiline and pretomanid, combined with HDTs, holds immense potential. Furthermore, advancements in genomics, proteomics, and biomarker identification are paving the way for personalized medicine, where treatment can be tailored to the specific characteristics of both the patient and the pathogen (Barik et al., 2025).

This integrated understanding has profound implications for intervention. The failure of BCG to consistently protect against adult pulmonary TB underscores the need for next-generation vaccines that can overcome the sophisticated immune evasion of diverse MTB strains by inducing a broader, more robust T-cell response. Underpinning all these efforts is fundamental research into bacterial virulence, such as the role of the PhoP regulator in controlling pathogenicity and stress tolerance, which continues to reveal new, potentially vulnerable therapeutic targets (Khan et al., 2024). By continuing to unravel the complex molecular dialogue between host and pathogen, we can develop the innovative tools needed to finally consign tuberculosis to the past.

2.6 Conclusions

MTB infection is a complex phenotype arising from a molecular dialogue between a highly adaptable pathogen and a genetically and physiologically diverse host. Bacterial determinants like the ESX-1 system and metabolic flexibility allow MTB to invade and persist, while host genetics, immune status, and comorbidities shape the battlefield. The true nature of TB pathogenesis lies not in the host or pathogen alone, but in their intricate interplay, where latency represents a delicate balance of containment and coexistence. Moving beyond a simplistic "pathogen vs. host" view to embrace this integrated perspective is essential for rationally designing the next generation of tools needed to finally end the TB pandemic.

While TB remains a formidable global health threat, a new era of innovative therapeutics and strategic approaches offers significant hope. The implementation of new drugs and shorter regimens for drug-resistant TB, combined with a deeper understanding of host-pathogen interactions, has improved treatment outcomes. However, the persistence of the disease and the continued emergence of resistance underscore that the fight is far from over. A sustained, multifaceted effort integrating basic science, clinical research, and public health initiatives is imperative to achieve better control of this ancient disease.

References

Abdallah AM, van Pittius NCG, Di Luca M. The type VII secretion systems of mycobacteria. In: Rubin E, editor. Mycobacterium tuberculosis. New York: Humana Press; 2021. p. 125–56.

Bailo R, Bhatt A, Aínsa JA. The challenge of new drug discovery for tuberculosis. Future Medicinal Chemistry 2015;7(13):1649–62.

Barik BS, Kar CD, Das S, Hussain T, Nayak S, Sahu AK, et al. Latent tuberculosis infection among household contacts, healthcare workers, and patients using QuantiFERON-TB Gold Plus and tuberculin skin test in a tertiary care hospital in Bhubaneswar: a cross-sectional study. Journal of Global Infectious Diseases 2025;17(1):52–9.

Behar SM, et al. Evasion of the innate and adaptive immune response by *Mycobacterium tuberculosis*. Journal of Cellular Physiology 2010;223(2):337–43.

Brennan PJ, Nikaido H. The envelope of mycobacteria. Annual Review of Biochemistry 1995;64: 29–63.

Brinkmann V, Zychlinsky A. Neutrophil extracellular traps: is immunity the second function of chromatin? The Journal of Cell Biology 2012;198(5):773–5.

Cavalcante-Silva LHA, Almeida FS, Andrade AG. A review of host-directed therapies for tuberculosis treatment. Tropical Medicine and Infectious Disease 2023;8(9):409.

Chen X, et al. The "double-edged sword" of regulatory T cells in human tuberculosis. Cellular & Molecular Immunology 2020;17(10):1027–39.

Choudhary J, Singhal A, Rao KV. Host DNA methylation in tuberculosis: new insights into disease pathogenesis. Frontiers in Immunology 2020;11:585466.

Conradie F, Diacon AH, Ngubane N, Howell P, Everitt D, Crook AM, et al. Treatment of highly drug-resistant pulmonary tuberculosis. The New England Journal of Medicine 2020;382(10): 893–902.

Cooper AM. Cell-mediated immune responses in tuberculosis. Annual Review of Immunology 2009;27:393–422.

Coscolla M, Gagneux S. Consequences of genomic diversity in *Mycobacterium tuberculosis*. Seminars in Immunology 2014;26(6):431–44.

Diacon AH, Pym A, Grobusch MP, Patientia RF, Rustomjee R, Ngoie M, et al. The diarylquinoline TMC207 for multidrug-resistant tuberculosis. The New England Journal of Medicine 2014;360(24):2397–405.

Egan CE, et al. The role of neutrophils in the immune response to *Mycobacterium tuberculosis*. Frontiers in Immunology 2021;12:646281.

Farzand R, Kolloli A, Subramanian V, Sharma S. The role of fatty acyl-CoA thioesterases in the pathogenesis of *Mycobacterium tuberculosis*. Frontiers in Microbiology 2022;13:865432.

Gauba V, Kaur J, Singh S. Epigenetic regulation of host immunity in *Mycobacterium tuberculosis* infection. Frontiers in Immunology 2021;12:682803.

Getahun H, Matteelli A, Chaisson RE, Raviglione M. Tackling tuberculosis in people with diabetes mellitus. Nature Reviews Endocrinology 2021;17(7):393–405.

Gopalaswamy R, Shanmugam S, Mondal R, Subbian S. Of tuberculosis and non-tuberculous mycobacterial infections—a comparative analysis of epidemiology, diagnosis and treatment. Journal of Biomedical Science 2020;27(1):74.

Gupta R, Raghunand TR, Bishai WR. The PhoPR two-component system of *Mycobacterium tuberculosis*. Tuberculosis (Edinburgh, Scotland) 2006;86(3-4):237–43.

Hamada Y, Getahun H, Tadesse BT, Ford N. HIV-associated tuberculosis. International Journal of STD and AIDS 2021;32(9):780–90.

Hanekom M, Gey van Pittius NCG, McEvoy C, Victor TC, Van Helden PD, Warren RM. *Mycobacterium tuberculosis* Beijing genotype: a template for success. Tuberculosis (Edinburgh, Scotland) 2011;91(6):510–23.

Kanabalan S, Gopal M, Bhat J. Current and emerging therapeutic strategies for tuberculosis. Antibiotics (Basel) 2021;10(12):1448.

Khan S, Singh S, Singh Y. Regulation of cAMP levels by PhoP is essential for *Mycobacterium tuberculosis* pathogenesis. Journal of Bacteriology 2024;216(4). e00570-23.

Kumar A, Saini V, Bharadwaj A. Methylglyoxal: a double-edged sword in *Mycobacterium tuberculosis* infection. Frontiers in Microbiology 2022;13:1046147.

Loddenkemper R, Hauer B, Diel R. Tuberculosis in children and adolescents. Deutsches Arzteblatt International 2016;113(7):101–7.

Mahomed H, Zimri N, Hussey GD. A review of the diagnosis and treatment of tuberculosis in children. South African Medical Journal 2023;113(9):718–24.

Menzies D, Al Jahdali H, Al Otaibi B. Recent developments in treatment of latent tuberculosis infection. The Indian Journal of Tuberculosis 2011;58(1):25–30.

Mokrousov I, Slavchev I, Solovieva N, Dogonadze M, Vyazovaya A, Valcheva V, et al. Molecular insight into *Mycobacterium tuberculosis* resistance to nitrofuranyl amides gained through metagenomics-like analysis of spontaneous mutants. Pharmaceuticals (Basel) 2022;15(9):1136.

Nau GJ, Richmond JF, Schlesinger A, Jennings EG, Lander ES, Young RA. Human macrophage activation programs induced by bacterial pathogens. Proceedings of the National Academy of Sciences of the United States of America 2000;97(7):3436–41.

Oehlers SH, Vega-López GA, Wharry CE, Tobin DM. Neutrophils in tuberculosis: friend or foe? Journal of Leukocyte Biology 2020;108(1):13–24.

Pandey AK, Sassetti CM. Mycobacterial persistence requires the utilization of host cholesterol. Proceedings of the National Academy of Sciences of the United States of America 2008;105(11):4376–80.

Papayannopoulos V. Neutrophil extracellular traps: a historical perspective. The Journal of Cell Biology 2018;217(9):3211–9.

Piras G, Monno R, Capparelli R, Iannaccone M. Host epigenetic changes induced by *Mycobacterium tuberculosis* infection. Frontiers in Microbiology 2019;10:2063.

Restrepo BI. Diabetes and tuberculosis. Microbiology Spectrum 2016;4(6). Available from 10.1128/microbiolspec.TNMI7-0023-2016.

Rustad TR, Minch KJ, Sherman DR. The non-replicating persister. Mycobacterium tuberculosis. New York: Springer; 2009. p. 237–60.

Sia JK, Rengarajan J. Immunology of *Mycobacterium tuberculosis* infections. Microbiology Spectrum 2019;7(4). GPP3-0022-2018.

Stanley SA, Raghavan S, Hwang WW, Cox JS. Acute infection and macrophage subversion by *Mycobacterium tuberculosis* require a specialized secretion system. Proceedings of the National Academy of Sciences of the United States of America 2003;100(22):13001–6.

Thwaites GE, Bang ND, Dung NH, Quy HT, Oanh DTT, Chau TTH, et al. Dexamethasone for the treatment of tuberculous meningitis in adolescents and adults. The New England Journal of Medicine 2004;351(17):1741–51.

Thye T, Owusu-Dabo E, Vannberg FO, van Crevel R, Curtis J, Sahiratmadja E, et al. Common variants at 11p13 associated with susceptibility to tuberculosis. Nature Genetics 2010;42(9):739–41.

Walters SB, Dubey VS, Kolesnikova I, Gengenbacher M, Rhee K. The *Mycobacterium tuberculosis* PhoPR two-component system regulates genes essential for virulence and complex lipid biosynthesis. Molecular Microbiology 2006;60(2):312–30.

Weiss G, Schaible UE. Macrophage defense mechanisms against intracellular *Mycobacterium tuberculosis*. Annual Review of Immunology 2015;33:517–48.

Wolf AJ, Desvignes L, Linas B, Banaiee N, Tamura T, Takatsu K, et al. Initiation of the adaptive immune response to *Mycobacterium tuberculosis* depends on antigen production in the local lymph node, not the lungs. The Journal of Experimental Medicine 2008;205(1):105–15.

World Health Organization. Global tuberculosis report 2023. Geneva: World Health Organization; 2023.

Yelamanchi SD, Surolia A. Metabolic virulence factors of *Mycobacterium tuberculosis*. Frontiers in Cellular and Infection Microbiology 2021;11:652907.

Zhang Y, Yew WW. Mechanisms of drug resistance in *Mycobacterium tuberculosis*. The International Journal of Tuberculosis and Lung Disease 2015;19(11):1300–16.

Zheng X, Chu T, Zhang H, Li Y, Wu J, Wang H. DNA methylation as a therapeutic target for tuberculosis. Epigenomics 2016;8(7):939–51.

CHAPTER 3

DNA structure and role of biomarkers

Neha Bisht and Chetna Dhand
Academic of Scientific and Innovative Research (AcSIR), CSIR-Advanced Materials and Processes Research Institute, Bhopal, Madhya Pradesh, India

3.1 Introduction

Tuberculosis (TB), caused by *Mycobacterium tuberculosis* (MTB), continues to be a serious public health concern (Yang et al., 2023). The challenges of quick identification, medication resistance, and pathogen–host interactions are the key reasons why this bacterium continue to be a major cause of morbidity and mortality even after decades of research. Its unique characteristics, such as its high guanine–cytosine (GC) content, repeated sequences like IS6110, and virulence genes like Early Secretory Antigenic Target 6 (ESAT-6) and culture filtrate protein 10 (CFP-10), demonstrate the bacteria's capacity to successfully elude the host immune system and create a chronic infection (Rahlwes et al., 2023). Recent developments in molecular diagnostic techniques have transformed the detection of MTB; gene sequences such insertion sequences (IS6110), 16S rRNA, hsp65, katG, and rpoB are potential candidates for assays that are sensitive and specific (Mousavi-Sagharchi et al., 2024). Protein biomarkers including lipoarabinomannan(LAM), ESAT-6, and CFP-10 have become significant in addition to genetic markers since they can distinguish between latent and active TB and may allow for less intrusive point-of-care testing. The intricate relationship between the host immune system and the MTB genome's evolution highlights the intricacy of disease pathophysiology and how it affects the precision medicine approach. Recent advancements including CRISPR-based diagnostic systems, bioinformatics analysis, and next-generation sequencing (NGS) are poised to transform the sector by offering real-time monitoring, individualized treatment plans, and quick medication resistance testing (Satam et al., 2023; He et al., 2025).

3.2 Structure of *Mycobacterium tuberculosis*

MTB is an obligate aerobe, non-motile, slender, and slightly curved bacillus that is 1–4 μm in length and about 0.4 μm in width. It does not have flagella or spores but has fimbriae and a capsule-like outer layer that enables it to adhere and survive in the host.

Advances in Biosensor Technology for Mycobacterium Tuberculosis Detection
DOI: https://doi.org/10.1016/B978-0-443-44589-7.00009-1

One of the most characteristic properties of MTB is its extremely slow rate of growth, which is mainly due to the presence of a remarkably high lipid content in its cell envelope, contributing to nearly 60% of the cell wall dry weight (Wang et al., 2022). A schematic illustration of the structural arrangement of MTB is given in Fig. 3.1. The cell envelope of MTB is a complex, multilayered structure. The innermost layer is made of peptidoglycan, which gives strength to the cell envelope and plays a role in cell division and immune recognition (Kalscheuer et al., 2019). Arabinogalactan is covalently bound to this layer. Arabinogalactan is a highly branched polysaccharide with terminal arabinan chains that are esterified to long-chain mycolic acids, forming the typical mycolyl–arabinogalactan–peptidoglycan complex (Farnia et al., 2025). Mycolic acids contribute to hydrophobicity, acid-fastness, and low permeability, thereby restricting the entry of dyes, antibiotics, and host immune factors. The glycolipids such as trehalose-6,6′-dimycolate, are embedded in the mycolic acid layer and represent an important virulence factor responsible for immune modulation and granuloma formation (Warrier, 2010). Immunogenic molecules are further shielded, and intracellular survival is improved by the outermost layer, which has a capsule-like structure composed of

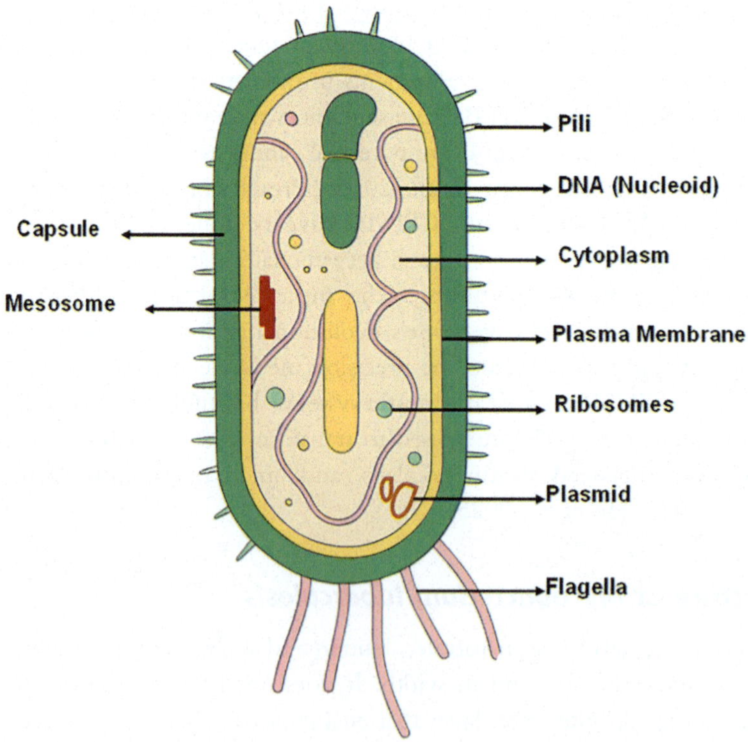

Figure 3.1 Structure of *Mycobacterium tuberculosis.*

proteins, lipids, and polysaccharides. Thus, the lipid and carbohydrate-rich envelope plays an important role in MTB pathogenicity and survival.

3.3 Genomic organization of *Mycobacterium tuberculosis*

The evolutionary optimization of MTB as an effective intracellular pathogen is reflected in its DNA. Understanding its persistence, pathogenicity, and resistance characteristics, all of which can be utilized to develop targeted treatments and vaccines, requires the sequencing of its genetic characteristics.

3.3.1 High guanine–cytosine content

The MTB genome is distinguished by its extraordinarily high GC content, which offers MTB the ability to survive and adapt (Gómez-González et al., 2023). Due to the thermal stability and oxidative resistance of GC-rich DNA, it can survive in very harsh conditions, like oxidative burst during macrophage infection (Kostyuk et al., 2015). The functional characteristic of high GC content significantly affects codon bias, gene expression efficiency, and mutation rates, in addition to influencing virulence factor translation and ensuring genetic stability. GC-rich genes are highly important for immune evasion and survival, such as those encoding regulatory proteins and PE/PPE proteins (Phelan et al., 2016). Nevertheless, this feature poses challenges for molecular diagnostics and genetic engineering since high-GC regions often create substantial secondary structures that obstruct PCR amplification and sequencing, requiring the use of long-read sequencing and better primer design.

3.3.2 Repetitive elements

Recurring insertion sequences (IS), particularly IS6110, are another intriguing aspect of the MTB genome (Comín et al., 2022). IS6110 DNA fingerprinting is based on the very variable number of IS6110 copies among the clinical isolates, which ranges from zero to more than 20 copies (Moström et al., 2002). By altering regulatory regions, deactivating coding genes, and resulting in chromosomal rearrangements, insertion elements enhance genomic flexibility and have a substantial effect on virulence, host adaptation, and treatment resistance (Vandecraen et al., 2017; Perdigão et al., 2014). With an average occurrence of 5–7 copies, some IS elements, such as IS1081, have been found to be very stable among MTBC isolates. Because of its stability, it has been discovered to be a useful marker for distinguishing MTBC from nontuberculous mycobacteria, particularly in isolates that lack IS6110 (Lyu et al., 2020). Its constant copy number has been used in molecular epidemiology studies for PCR-based assay and quality control. IS1547 has a polymorphisms pattern and controls insertional inactivation and downstream genes to support genomic evolution, despite its limited exploration (Fang et al., 1999).

3.3.3 Virulence-associated genes

MTB pathogenicity is generally determined by genomic regions known as Regions of Difference, which code for various virulence proteins involved in host invasion and manipulation of the immune system. Of these, RD1 is particularly important because it codes for ESAT-6 and CFP-10 through the ESX-1 secretion system, which is involved in cytosolic evasion, phagosomal membrane disruption, and manipulation of the host immune response-all of which enhance survival and transmission (Forrellad et al., 2013). Apart from RD1, immune evasion and antigenic variation are facilitated by the PE/PPE multigene families, which make up around 10% of the genome. Additionally, they contribute to strain-specific pathogenicity through their involvement in T-cell regulation and cell wall biosynthesis (Akhter et al., 2012). Granuloma development and immunological suppression are caused by genes related to lipid metabolism that encode sulfolipids and phthiocerol dimycocerosates (PDIMs) (Augenstreich et al., 2020).

3.4 Key genetic biomarkers for *Mycobacterium tuberculosis* detection

Rapid and accurate detection of MTB remains critical for preventing TB transmission and fatalities. Although they are often used, traditional diagnostic techniques including smear microscopy, culture, and chest radiography have poor sensitivity, specificity, and turnaround times. Molecular diagnostics based on MTB-specific genomic targets have overcome these limitations and developed into powerful tools for fast, sensitive, and specific pathogen identification. Among the most thoroughly studied and therapeutically significant targets are IS6110, 16S rRNA, and functional genes including hsp65, katG, rpoB, and Rv3616c.

3.4.1 IS6110 as a diagnostic marker

IS6110, a member of the IS3 family of insertion elements, is one of the most stable MTB markers for detection. It is found in multiple copies (1–20) in most strains of MTBC and increases the sensitivity of PCR-based tests (Gonzalo-Asensio et al., 2018; McEvoy et al., 2007). Its high specificity enables it to clearly differentiate MTB from nontuberculous mycobacteria (NTM), and hence it is the standard for most tests (Githui et al., 1999). However, its application is hampered by low or null copy-number variants that exist in some regions of Southeast Asia and Africa. In these instances, additional markers are employed. Although it has a serious limitation, IS6110 is a critical element in PCR-based and biosensor technologies, such as electrochemical and fluorescent DNA sensors.

3.4.2 Function of 16S rRNA in *Mycobacterium tuberculosis* Identification

The 16S ribosomal RNA gene is very conserved among bacteria, but there are hypervariable regions that help in the classification of species. In MTB, the 16S rRNA-targeted test is useful for both detection and phylogenetic classification. The benefit lies in its higher copy number compared to genomic DNA, making it more sensitive, especially in low bacterial load cases of extrapulmonary or pediatric TB (Mousavi-Sagharchi et al., 2024; Tawfick et al., 2023; Choi et al., 2015). Techniques such as reverse transcription PCR (RT-PCR) and quantitative PCR (qPCR) have been shown to be of immense use. While 16S rRNA sequencing is a very useful technique for reference purposes, it is not very practical in the clinical setting beacause of cost and technical issues.

3.4.3 hsp65, katG, rpoB, and Rv3616c as diagnostic targets

There are a number of protein-coding genes that can be used as supplementary diagnostic and resistance markers. The hsp65 gene, which encodes a 65-kDa heat shock protein, is conserved but sufficiently diverse to distinguish the MTB from the closely related NTM (Kim and Shin, 2018). The katG gene encodes a catalase–peroxidase that activates isoniazid (INH), an important first-line drug. Mutations at codon 315 are strongly linked with INH resistance, making katG a crucial marker in both diagnostic and drug-susceptibility tests (Torres et al., 2015). The rpoB gene, which encompasses the β-subunit of RNA polymerase, is a vital marker for rifampin resistance. Most resistance mutations occur in the 81-bp rifampin resistance-determining region (RRDR), which forms the basis for tests such as the GeneXpert MTB/ RIF and line probe assay (Mao et al., 2025). Another interesting biomarker, Rv3616c of the RD-1 chromosomal locus, has been associated with MTB pathogenicity and immunomodulation (Ruhwald et al., 2017; Millington et al., 2011).

3.5 Protein biomarkers and their role in *Mycobacterium tuberculosis* diagnosis

Recently, protein biomarkers have been considered promising alternatives for enhancing TB diagnosis because they can simulate the host's physiological and immunological reaction to MTB infection or simply indicate the bacterial protein's presence. These biomarkers, of either MTB or the host's immune reaction, can be detected in different bodily fluids such as urine, serum, and sputum, making them minimally invasive or noninvasive techniques for diagnosis. Among the different candidates being considered,

LAM, ESAT-6, and CFP-10 have particularly caught the attention of researchers because of their potential to diagnose both active TB (ATB) and latent TB infection (LTBI) (Correia-Neves et al., 2019; Jafari et al., 2009).

3.5.1 Lipoarabinomannan as a diagnostic marker

LAM is a major and specific glycolipid component of the outer mycobacterial cell wall (Hamasur et al., 2001). It makes an interesting diagnostic candidate owing to its amphipathic nature and high concentration. LAM is shed by metabolically active or decomposing pathogens and can be identified in urine, particularly in HIV coinfected patients with TB with a high bacterial load and poor renal function (Flores et al., 2021). Because of its ease of use, quick turnaround time, and low cost, the urine LAM assay, specifically the Alere Determine TB LAM Ag test, has been WHO-approved for use in HIV-infected patients with low CD4 counts (Mwaura and Engel, 2021). However, sensitivity remains low in HIV-negative patients, as well as in drug-susceptible or extrapulmonary TB. Monoclonal antibodies specific for certain LAM epitopes and combining LAM testing with other indicators are two approaches for improving assay performance (Corrigan et al., 2023). More advanced approaches, including as nanoparticle-based assays and electrochemical biosensors, are being actively pursued to improve sensitivity and specificity. Although LAM is already beneficial in high-burden environments, further improvement is required for wider application.

3.5.2 Early secretory antigenic target 6

ESAT-6 is a 6-kDa secreted MTB protein encoded within the RD1 locus. It is absent in most NTMs and all BCG vaccine strains, making it highly specific for MTB identification (Gcebe et al., 2016). As an immunodominant antigen, ESAT-6 is detected early by the host immune system and is often used in IGRAs, such as QuantiFERON-TB Gold and T-SPOT.TB, which measure T-cell responses to MTB antigens (Passos et al., 2024). The tests are exceedingly specific, but they cannot differentiate between active and latent tuberculosis. Aside from immunoassays, ESAT-6 is being tested for direct antigen detection in serum, sputum, and other specimens using multiplex platforms and biosensor-based assays (Chauke et al., 2024). Due to its significance in MTB pathogenicity and immunological evasion, it serves as both a diagnostic and therapeutic target. Point-of-care (POC) devices based on ESAT-6-specific antibodies or aptamers are being developed to provide diagnostics to low-resource environments.

3.5.3 Culture filtrate protein 10

CFP-10, which is also expressed by the RD1 region, is released alongside ESAT-6 and has been shown to synergistically affect MTB pathogenesis and host immune perception

(Guinn et al., 2004). It is employed in IGRAs and, like ESAT-6, produces robust T-cell responses. It has been found in clinical samples such as blood, respiratory secretions, etc., and research is indicating its potential to be incorporated in antigen-detection tests (Liu et al., 2021; Beamer et al., 2008). The antigen offers better sensitivity and specificity to rapid tests and is also capable of differentiating MTB from NTM disease when it is either alone or used with ESAT-6. With the help of multiplex detection technology such as lateral flow technology or microfluidic biosensors, this antigen is being incorporated for multi-analyte tests (Weldingh and Andersen, 2008).

3.6 Role of MTB genomic variability in host-pathogen interactions

The process of coordination in host and pathogen interactions, immunological recognition, the adaptability of the pathogen, and disease outcomes all depend on the availability of MTB DNA. Genetic mutation and diversity in the genome of MTB have been recognized as significant influences on the process of immunological regulation and disease outcomes, as it undergoes a process of co-evolution with its host. Innate and acquired immunological responses and pathogenicity depending on DNA variations are all part of the complex association of MTB variations and host immunology. The role of the host immune system in recognition of MTB infection is perhaps one of the most important parts of this complex association. Pathogen-associated molecular patterns (PAMPs) of MTB, such as its DNA, RNA, lipids, and proteins, are known by toll-like receptors, NOD-like receptors, and DNA-recognition proteins (Sun et al., 2013; Van der Roest et al., 2025). TLR9 senses the unmethylated CpG motifs in the DNA of MTB, which initiates a signal transduction cascade culminating in the production of pro-inflammatory cytokines and interferon type I (Kawai and Akira, 2009). MTB antigens, such as ESAT-6 and CFP-10, also activate a cascade of T-cell responses to combat the infection. In hiding the DNA patterns and regulating the production of proteins to avoid recognition, a method of evading the host's recognition has been created by MTB. MTB DNA stimulates cytosolic DNA sensors, including cyclic GMP-AMP synthase (cGAS), activating the STING pathway and triggering interferon-stimulated genes, which may indicate bacterial clearance or immune-mediated illness (Sun et al., 2013; Watson et al., 2015). Bacterial replication rates, dormancy regulation, and immunological interactions can all be influenced by single nucleotide polymorphisms, insertions, deletions, and gene rearrangements (Van der Roest et al., 2025). MTB can live inside macrophages and granulomas because of genetic alterations in regulatory genes such phoP, sigH, and dosR, which control virulence gene expression and stress responses (Pacl et al., 2018). While genetic variants within ESX-1 secretion system genes may lead to immunological modulation or hypervirulence phenotypes, genetic variety within the PE/PPE gene family has a comparable role in modifying antigenic diversity and immune control (Zhang et al., 2025; Chirakos et al., 2020).

The emergence of drug resistance resulting from mutations in the rpoB, katG, and isoniazid (inhA) genes exacerbates these issues by causing immunodepletion, severe tissue damage, and persistent infection (Solo et al., 2020).

3.7 Challenges in *Mycobacterium tuberculosis* genomic studies and future perspectives

Our knowledge of MTB genesis, transmission, and medication resistance has significantly increased with the advancement of genomic technology. The genetic diversity and survival patterns of the causal agent have been better understood by to techniques such as whole genome sequencing (WGS), next-generation sequencing (NGS), and others. However, it is difficult to apply these results to treatment modalities for clinical and public health applications due to limitations in diagnosis, complexity of host–pathogen interactions, and lack of genetic modalities. Genetic diversity among MTB strains is a major impediment to the detection of the pathogen at the molecular level. Although GeneXpert MTB/RIF and line probe assays can identify general resistance mutations (rpoB, katG, and inhA), heteroresistance and low-level or lineage-specific mutations can result in false-negative result. Traditional sequencing frequently fails to detect low-frequency variations, whereas WGS can detect them but is limited in dense locations because of cost, infrastructural limitations, and the lack of common interpretive procedures. Host–pathogen interactions further complicate tuberculosis control. MTB uses immune evasion mechanisms such phagosome–lysosome interference and cytokine regulation to survive and reactivate. Host factors such as genetics, immunological state, HIV co-infection, and comorbidities all have an impact on illness outcomes. Poor experimental models limit the translation of susceptibility pathways identified in host genomic investigations (e.g., TLR2, IL-12, IFN-γ) into therapeutics. Organoids and 3D granulomas are sophisticated systems that show promise, but they require optimization for high-throughput applications. Future progress will be driven by novel, low-cost tools. CRISPR-based testing, nanopore sequencing, and digital PCR may provide rapid, field-deployable diagnostics. Integration of genomic information and artificial intelligence could help in the prediction of resistance, treatment outcomes, and transmission. Personalized medicine approaches that include host–pathogen data could help in improving drugs and reduce treatment failure. Genomic surveillance could also help in the detection of drug-resistant strains. Projects such as CRyPTIC and WHO demonstrate that there is some progress being made, but more comprehensive networks and sharing of information are required to provide useful biomarkers for diagnosis, treatment follow-up, and prevention of recurrence. Building capacity in infrastructure and access to technology is required to ensure that these advances are also made available to the most vulnerable groups of patients with TB.

3.8 Conclusion

TB is a serious worldwide health issue that calls for coordinated efforts to find remedies ranging from genomics to molecular diagnostics, nanotechnology, and biotechnology. Recent advances in high-throughput sequencing, transcriptomics, and proteomics have revealed key genetic elements, immunemodulatory mechanisms, and pathogen-specific markers (from circulating microRNAs to volatile organic compounds) that can be harnessed for their early and accurate detection. Electrochemical biosensors and point-of-care devices, when integrated with functional nanomaterials, hold promise for rapid, accurate, and affordable solutions, particularly in resource-poor settings. From a translation perspective, the following biosensing technologies are on their way to clinical trials for real-time diagnosis and epidemiological surveillance: biosensors that utilize smartphone and handheld portable devices, CRISPR-based biosensing technology, and lab-on-a-chip technology. Whole-genome sequencing facilitates better drug- susceptibility testing and precision therapy for MDR-TB and XDR-TB. Advances in drug delivery systems, such as biomimetic nanocarriers and pulmonary aerosol-based delivery systems, are being developed to improve target specificity and bioavailability, while epitope-based vaccine candidates are being developed through reverse vaccinology approaches.

In the future, the fusion of multi-omics, single-cell and spatial transcriptomics, and AI-assisted analysis will play a crucial role in enhancing biomarker panels and the discovery of novel therapeutic approaches. Miniaturized AI-integrated diagnostic tools and CRISPR-based isothermal tests suggest the coming age of decentralized ultra-fast TB diagnostics. These dreams will be realized through continued interdisciplinary collaboration, support from the regulatory environment, and equal access to innovation. Finally, the convergence of precision medicine, digital health, and bioinspired engineering holds the promise of revolutionizing MTB control strategies and moving the world one step closer to TB elimination.

References

Akhter Y, Ehebauer MT, Mukhopadhyay S, Hasnain SE. The PE/PPE multigene family codes for virulence factors and is a possible source of mycobacterial antigenic variation: perhaps more? Biochimie 2012;94(1):110–6.

Augenstreich J, Haanappel E, Sayes F, Simeone R, Guillet V, Mazeres S, et al. Phthiocerol dimycocerosates from Mycobacterium tuberculosis increase the membrane activity of bacterial effectors and host receptors. Frontiers in Cellular and Infection Microbiology 2020;10:420.

Beamer GL, Flaherty DK, Vesosky B, Turner J. Peripheral blood gamma interferon release assays predict lung responses and Mycobacterium tuberculosis disease outcome in mice. Clinical and Vaccine Immunology: CVI 2008;15(3):474–83. Available from https://doi.org/10.1128/cvi.00408-07.

Chauke SH, Nzuza S, Ombinda-Lemboumba S, Abrahamse H, Dube FS, Mthunzi-Kufa P. Advances in the detection and diagnosis of tuberculosis using optical-based devices. Photodiagnosis and

Photodynamic Therapy 2024;45:103906. Available from https://doi.org/10.1016/j.pdpdt.2023.103906.

Chirakos AE, Balaram A, Conrad W, Champion PA. Modeling Tubercular ESX-1 Secretion Using Mycobacterium marinum. Microbiology and Molecular Biology Reviews 2020;84(4). Available from https://doi.org/10.1128/mmbr.00082-19, 10.1128/mmbr. 00082-19.

Choi Y, Hong S, Jeon B, Wang H, Lee G, Cho S, et al. Conventional and real-time PCR targeting 16S ribosomal RNA for the detection of Mycobacterium tuberculosis complex. The International Journal of Tuberculosis and Lung Disease 2015;19(9):1102–8.

Comín J, Otal I, Samper S. In-depth analysis of IS 6110 genomic variability in the Mycobacterium tuberculosis complex. Frontiers in Microbiology 2022;13:767912.

Correia-Neves M, Fröberg G, Korshun L, Viegas S, Vaz P, Ramanlal N, et al. Biomarkers for tuberculosis: the case for lipoarabinomannan. ERJ open research 2019;5(1).

Corrigan DT, Ishida E, Chatterjee D, Lowary TL, Achkar JM. Monoclonal antibodies to lipoarabinomannan/arabinomannan–characteristics and implications for tuberculosis research and diagnostics. Trends in Microbiology 2023;31(1):22–35.

Fang Z, Doig C, Morrison N, Watt B, Forbes K. Characterization of IS 1547, a new member of the IS 900 family in the Mycobacterium tuberculosis complex, and its association with IS 6110. Journal of Bacteriology 1999;181(3):1021–4.

Farnia P, Velayati AA, Ghanavi J, Farnia P. Contribution of Proteins to Membrane and Cell Wall Structures in Mycobacterium tuberculosis. Proteins in Mycobacterium Tuberculosis: Functions and Therapeutic Advances. Springer; 2025. p. 95–126.

Flores J, Cancino JC, Chavez-Galan L. Lipoarabinomannan as a point-of-care assay for diagnosis of tuberculosis: how far are we to use it? Frontiers in microbiology 2021;12:638047.

Forrellad MA, Klepp LI, Gioffré A, Sabio y Garcia J, Morbidoni HR, Santangelo MDIP, et al. Virulence factors of the Mycobacterium tuberculosis complex. Virulence 2013;4(1):3–66.

Gcebe N, Michel A, Gey van Pittius NC, Rutten V. Comparative genomics and proteomic analysis of four non-tuberculous Mycobacterium species and Mycobacterium tuberculosis complex: occurrence of shared immunogenic proteins. Frontiers in microbiology 2016;7:795.

Githui WA, Wilson SM, Drobniewski FA. Specificity of IS 6110-based DNA fingerprinting and diagnostic techniques for Mycobacterium tuberculosis complex. Journal of Clinical Microbiology 1999;37(4):1224–6.

Gonzalo-Asensio J, Pérez I, Aguiló N, Uranga S, Picó A, Lampreave C, et al. New insights into the transposition mechanisms of IS 6110 and its dynamic distribution between Mycobacterium tuberculosis complex lineages. PLoS Genetics 2018;14(4):e1007282.

Guinn KM, Hickey MJ, Mathur SK, Zakel KL, Grotzke JE, Lewinsohn DM, et al. Individual RD1-region genes are required for export of ESAT-6/CFP-10 and for virulence of Mycobacterium tuberculosis. Molecular Microbiology 2004;51(2):359–70. Available from https://doi.org/10.1046/j.1365-2958.2003.03844.x.

Gómez-González PJ, Grabowska AD, Tientcheu LD, Tsolaki AG, Hibberd ML, Campino S, et al. Functional genetic variation in pe/ppe genes contributes to diversity in Mycobacterium tuberculosis lineages and potential interactions with the human host. Frontiers in microbiology 2023;14:1244319.

Hamasur B, Bruchfeld J, Haile M, Pawlowski A, Bjorvatn B, Källenius G, et al. Rapid diagnosis of tuberculosis by detection of mycobacterial lipoarabinomannan in urine. Journal of Microbiological Methods 2001;45(1):41–52.

He Y, Tu X, Xue Y, Chen Y, Ye B, Li X, et al. CRISPR screening redefines therapeutic target identification and drug discovery with precision and scalability. Journal of Pharmaceutical Analysis 2025;16(2):101357. Available from https://doi.org/10.1016/j.jpha.2025.101357.

Jafari C, Thijsen S, Sotgiu G, Goletti D, Benítez JAD, Losi M, et al. Bronchoalveolar lavage enzyme-linked immunospot for a rapid diagnosis of tuberculosis: a Tuberculosis Network European Trialsgroup study. American Journal of Respiratory and Critical Care Medicine 2009;180(7):666–73.

Kalscheuer R, Palacios A, Anso I, Cifuente J, Anguita J, Jacobs WR, et al. The Mycobacterium tuberculosis capsule: a cell structure with key implications in pathogenesis. Biochemical Journal 2019;476(14):1995–2016.

Kawai T, Akira S. The roles of TLRs, RLRs and NLRs in pathogen recognition. International Immunology 2009;21(4):317–37. Available from https://doi.org/10.1093/intimm/dxp017.

Kim SH, Shin JH. Identification of nontuberculous mycobacteria using multilocous sequence analysis of 16S rRNA, hsp65, and rpoB. Journal of Clinical Laboratory Analysis 2018;32(1):e22184.

Kostyuk S, Smirnova T, Kameneva L, Porokhovnik L, Speranskij A, Ershova E, et al. GC-Rich extracellular DNA induces oxidative stress, double-strand DNA breaks, and DNA damage response in human adipose-derived mesenchymal stem cells. Oxidative Medicine and Cellular Longevity 2015;2015(1):782123.

Liu S, Wu M, Wu EAS, Geng S, Li Z, Li M, et al. Factors associated with differential T cell responses to antigens ESAT-6 and CFP-10 in pulmonary tuberculosis patients. Medicine 2021;100(8):e24615. Available from https://doi.org/10.1097/md.0000000000024615.

Lyu L, Li Z, Pan L, Jia H, Sun Q, Liu Q, et al. Evaluation of digital PCR assay in detection of M. tuberculosis IS6110 and IS1081 in tuberculosis patients plasma. BMC Infectious Diseases 2020;20(1):657.

Mao ZQ, Zhang QL, Zheng H, Liu ZQ, Li FM, Xu LS, et al. RpoB mutation patterns in Rifampicin-resistant tuberculosis: a Jiangxi Province study, 2021–2023. Scientific Reports 2025;15(1):27988. Available from https://doi.org/10.1038/s41598-025-11949-0.

McEvoy CR, van Pittius NCG, Victor TC, van Helden PD, Warren RM. The role of IS6110 in the evolution of Mycobacterium tuberculosis. Tuberculosis 2007;87(5):393–404.

Millington KA, Fortune SM, Low J, Garces A, Hingley-Wilson SM, Wickremasinghe M, et al. Rv3615c is a highly immunodominant RD1 (Region of Difference 1)-dependent secreted antigen specific for Mycobacterium tuberculosis infection. Proceedings of the National Academy of Sciences 2011;108(14):5730–5.

Moström P, Gordon M, Sola C, Ridell M, Rastogi N. Methods used in the molecular epidemiology of tuberculosis. Clinical microbiology and infection 2002;8(11):694–704.

Mousavi-Sagharchi SMA, Afrazeh E, Seyyedian-Nikjeh SF, Meskini M, Doroud D, Siadat SD. New insight in molecular detection of Mycobacterium tuberculosis. Amb Express 2024;14(1):74.

Mwaura M, Engel N. Constructing confidence: User perspectives on AlereLAM testing for tuberculosis. International Journal of Infectious Diseases 2021;112:237–42. Available from https://doi.org/10.1016/j.ijid.2021.09.018.

Pacl HT, Reddy VP, Saini V, Chinta KC, Steyn AJC. Host-pathogen redox dynamics modulate Mycobacterium tuberculosis pathogenesis. Pathogens and Disease 2018;76(5). Available from https://doi.org/10.1093/femspd/fty036.

Passos BBS, Araújo-Pereira M, Vinhaes CL, Amaral EP, Andrade BB. The role of ESAT-6 in tuberculosis immunopathology. Frontiers in Immunology 2024;15:1383098. Available from https://doi.org/10.3389/fimmu.2024.1383098.

Perdigão J, Silva H, Machado D, Macedo R, Maltez F, Silva C, et al. Unraveling Mycobacterium tuberculosis genomic diversity and evolution in Lisbon, Portugal, a highly drug resistant setting. BMC Genomics 2014;15(1):991.

Phelan JE, Coll F, Bergval I, Anthony RM, Warren R, Sampson SL, et al. Recombination in pe/ppe genes contributes to genetic variation in Mycobacterium tuberculosis lineages. BMC Genomics 2016;17(1):151.

Rahlwes KC, Dias BR, Campos PC, Alvarez-Arguedas S, Shiloh MU. Pathogenicity and virulence of Mycobacterium tuberculosis. Virulence 2023;14(1):2150449.

Ruhwald M, de Thurah L, Kuchaka D, Zaher MR, Salman AM, Abdel-Ghaffar A-R, et al. Introducing the ESAT-6 free IGRA, a companion diagnostic for TB vaccines based on ESAT-6. Scientific Reports 2017;7(1):45969. Available from https://doi.org/10.1038/srep45969.

Satam H, Joshi K, Mangrolia U, Waghoo S, Zaidi G, Rawool S, et al. Next-generation sequencing technology: current trends and advancements. Biology 2023;12(7):997.

Solo ES, Nakajima C, Kaile T, Bwalya P, Mbulo G, Fukushima Y, et al. Mutations in rpoB and katG genes and the inhA operon in multidrug-resistant *Mycobacterium tuberculosis* isolates from Zambia. Journal of Global Antimicrobial Resistance 2020;22:302–7. Available from https://doi.org/10.1016/j.jgar.2020.02.026.

Sun L, Wu J, Du F, Chen X, Chen ZJ. Cyclic GMP-AMP synthase is a cytosolic DNA sensor that activates the type I interferon pathway. Science (New York, N.Y.) 2013;339(6121):786–91. Available from https://doi.org/10.1126/science.1232458.

Tawfick MM, Badawy MSE, Taleb MH, Menofy NGE. Tuberculosis diagnosis and detection of drug resistance: a comprehensive updated review. Journal of Pure & Applied Microbiology 2023;17(4).

Torres JN, Paul LV, Rodwell TC, Victor TC, Amallraja AM, Elghraoui A, et al. Novel katG mutations causing isoniazid resistance in clinical *M. tuberculosis* isolates. Emerging microbes & infections 2015;4(1):1–9.

Van der Roest BR, Bootsma MCJ, Fischer EAJ, Gröschel MI, Anthony RM, de Zwaan R, et al. Phylodynamic assessment of SNP distances from whole genome sequencing for determining Mycobacterium tuberculosis transmission. Scientific Reports 2025;15(1):10694. Available from https://doi.org/10.1038/s41598-025-94646-2.

Vandecraen J, Chandler M, Aertsen A, Van Houdt R. The impact of insertion sequences on bacterial genome plasticity and adaptability. Critical Reviews in Microbiology 2017;43(6):709–30.

Wang Y, Shi Q, Chen Q, Zhou X, Yuan H, Jia X, et al. Emerging advances in identifying signal transmission molecules involved in the interaction between *Mycobacterium tuberculosis* and the host. Frontiers in Cellular and Infection Microbiology 2022;12:956311.

Warrier T. Novel small molecules targeting Ag85C, mycolyl transferase of Mycobacterium tuberculosis. Berlin: Humboldt-Univ., Diss.; 2010. Available from https://doi.org/10.18452/16160.

Watson RO, Bell SL, MacDuff DA, Kimmey JM, Diner EJ, Olivas J, et al. The cytosolic sensor cGAS detects *Mycobacterium tuberculosis* DNA to induce type I interferons and activate autophagy. Cell Host & Microbe 2015;17(6):811–9. Available from https://doi.org/10.1016/j.chom.2015.05.004.

Weldingh K, Andersen P. ESAT-6/CFP10 skin test predicts disease in *M. tuberculosis*-infected guinea pigs. PLoS One 2008;3(4):e1978. Available from https://doi.org/10.1371/journal.pone.0001978.

Yang J, Zhang L, Qiao W, Luo Y. *Mycobacterium tuberculosis*: pathogenesis and therapeutic targets. MedComm 2023;4(5):e353.

Zhang Z, Dong L, Li X, Deng T, Wang Q. The PE/PPE family proteins of *Mycobacterium tuberculosis*: evolution, function, and prospects for tuberculosis control. Frontiers in Immunology 2025;16:1606311. Available from https://doi.org/10.3389/fimmu.2025.1606311.

CHAPTER 4

Conventional diagnostic approaches for *Mycobacterium tuberculosis* detection

Mansi Chaturvedi and Chetna Dhand
Academic of Scientific and Innovative Research (AcSIR), CSIR-Advanced Materials and Processes Research Institute, Bhopal, Madhya Pradesh, India

4.1 Introduction

Tuberculosis (TB) infections may remain infectious despite the absence of symptoms, thereby sustaining transmission and contributing to persistence within communities {Srivastava, 2015 #370}. Therefore, early detection of *Mycobacterium tuberculosis* (MTB) infection is crucial for preventing continued presence of the disease and for improving the likelihood of successful treatment outcomes {Rangaka, 2015 #371}. Early diagnosis is essential not only for enhancing the well-being of patients already affected by the disease, but also for reducing the risk of transmission to others. While identifying individuals with suspected TB is important, early diagnosis is even more critical for effective disease control. However, despite the availability of diagnostic options, numerous barriers continue to hinder the timely diagnosis of TB, particularly in resource-limited settings. Major impediments include the costs associated with traveling to diagnostic facilities and limited access to health care services, which often result in delays in confirming TB infection in affected patients. Commonly used diagnostic methods for TB include sputum smear microscopy (SSM), culture-based techniques, and radiographic assessments (Fig. 4.1) {Ryu, 2015 #372}. SSM remains the most widely used and cost-effective diagnostic approach, employing Ziehl–Neelsen (ZN) and fluorescence staining techniques because of their simplicity and ease of implementation {Datta, 2017 #373}. However, these methods exhibit relatively low sensitivity, particularly in cases of extrapulmonary TB and in patients with low bacillary burden. Culture-based diagnostics are regarded as the gold standard for identifying of MTB due to their higher sensitivity compared with SSM {Chopra, 2020 #374}. In addition, culture methods allow subsequent drug susceptibility testing. Both solid culture systems (e.g., Löwenstein–Jensen medium) and liquid culture techniques (e.g., MGIT) are effective for MTB detection; however, the slow growth rate of the organism renders culture-based diagnostics time-consuming, often requiring several weeks to obtain definitive results. Radiographic techniques, such as chest X-rays (CXR), can reveal lung abnormalities suggestive of TB, but these findings are not specific to the disease. Overall, the

Advances in Biosensor Technology for Mycobacterium Tuberculosis Detection
DOI: https://doi.org/10.1016/B978-0-443-44589-7.00011-X

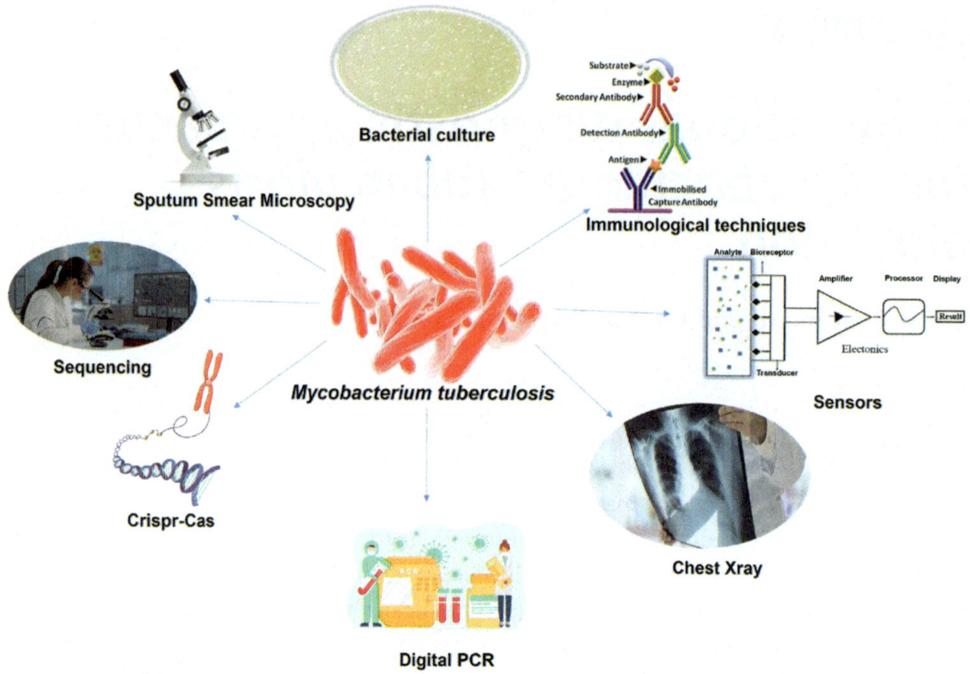

Figure 4.1 Summarize the common conventional diagnostic methods for *Mycobacterium tuberculosis* (MTB) detection.

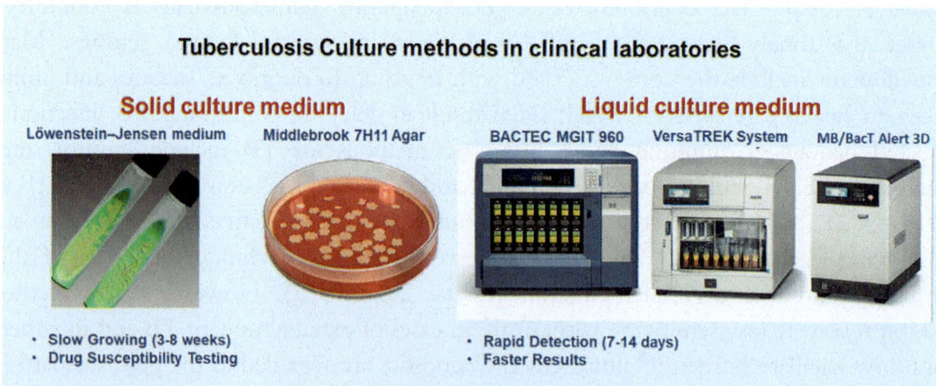

Figure 4.2 Schematic overview of tuberculosis (TB) diagnostic approaches highlighting solid culture and liquid culture medium for *Mycobacterium tuberculosis* (MTB) detection.

limitations of traditional TB diagnostic methods, including low sensitivity in smear-negative cases, prolonged turnaround times, and dependence on specialized laboratory infrastructure, underscore the urgent need for diagnostic approaches that are faster, more sensitive, and less reliant on centralized laboratory facilities (Fig. 4.2).

4.2 Culture method

One of the principal diagnostic techniques for detectingMTB in clinical samples is culture-based diagnosis, which is regarded as the gold standard because it enables sensitive detection, accurate species identification, and phenotypic drug susceptibility testing (Gholoobi et al., 2014). Despite the slow growth rate of MTB, culture methods are capable of detecting as few as 10^2 bacilli per mL of sputum, making them highly reliable confirmatory tools. In clinical laboratories, both solid and liquid culture systems are routinely employed. MTB is commonly cultivated on solid media, which not only support bacterial growth but also facilitate species identification and drug susceptibility testing, thereby guiding clinicians in selecting appropriate antibacterial therapy (Kenaope et al., 2020). Hard-to-prepare solid media, including agar-based formulations such as Middlebrook 7H11 (Adúriz et al., 1995) and egg-based media such as Lowenstein–Jensen (Yang et al., 2021) are commonly used. In contrast, liquid culture systems—such as the BACTEC MGIT 960, VersaTREK, and MB/BacT Alert 3D—as well as liquid broth media such as Middlebrook 7H12 (Siddiqi et al., 1988) allow more rapid detection of MTB, often within a few days, which is significantly faster than growth on conventional solid media. In addition, culture media are supplemented with antibiotics to enhance selectivity for mycobacteria and suppress the growth of contaminating microorganisms. (Fig. 4.3).

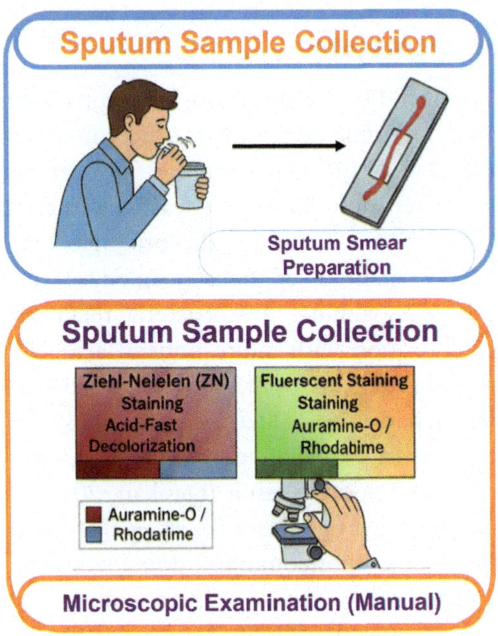

Figure 4.3 Schematic representation of sputum smear microscopy (SSM).

While solid media typically require 3–8 weeks for visible colony development, liquid broth cultures significantly reduce detection time, with an average of approximately 10 days for smear-positive specimens and 20 days for smear-negative cases. In contrast, solid media offer distinct advantages, including the ability to assess colony morphology, detect mixed cultures, and identify strains that grow only on solid substrates (Russell, 2001). Although culture is a highly sensitive and important diagnostic method, it has significant limitations, including the need for specialized laboratory infrastructure and prolonged turnaround times, which may extend up to 2–4 weeks to obtain conclusive results (Rahman et al., 2011; de Steenwinkel et al., 2012). The BACTEC MGIT 960 system detects the growth of MTB by monitoring oxygen-quenched fluorescence, which increases as bacterial proliferation occurs within the culture tube. Studies have demonstrated that the MGIT 960 system enables faster detection and earlier TB diagnosis than conventional Löwenstein–Jensen solid culture media (Hasan et al., 2013). Similarly, the VersaTREK system detects mycobacterial growth by monitoring pressure changes above the broth surface that result from microbial metabolic activity (Espasa et al., 2012). Conversely, the MB/BacT Alert 3D system employs a colorimetric CO_2 sensor to detect bacterial growth, providing another reliable liquid culture diagnostic option (Piersimoni et al., 2001). Although liquid culture procedures have been substantially improved, the intrinsically slow growth rate of the MTB complex means that positive cultures typically require at least one week for detection, whereas confirmation of negative results may take up to eight weeks (Lee et al., 2003).

4.3 Sputum smear microscopy

For pulmonary tuberculosis (PTB), SSM remains the primary diagnostic method and is the most commonly preferred initial test because of its robustness and rapid turnaround time (Steingart et al., 2006). In many remote, resource-constrained settings in developing countries, it remains the only feasible diagnostic method available (Ben-Selma et al., 2009). SSM is based on the principle of acid-fast staining, most commonly using the ZN method or fluorescent dyes such as auramine O or auramine–rhodamine. In ZN staining, mycobacteria retain the primary dye, carbol fuchsin, even after decolorization with dilute mineral acid. This property, known as acid-fastness, allows mycobacteria to resist acid–alcohol treatment and to be distinguished microscopically (Acharya et al., 2020). Nevertheless, ZN smear microscopy has low sensitivity, as a high bacillary load of approximately 5,000 to 10,000 CFU/mL of sputum is required to obtain a positive result (Desikan, 2012).

This limitation contributes to a significant false-negative rate and increases the risk of misdiagnosis. To address these shortcomings, several advances in microscopy-based diagnostic approaches have been developed to enhance detection performance. Automated SSM systems, such as the ZEISS Axio Scan, have demonstrated improved efficiency, achieving a sensitivity of 97.06% and a specificity of 86.44% for the automated detection and enumeration of acid-fast bacilli (Zingue et al., 2018; Zingue et al., 2018). In addition, laser confocal microscopy combined with specific fluorescent antibody labeling has further enhanced the identification of MTB within lung tissues, particularly in cases where ZN staining yields weak bacillary signals (Erokhina et al., 2019). Moreover, digital pathology tools such as the Pat-Scan program enable automated quantification of bacilli in paraffin-embedded ZN-stained tissue sections, contributing to reduced diagnostic turnaround time (Sua et al., 2021). Despite these advances, SSM still has significant limitations. It cannot distinguish between viable and nonviable bacilli, nor can it differentiate MTBs from nontuberculous mycobacteria. However, ZN staining can provide useful insights into the acid-fastness of bacilli, particularly in patients undergoing antiTB therapy, a feature that is generally lacking in fluorescence-based SSM techniques.

4.4 Chest radiography

X-rays are a form of electromagnetic radiation produced by an X-ray tube and used in radiography to visualize internal anatomical structures (Kotzé et al., 2012; Hessenbruch, 2002). Chest radiography, commonly referred to as CXR, is a rapid and widely used imaging modality for detecting pulmonary abnormalities. Although CXR demonstrates high sensitivity for identifying lung changes suggestive of PTB, its specificity is relatively low. Consequently, while bacteriological methods remain the cornerstone of modern TB diagnosis, CXR continues to play an important role in screening individuals who require microbiological testing and in evaluating cases in which bacteriological results are inconclusive. However, limited access to high-quality radiographic facilities in many regions remains a significant challenge (Schalekamp et al., 2014).

4.5 Pure protein derivative test/tuberculin skin test

The purified protein derivative (PPD) skin test, also known as the TST or Mantoux test is based on a delayed-type (type IV) hypersensitivity reaction to tuberculin antigens (Starke, 1993). First introduced by Robert Koch in 1890 (Raj, 2021) and later refined by Charles Mantoux (Nayak and Acharjya, 2012) through the intradermal administration technique, the test continues to be widely used in many countries, including India, as an initial screening tool for exposure to MTB. In this test, PPD is administered intradermally, and the resulting induration at the injection site is measured after 48–72 hours (Hasan et al., 2024). Generally, an induration of ≥5 mm is considered a positive result. The tuberculin used is a PPD extracted from MTB, typically standardized as PPD-S. The reaction is mediated by immune cell infiltration, leading to localized inflammation, with the peak response occurring approximately 24 hours after injection.

The procedure requires at least two patient visits: one for intradermal administration of the test and a second for interpretation of the results by trained health care professionals. Although the TST is simple to perform, it has several limitations. False-positive results may occur in individuals who have received the Bacille Calmette-Guérin (BCG) vaccine or those exposed to nontuberculous mycobacteria, whereas immunocompromised individuals often exhibit reduced test sensitivity (Acharya et al., 2020; Bianchi et al., 2009). Although the PPD test plays an important supportive role in the diagnosis of TB, particularly in pediatric populations, it cannot reliably differentiate between prior BCG vaccination and true MTB infection, nor does it yield dependable results in immunosuppressed individuals.

4.6 Interferon-γ release assay

Interferon-gamma release assays (IGRAs) were approved by the U.S. Food and Drug Administration in 2001 for the diagnosis of MTB infection (Parrey et al., 2023). IGRAs detect T-cell–mediated interferon-γ (IFN-γ) production following stimulation of whole blood with MTB–specific antigens, including ESAT-6, CFP-10, and TB7.7 (included in some assay formats). Two principal commercial formats are available: the enzyme-linked immunospot (ELISPOT) assay, exemplified by the T-SPOT® TB test, which enumerates antigen-specific IFN-γ–producing T cells; and the enzyme-linked immunosorbent assay (ELISA), such as QuantiFERON®-TB (QFT-Gold and QFT-Gold In-Tube), which measures the concentration of IFN-γ released into plasma. Because BCG vaccine strains lack ESAT-6 and CFP-10, prior BCG vaccination does not influence IGRA results. However, IGRAs are unable to distinguish between latent tuberculosis infection (LTBI) and active disease. Multiple systematic reviews have shown that IGRAs do not demonstrate superior diagnostic accuracy than the TST for detecting latent or active TB in HIV-positive or severely immunocompromised individuals. Consequently, the World Health Organization recommends against the use of IGRAs for the diagnosis of active pulmonary TB in low- and middle-income countries with a high TB burden (Wang et al., 2013). Although IGRAs provide higher specificity than the TST and serve as valuable blood-based alternatives for detecting MTB exposure, their use is limited by the need for venous blood collection, higher costs, laboratory infrastructure requirements, and reduced diagnostic performance in children, HIV-infected individuals, and other immunocompromised populations (Smith et al., 2011).

4.7 Xpert MTB/RIF

The GeneXpert MTB/RIF assay is one of the most widely implemented molecular diagnostic platforms for TB. It employs a seminested, real-time fluorescent polymerase chain reaction (PCR) assay to simultaneously detect MTB complex DNA and mutations associated with rifampicin resistance. The next-generation Xpert MTB/RIF Ultra assay incorporates two high-copy genetic targets and a larger reaction chamber, resulting in significantly enhanced analytical sensitivity, particularly in paucibacillary and smear-negative cases (World Health Organization, 2017). As a result, the limit of detection is markedly improved from approximately 112.6 CFU/mL for the Xpert MTB/RIF assay to 15.6 CFU/mL for the Xpert MTB/RIF Ultra. This cartridge-based system can analyze either raw or concentrated sputum samples using automated technology and delivers results within approximately 2 hours (Bodmer and Ströhle, 2012). In 2010, the World Health Organization endorsed Xpert MTB/RIF for the rapid diagnosis of TB and rifampicin resistance, with highest priority recommended for people living with HIV and individuals suspected of multidrug-resistant TB (Weyer et al., 2013). Despite its substantial diagnostic advantages, widespread implementation of this technology is challenged by requirements for trained laboratory personnel, reliable power supply, controlled operating conditions, and ongoing maintenance support, which can limit accessibility in primary health care settings and resource-constrained environments (Gidado et al., 2019).

4.8 Nucleic acid amplification testing

Another important direct diagnostic approach is nucleic acid amplification testing (NAAT), which enables the rapid and precise detection of MTB in clinical specimens (Baron et al., 2018). Although NAATs are fast and highly sensitive, they are not considered replacements for clinical evaluation, smear microscopy, or culture-based diagnosis of TB (Al-Zamel, 2009). A key advantage of NAAT is its ability to detect mycobacterial DNA even in the absence of viable organisms, making it particularly useful in cases in which prior antimicrobial therapy has reduced bacterial viability (Whiley et al., 2008). However, NAATs also have limitations, including relatively complex protocols, longer turnaround times than newer rapid assays, and the need for larger reaction volumes and specialized laboratory expertise (Lee et al., 2019). Loop-mediated isothermal amplification (LAMP) is an advanced NAAT technique that employs a strand-displacing DNA polymerase and multiple specifically designed primers to amplify target DNA under isothermal conditions. The single-step LAMP (SS-LAMP) assay utilizes six primers targeting eight regions of the Mycobacterium tuberculosis complex (MTBC) specific insertion sequence IS6110, enabling direct detection of MTBC DNA in liquefied sputum samples (Bentaleb et al., 2016). In a study involving 157 suspected TB sputum specimens from Morocco, SS-LAMP demonstrated a sensitivity of 82.93% and a specificity of 99.14% compared with conventional Löwenstein–Jensen culture. Owing to its rapid turnaround time, high accuracy, and minimal equipment requirements, the LAMP method is particularly well suited for implementation in resource-limited settings.

4.9 Digital PCR

Digital PCR (dPCR) is an advanced nucleic acid quantification technique that enables absolute quantification of target DNA molecules without the need for a standard calibration curve, even when only minimal amounts of target DNA are present (Kuypers and Jerome, 2017). This method offers exceptionally high analytical sensitivity and accuracy, with the capability to detect single-copy DNA molecules (Nyaruaba et al., 2019). dPCR is applicable to a wide range of clinical sample types, including sputum, blood, formalin-fixed, paraffin-embedded tissues, and exhaled breath, and can also support drug susceptibility testing (Surat et al., 2014; Ushio et al., 2016; Patterson et al., 2018; Luo et al., 2019; Cho et al., 2020; Cao et al., 2020). Detection of MTB using dPCR commonly targets the insertion sequence IS6110; however, multiplexing with IS1081 has been shown to further enhance sensitivity, particularly improving diagnostic accuracy in smear-negative TB cases compared with conventional quantitative PCR (qPCR) assays (Lyu et al., 2020). Overall, dPCR demonstrates strong diagnostic performance in pulmonary and extrapulmonary TB, as well as in latent TB infection. Nevertheless, its specificity may be affected by technical challenges and limited expertise in low-resource laboratory settings (Nyaruaba et al., 2019).

4.10 Immuno-PCR

Immuno-PCR (I-PCR) exploits the high specificity of antibody–antigen interactions to amplify tuberculosis-associated antigens or circulating antibodies, achieving extremely high analytical sensitivity that allows detection in very small clinical samples and amplification of minute quantities of target antigens. Recent developments, including automated one-step I-PCR formats, have substantially reduced processing time, while the incorporation of magnetic beads (MBs) and gold nanoparticles (GNPs) has minimized background interference and improved assay efficiency, demonstrating strong clinical potential (Mehta et al., 2017). For example, the MB-GNP I-PCR assay enables early detection of the ESAT-6 antigen (Singh et al., 2018), while the GNP-RT-I-PCR system targeting CFP-10 has demonstrated sensitivities of 83.7% and 76.2% and specificities exceeding 93% in pulmonary and extrapulmonary TB, respectively (Dahiya et al., 2020). Similarly, the MB-AuNP-I-PCR system targeting MPT64 and CFP-10 has shown superior performance, particularly in smear-negative and extrapulmonary TB cases, compared with Magneto-ELISA and the GeneXpert MTB/RIF assay (Dahiya et al., 2020). Other nucleic acid-based immunoassays also contribute to rapid diagnosis and drug-resistance detection. PCR-ELISA can identify mutations associated with rifampicin and isoniazid resistance by targeting the rpoB, katG, and inhA genes (Zhou et al., 2020). Additionally, culture-independent, ultrasensitive ELISA employing thionicotinamide adenine dinucleotide cycling exhibits high analytical sensitivity (Watabe et al., 2014). As the MPT64 antigen is specifically released by viable MTB under heating conditions, this approach enables selective detection of live bacilli, achieving diagnostic performance comparable to smear microscopy and Xpert MTB/RIF (Wang et al., 2020; Cao et al., 2021).

4.11 Lateral flow urine lipoarabinomannan assay

Lipoarabinomannan (LAM) is a major lipopolysaccharide component of the MTB cell wall and functions as an important virulence factor, containing pathogen-specific structural epitopes (Sarkar et al., 2014). LAM-based diagnostics, particularly lateral flow LAM (LF-LAM) assays, demonstrate considerably higher sensitivity in HIV-positive individuals with low $CD4^+$ T-cell counts, owing to increased antigen burden in advanced immunosuppression. Consequently, the World Health Organization has endorsed LF-LAM as an adjunctive diagnostic tool for tuberculosis screening and diagnosis among HIV-infected populations (World Health Organization, 2015). However, despite their demonstrated utility in high-risk groups, LAM-based assays exhibit limited sensitivity in HIV-negative individuals and those in the early stages of HIV infection. Therefore, extensive clinical validation across larger and more diverse patient cohorts is required to define their diagnostic accuracy clearlyand establish the broader contexts in which these assays can be reliably applied in routine clinical practice (Sigal et al., 2018).

4.12 Conclusion and future prospects

Conventional diagnostic approaches have long formed the backbone of tuberculosis (TB) detection and control strategies worldwide. Techniques such as SSM, culture-based methods, chest radiography, and immunological tests continue to be widely employed because of their established reliability, relative affordability, and availability across diverse health care settings. Culture-based diagnosis remains the reference standard, offering high specificity and enabling drug susceptibility testing, which is essential for effective TB management.

Despite their importance, these conventional methods exhibit significant limitations that restrict their effectiveness in contemporary TB control efforts. Challenges such as low sensitivity in smear-negative and extrapulmonary TB cases, prolonged turnaround times associated with culture techniques, and dependence on skilled personnel and laboratory infrastructure often delay diagnosis and treatment initiation. These limitations are particularly problematic in high-burden and resource-limited regions, where delayed detection contributes to ongoing transmission, disease progression, and increased mortality. Consequently, while conventional diagnostic tools remain indispensable, they are insufficient on their own to meet the growing demand for rapid, accurate, and accessible TB diagnostics.

Looking ahead, the future of TB diagnosis lies in addressing these gaps through technological innovation and the strategic integration of emerging diagnostic platforms. Advances in molecular diagnostics, immunoassays, and biosensor-based technologies offer promising alternatives with improved sensitivity, reduced diagnostic time, and an enhanced capability to detect drug-resistant strains. The development of cost-effective, portable, and user-friendly point-of-care diagnostic tools is particularly critical for expanding access to timely diagnosis in decentralized and underserved health care settings.

Furthermore, combining conventional diagnostic approaches with next-generation technologies through integrated diagnostic algorithms may significantly enhance overall diagnostic accuracy and efficiency. Emerging fields such as nanotechnology, artificial intelligence-assisted imaging, and host biomarker discovery are expected to play a transformative role in shaping future TB diagnostic strategies. In summary, sustained improvement of conventional methods, coupled with the adoption of innovative diagnostic solutions, will be essential for early detection, effective treatment, and the long-term goal of tuberculosis elimination.

References

Acharya B, Acharya A, Gautam S, Ghimire SP, Mishra G, Parajuli N, et al. Advances in diagnosis of Tuberculosis: an update into molecular diagnosis of *Mycobacterium tuberculosis*. Molecular Biology Reports 2020;47(5):4065–75. Available from https://doi.org/10.1007/s11033-020-05413-7.

Adúriz JJ, Juste RA, Cortabarria N. Lack of mycobactin dependence of mycobacteria isolated on Middlebrook 7H11 from clinical cases of ovine paratuberculosis. Veterinary Microbiology 1995;45(2-3):211–7. Available from https://doi.org/10.1016/0378-1135(95)00037-b.

Al-Zamel FA. Detection and diagnosis of mycobac terium tuberculosis. Expert Review of Anti-Infective Therapy 2009;7(9):1099–108. Available from https://doi.org/10.1586/ERI.09.92.

Baron EJ, Tenover FC, Gnanashanmugam D. Direct Detection of *Mycobacterium tuberculosis* in Clinical Specimens Using Nucleic Acid Amplification Tests. Clinical Microbiology Newsletter 2018;40(13):107–12. Available from https://doi.org/10.1016/j.clinmicnews.2018.06.003.

B-Selma W, Ben-Kahla I, Marzouk M, Ferjeni A, Ghezal S, Ben-Said M, et al. Rapid detection of *Mycobacterium tuberculosis* in sputum by Patho-TB kit in comparison with direct microscopy and culture. Diagnostic Microbiology and Infectious Disease 2009;65(3):232–5. Available from https://doi.org/10.1016/j.diagmicrobio.2009.07.021.

Bentaleb EM, Abid M, El Messaoudi MD, Lakssir B, Ressami EM, Amzazi S, et al. Development and evaluation of an in-house single step loop-mediated isothermal amplification (SS-LAMP) assay for the detection of *Mycobacterium tuberculosis* complex in sputum samples from Moroccan patients. BMC Infectious Diseases 2016;16(1). Available from https://doi.org/10.1186/s12879-016-1864-9.

Bianchi L, Galli L, Moriondo M, Veneruso G, Becciolini L, Azzari C, et al. Interferon-Gamma Release Assay Improves the Diagnosis of Tuberculosis in Children. Pediatric Infectious Disease Journal 2009;28(6):510–4. Available from https://doi.org/10.1097/inf.0b013e31819abf6b.

Bodmer T, Ströhle A. Diagnosing Pulmonary Tuberculosis with the Xpert MTB/RIF Test. Journal of Visualized Experiments 2012(62). Available from https://doi.org/10.3791/3547.

Cao XJ, Li YP, Wang JY, Zhou J, Guo XG. MPT64 assays for the rapid detection of *Mycobacterium tuberculosis*. BMC Infectious Diseases 2021;21(1). Available from https://doi.org/10.1186/s12879-021-06022-w.

Cao Z, Wu W, Wei H, Gao C, Zhang L, Wu C, et al. Using droplet digital PCR in the detection of *Mycobacterium tuberculosis* DNA in FFPE samples. International Journal of Infectious Diseases 2020;99:77–83. Available from https://doi.org/10.1016/j.ijid.2020.07.045.

Cho SM, Shin S, Kim Y, Song W, Hong SG, Jeong SH, et al. A novel approach for tuberculosis diagnosis using exosomal DNA and droplet digital PCR. Clinical Microbiology and Infection 2020;26(7):942.e1. Available from https://doi.org/10.1016/j.cmi.2019.11.012.

Dahiya B, Prasad T, Singh V, Khan A, Kamra E, Mor P, et al. Diagnosis of Tuberculosis by Nanoparticle-Based Immuno-PCR Assay Based on Mycobacterial MPT64 and CFP-10 Detection. Nanomedicine: Nanotechnology, Biology, and Medicine 2020;15(26):2609–24. Available from https://doi.org/10.2217/nnm-2020-0258.

Desikan KV. Elimination of leprosy & possibility of eradication – the Indian scenario. Indian Journal of Medical Research 2012;135(1):3–5. Available from https://doi.org/10.4103/0971-5916.93415.

Erokhina MV, Lepekha LN, Voronezhskaya EE, Nezlin LP, Avdienko VG, Ergeshov AE. Application of laser scanning confocal microscopy for the visualization of M. tuberculosis in lung tissue samples with weak Ziehl–Neelsen staining. Journal of Clinical Medicine 2019;8(8). Available from https://doi.org/10.3390/jcm8081185.

Espasa M, Salvadó M, Vicente E, Tudó G, Alcaide F, Coll P, et al. Evaluation of the VersaTREK System Compared to the Bactec MGIT 960 System for First-Line Drug Susceptibility Testing of *Mycobacterium tuberculosis*. Journal of Clinical Microbiology 2012;50(2):488–91. Available from https://doi.org/10.1128/jcm.06432-11.

Gholoobi A, Masoudi-Kazemabad A, Meshkat M, Meshkat Z. Comparison of Culture and PCR Methods for Diagnosis of *Mycobacterium tuberculosis* in Different Clinical Specimens. Jundishapur Journal of Microbiology 2014;7(2). Available from https://doi.org/10.5812/jjm.8939.

Gidado M, Nwokoye N, Ogbudebe C, Nsa B, Nwadike P, Ajiboye P, et al. Assessment of GeneXpert MTB/RIF performance by type and level of health-care facilities in Nigeria. Nigerian Medical Journal 2019;60(1):33. Available from https://doi.org/10.4103/nmj.nmj_12_19.

Hasan M, Munshi SK, Banu Momi MS, Rahman F, Noor R. Evaluation of the effectiveness of BACTEC MGIT 960 for the detection of mycobacteria in Bangladesh. International Journal of Mycobacteriology 2013;2(4):214–9. Available from https://doi.org/10.1016/j.ijmyco.2013.09.001.

Hasan MR, Anzar N, Sharma P, Singh S, Hassan H, Rawat C, et al. *Mycobacterium tuberculosis* diagnosis from conventional to biosensor-a systematic review. International Journal of Environmental Analytical Chemistry 2024;104(18):6519–34. Available from https://doi.org/10.1080/03067319.2022.2147427.

Hessenbruch A. A brief history of x-rays. Endeavour 2002;26(4):137–41. Available from https://doi.org/10.1016/s0160-9327(02)01465-5.

Kenaope L, Ferreira H, Seedat F, Otwombe K, Martinson NA, Variava E. Sputum culture and drug sensitivity testing outcome among X-pert *Mycobacterium tuberculosis*/rifampicin-positive, rifampicin-resistant sputum: A retrospective study — Not all rifampicin resistance is multi-drug resistant. Journal of Global Antimicrobial Resistance 2020;21:434–8. Available from https://doi.org/10.1016/j.jgar.2019.11.008.

Kotzé SH, Mole CG, Greyling LM. The translucent cadaver: an evaluation of the use of full body digital X-ray images and drawings in surface anatomy education. Anatomical sciences education 2012;5(5):287–94. Available from https://doi.org/10.1002/ase.1277.

Kuypers J, Jerome KR. Applications of digital PCR for clinical microbiology. Journal of Clinical Microbiology 2017;55(6):1621–8. Available from https://doi.org/10.1128/JCM.00211-17.

Lee JJ, Suo J, Lin CB, Wang JD, Lin TY, Tsai YC. Comparative evaluation of the BACTEC MGIT 960 system with solid medium for isolation of mycobacteria. International Journal of Tuberculosis and Lung Disease 2003;7(6):569–74.

Lee SH, Park SM, Kim BN, Kwon OS, Rho WY, Jun BH. Emerging ultrafast nucleic acid amplification technologies for next-generation molecular diagnostics. Biosensors and Bioelectronics 2019;141. Available from https://doi.org/10.1016/j.bios.2019.111448.

Luo J, Luo M, Li J, Yu J, Yang H, Yi X, et al. Rapid direct drug susceptibility testing of *Mycobacterium tuberculosis* based on culture droplet digital polymerase chain reaction. The International Journal of Tuberculosis and Lung Disease 2019;23(2):219–25. Available from https://doi.org/10.5588/ijtld.18.0182.

Lyu L, Li Z, Pan L, Jia H, Sun Q, Liu Q, et al. Evaluation of digital PCR assay in detection of M.tuberculosis IS6110 and IS1081 in tuberculosis patients plasma. BMC Infectious Diseases 2020;20(1). Available from https://doi.org/10.1186/s12879-020-05375-y.

Mehta PK, Dahiya B, Sharma S, Singh N, Dharra R, Thakur Z, et al. Immuno-PCR, a new technique for the serodiagnosis of tuberculosis. Journal of Microbiological Methods 2017;139:218–29. Available from https://doi.org/10.1016/j.mimet.2017.05.009.

Nayak S, Acharjya B. Mantoux test and its interpretation. Indian Dermatology Online Journal 2012;3(1):2. Available from https://doi.org/10.4103/2229-5178.93479.

Nyaruaba R, Mwaliko C, Kering KK, Wei H. Droplet digital PCR applications in the tuberculosis world. Tuberculosis 2019;117:85–92. Available from https://doi.org/10.1016/j.tube.2019.07.001.

Parrey AH, Koka M, Assad M, Ismail M. Tuberculid caused by atypical mycobacteria in an immunocompetent patient: a rare case report. Infekcionnye bolezni 2023;21(4):106–8. Available from 10.20953/1729-9225-2023-4-106-108.

Patterson B, Morrow C, Singh V, Moosa A, Gqada M, Woodward J, et al. Detection of *Mycobacterium tuberculosis* bacilli in bio-aerosols from untreated TB patients. Gates open research 2018;1:11.

Piersimoni C, Scarparo C, Callegaro A, Tosi CP, Nista D, Bornigia S, et al. Comparison of MB/BacT ALERT 3D System with Radiometric BACTEC System and Löwenstein-Jensen Medium for Recovery and Identification of Mycobacteria from Clinical Specimens: a Multicenter Study. Journal of Clinical Microbiology 2001;39(2):651–7. Available from https://doi.org/10.1128/jcm.39.2.651-657.2001.

Rahman F, Munshi SK, Kamal SMM, Rahman ASMM, Rahman MM, Noor R. Comparison of different microscopic methods with conventional TB culture. Stamford Journal of Microbiology 2011;1(1):46–50. Available from https://doi.org/10.3329/sjm.v1i1.9133.

Raj GM. Antimycobacterial Drugs. In: Paul A, Anandabaskar N, Mathaiyan J, Raj GM, editors. Introduction to Basics of Pharmacology and Toxicology. Singapore: Springer; 2021. p. 869–904. Available from https://doi.org/10.1007/978-981-33-6009-9_56.

Russell DG. *Mycobacterium tuberculosis*: here today, and here tomorrow. Nature Reviews. Molecular Cell Biology 2001;2(8):569–78. Available from https://doi.org/10.1038/35085034.

Sarkar P, Biswas D, Sindhwani G, Rawat J, Kotwal A, Kakati B. Application of lipoarabinomannan antigen in tuberculosis diagnostics: current evidence. Postgraduate Medical Journal 2014;90(1061):155–63. Available from https://doi.org/10.1136/postgradmedj-2013-132053.

Schalekamp S, van Ginneken B, Karssemeijer N, Schaefer-Prokop CM. Chest radiography: new technological developments and their applications. Seminars in Respiratory and Critical Care Medicine 2014;35(01):003–16. Available from https://doi.org/10.1055/s-0033-1363447.

Siddiqi SH, Libonati JP, Carter ME, Hooper NM, Baker JF, Hwangbo CC, et al. Enhancement of mycobacterial growth in middlebrook 7H12 medium by polyoxyethylene stearate. Current Microbiology 1988;17(2):105–10. Available from https://doi.org/10.1007/bf01568794.

Sigal GB, Pinter A, Lowary TL, Kawasaki M, Li A, Mathew A, et al. A novel sensitive immunoassay targeting the 5-methylthio-D-xylofuranose–lipoarabinomannan epitope meets the WHO's performance target for tuberculosis diagnosis. Journal of Clinical Microbiology 2018;56(12). Available from https://doi.org/10.1128/JCM.01338-18.

Singh N, Dahiya B, Radhakrishnan VS, Prasad T, Mehta PK. Detection of *Mycobacterium tuberculosis* purified ESAT-6 (Rv3875) by magnetic bead-coupled gold nanoparticle-based immuno-PCR assay. International Journal of Nanomedicine 2018;13:8523–35. Available from https://doi.org/10.2147/IJN.S181052.

Smith R, Cattamanchi A, Steingart KR, Denkinger C, Dheda K, Winthrop KL, et al. Interferon-gamma release assays for diagnosis of latent tuberculosis infection: evidence in immune-mediated inflammatory disorders. Current Opinion in Rheumatology 2011;23(4):377–84. Available from https://doi.org/10.1097/BOR.0b013e3283474d62.

Starke JR. The tuberculin skin test. Pediatric Annals 1993;22(10):612–20. Available from https://doi.org/10.3928/0090-4481-19931001-10.

de Steenwinkel JEM, de Knegt GJ, ten Kate MT, Verbrugh HA, Ottenhoff THM, Bakker-Woudenberg IAJM. Dynamics of interferon-gamma release assay and cytokine profiles in blood and respiratory tract specimens from mice with tuberculosis and the effect of therapy. Journal of Clinical Microbiology & Infectious Diseases 2012;31(6):1195–201. Available from https://doi.org/10.1007/s10096-011-1428-2.

Steingart KR, Ng V, Henry M, Hopewell PC, Ramsay A, Cunningham J, et al. Sputum processing methods to improve the sensitivity of smear microscopy for tuberculosis: a systematic review. Lancet Infectious Diseases 2006;6(10):664–74. Available from https://doi.org/10.1016/S1473-3099(06)70602-8.

Sua LF, Bolaños JE, Maya J, Sánchez A, Medina G, Zúñiga-Restrepo V, et al. Detection of mycobacteria in paraffin-embedded Ziehl–Neelsen-Stained tissues using digital pathology. Tuberculosis 2021;126:102025. Available from https://doi.org/10.1016/j.tube.2020.102025.

Surat G, Wallace WA, Laurenson IF, Seagar AL. Rapid real-time PCR for detection of *Mycobacterium tuberculosis* complex DNA in formalin-fixed paraffin embedded tissues: 16% of histological 'sarcoid' may contain such DNA. Journal of Clinical Pathology 2014;67(12):1084–7. Available from https://doi.org/10.1136/jclinpath-2014-202307.

Ushio R, Yamamoto M, Nakashima K, Watanabe H, Nagai K, Shibata Y, et al. Digital PCR assay detection of circulating *Mycobacterium tuberculosis* DNA in pulmonary tuberculosis patient plasma. Tuberculosis 2016;99:47–53. Available from https://doi.org/10.1016/j.tube.2016.04.004.

Wang SQ, Inci F, De Libero G, Singhal A, Demirci U. Point-of-care assays for tuberculosis: role of nanotechnology/microfluidics. Biotechnology Advances 2013;31(4):438–49. Available from https://doi.org/10.1016/j.biotechadv.2013.01.006.

Wang WH, Takeuchi R, Jain SH, Jiang YH, Watanuki S, Ohtaki Y, et al. A novel, rapid (within hours) culture-free diagnostic method for detecting live *Mycobacterium tuberculosis* with high sensitivity. EBioMedicine 2020;60. Available from https://doi.org/10.1016/j.ebiom.2020.103007.

Watabe S, Kodama H, Kaneda M, Morikawa M, Nakaishi K, Yoshimura T, et al. Ultrasensitive enzyme-linked immunosorbent assay (ELISA) of proteins by combination with the thio-NAD cycling method. Biophysics 2014;10(0):49–54. Available from https://doi.org/10.2142/biophysics.10.49.

Weyer K, Mirzayev F, Migliori GB, Van Gemert W, D'Ambrosio L, Zignol M, et al. Rapid molecular TB diagnosis: evidence, policy making and global implementation of Xpert MTB/RIF. Respiratory Journal 2013;42(1):252–71. Available from https://doi.org/10.1183/09031936.00157212.

Whiley DM, Lambert SB, Bialasiewicz S, Goire N, Nissen MD, Sloots TP. False-negative results in nucleic acid amplification tests—do we need to routinely use two genetic targets in all assays to overcome problems caused by sequence variation? Critical Reviews in Microbiology 2008;34(2):71–6. Available from https://doi.org/10.1080/10408410801960913.

World Health Organization. The Use of Lateral Flow Urine Lipoarabinomannan Assay (LF-LAM) for the Diagnosis and Screening of Active Tuberculosis in People Living with HIV: Policy Guidance. Geneva: World Health Organization; 2015.

World Health Organization. WHO Meeting Report of a Technical Expert Consultation: Non-Inferiority Analysis of Xpert MTB/RIF Ultra Compared to Xpert MTB/RIF. Geneva: World Health Organization; 2017.

Yang C, Luo X, Fan L, Sha W, Xiao H, Cui H. Performance of interferon-gamma release assays in the diagnosis of nontuberculous mycobacterial diseases—a retrospective survey from 2011 to 2019. Frontiers in Cellular and Infection Microbiology 2021;10. Available from https://doi.org/10.3389/fcimb.2020.571230.

Zhou Y-C, He S-M, Wen Z-L, Zhao J-W, Song Y-Z, Zhang Y, et al. A rapid and accurate detection approach for multidrug-resistant tuberculosis based on PCR-ELISA microplate hybridization assay. Laboratory Medicine 2020;51(6):606–13. Available from https://doi.org/10.1093/labmed/lmaa016.

Zingue D, Weber P, Soltani F, Raoult D, Drancourt M. Automatic microscopic detection of mycobacteria in sputum: a proof-of-concept. Scientific Reports 2018;8(1). Available from https://doi.org/10.1038/s41598-018-29660-8.

CHAPTER 5

Biosensor as advanced detection approach for *Mycobacterium tuberculosis*

Neha Bisht and Chetna Dhand
Academic of Scientific and Innovative Research (AcSIR), CSIR-Advanced Materials and Processes Research Institute, Bhopal, Madhya Pradesh, India

5.1 Biosensing approach for tuberculosis detection

Biosensing devices for tuberculosis (TB) diagnosis have emerged as a paradigm shift in the diagnostic approaches as they have the potential to offer accurate, rapid, and often point-of-care (PoC) alternatives to traditional approaches of culture, smear microscopy, and PCR for TB diagnosis. These analytical tools are often employed for rapid, sensitive, and real-time analysis of biomolecules that act as disease biomarkers or therapeutic targets (Joshi et al., 2022). Fig. 5.1 illustrates the basic components of a typical biosensor employed for the detection of *Mycobacterium tuberculosis* (MTB). Biosensors are capable of detecting chemical or biological events through the binding of their specific biological recognition elements, such as enzymes, proteins, nucleic acids, microorganisms, antigens, antibodies, and aptamers, and the generated signal indicates the presence of the target analyte in the sample (Lino et al., 2022). Multiple signal-sensing transducer elements include piezoelectric (mass-based), optical (absorbance, luminescence, refractive, reflective, fluorescence), magnetic (magnetic property-based), thermometric (heat-based), and electrochemical (current, potential, voltage, conductivity, impedance) transducers (Nguyen et al., 2019).

When MTB biomarkers are selectively attached to biological recognition elements such as antibodies, nucleic acids, or aptamers, biosensors turn them into detectable signals using electrochemical, optical, piezoelectric, magnetic, or microfluidic systems. Electrochemical biosensors have sparked widespread interest because of their great sensitivity, low cost, mobility, and compatibility with compact devices. These sensors detect TB-specific biomarkers such as insertion sequence (IS6110) DNA (Yuanfeng et al., 2020; Bai et al., 2019), culture filtrate protein-10 (CFP-10) (Li et al., 2019; Ribeiro et al., 2024), and (early secreted antigenic target, 6) ESAT-6 proteins (Diouani et al., 2017; Li et al., 2018) by improving electron transport and signal intensity using nanostructured materials. All these developments have contributed to the emergence of technologies that are appropriate for the early and accurate diagnosis of TB, with

Advances in Biosensor Technology for Mycobacterium Tuberculosis Detection
DOI: https://doi.org/10.1016/B978-0-443-44589-7.00008-X

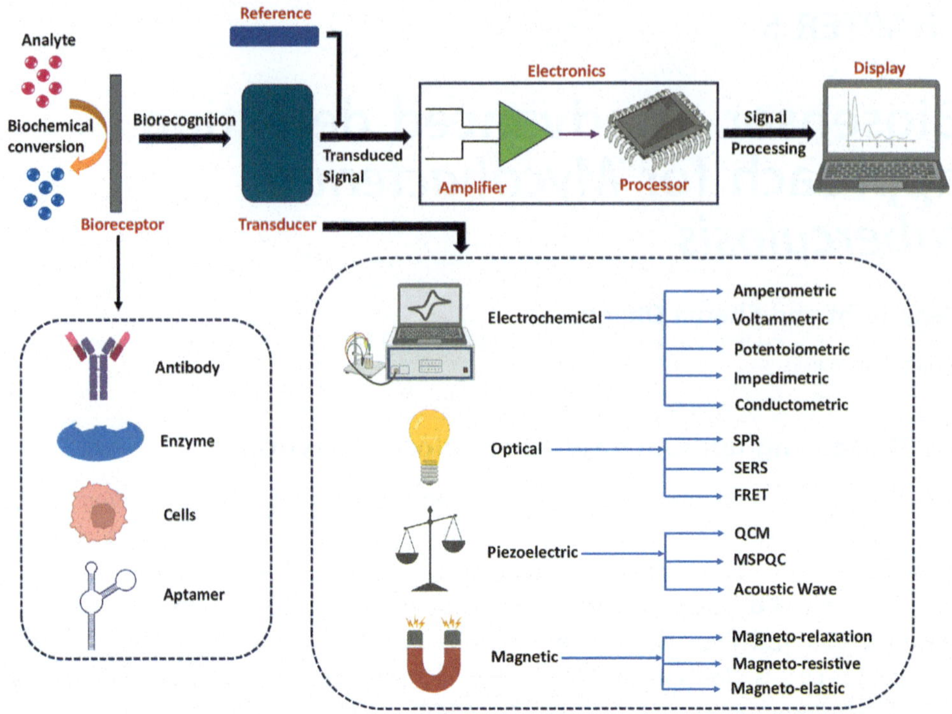

Figure 5.1 Schematic illustration of the basic structure and elements of a typical biosensor.

detection limits that go as low as the attomolar level. Electrochemical techniques are combined with other techniques such as surface plasmon resonance (SPR), fluorescence, and colorimetry in optical biosensors. In the case of TB markers, these techniques offer useful quantitative or graphical output. With aptamers or peptides as functionalization agents, SPR biosensors are very efficient at detecting MTB exosomes and antigens, even without labeling (Maphanga et al., 2023). Visible detection is enabled by colorimetric biosensors, which is a highly desirable feature, especially in resource-limited settings (Kooti et al., 2023). Fluorescence biosensors using quantum dots or FRET are highly sensitive for detecting nucleic acids (Liang et al., 2021). Optical biosensors can be made more convenient to use in the field by incorporating them into portable electronics such as cellular phones. Piezoelectric biosensors, especially quartz crystal microbalance biosensors, introduce a mechanical element into label-free biosensing by detecting the changes in frequency that result from molecular binding (Montoya et al., 2017; Kaewphinit et al., 2012). These biosensors are highly effective for the detection of DNA sequences and MTB-specific antigens, and they also give real-time results. Despite its high sensitivity, environmental instability and nonspecific binding are challenges that need to be overcome to ensure reliable performance. To improve selectivity and facilitate sample processing, magnetic biosensors rely on TB-specific ligand-modified magnetic nanoparticles (MNPs) to capture and concentrate target molecules (Gordillo-Marroquín et al., 2018; Costa et al., 2014). Magnetic relaxation switching or magnetoresistive sensing is commonly used in these biosensors to detect MTB DNA and antigens with high sensitivity and specificity. Because magnetic separation makes the process simple and is easier to integrate with other analysis techniques, it is a very attractive option for automated diagnosis. Microfluidic platforms are diagnostic devices that integrate the functions of a variety of biosensors by combining sampling, reaction, and analysis on a very small lab-on-a-chip device. Such devices can be combined with optical, magnetic, or electrochemical biosensing components. Paper-based microfluidic analytical devices (μPADs) have emerged as disposable, low-cost tools for TB detection in formats based on lateral flow or colorimetry. In resource-limited settings, where early detection is key, a combination of these technologies offers a flexible, sensitive, and scalable approach to decentralized TB testing.

5.2 Electrochemical biosensors for *Mycobacterium Tuberculosis* detection

Recently, electrochemical biosensors have attracted significant attention owing to their low cost, simplicity, portability, ease of use, and applicability for PoC and real-time testing. The flexibility of electrochemical biosensors in enzyme, nucleic acid, and microbial detection has made them popular for the detection of MTB and other biological targets. A normal electrochemical biosensor consists of three electrodes, as shown in Fig. 5.2, namely the working electrode (WE), the reference electrode (RE), and the counter electrode (CE) (Chaturvedi et al., 2024). The WE is the main site for the immobilization of the biorecognition elements and is typically made of glassy

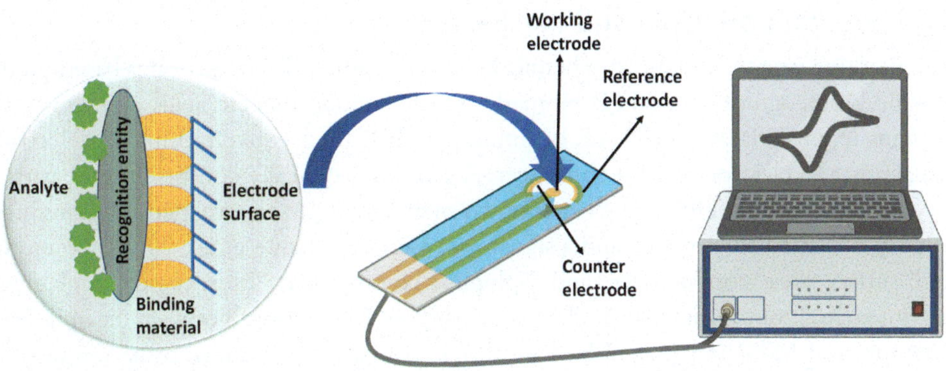

Figure 5.2 Schematic illustration of a typical electrochemical biosensor.

carbon, screen-printed carbon, carbon paste, indium tin oxide (ITO), gold, or platinum. The RE, typically Ag/AgCl or calomel, is used to maintain the reference potential, while the CE, typically platinum or graphite, completes the circuit. Biosensors based on these components operate by detecting the change in the electrical signal, such as current, potential, impedance, resistance, capacitance, and conductance, which is produced as a consequence of the specific biological interactions at the functionalized electrode interface. Electrochemical biosensors can be categorized on the basis of the type of electrical signal that is measured, including amperometric, voltammetric, potentiometric, impedimetric, and conductometric biosensors (Manoharan Nair Sudha Kumari and Thankappan Suryabai, 2024). Amperometric biosensors function at a fixed potential difference and are sensitive to the redox current, which is highly sensitive and linear; interestingly, pulsed amperometric detection (PAD) has been used to detect MTB-related carbohydrates and urinary markers such as mannitol and lactulose (Generoso et al., 2003). Voltammetric biosensors, on the other hand, scan a potential window to measure current changes, where techniques such as cyclic voltammetry (CV), square wave voltammetry (SWV), and differential pulse voltammetry (DPV) provide characteristic current peaks that are directly proportional to the concentration of the analyte. (Singh et al., 2021). Potentiometric biosensors operate based on the potential difference between electrodes under zero-current conditions, hence energy-saving and ideal for ionic or charged biomolecules (la Grasta et al., 2023). Impedimetric biosensors, often using electrochemical impedance spectroscopy (EIS), track impedance changes at the electrode interface caused by recognition events such as antigen–antibody binding or nucleic acid hybridization, enabling sensitive and label-free detection (Sandoval Bojórquez et al., 2023). Finally, conductometric biosensors monitor solution conductivity changes induced by biochemical interactions, offering simplicity and cost-effectiveness, although often with lower selectivity compared to other techniques (Shanbhag et al., 2023).

5.2.1 Working principle and mechanism

The working principle of electrochemical biosensors for MTB detection is based on the specific biorecognition of MTB biomarkers that include nucleic acids (e.g., IS6110), proteins (e.g., CFP-10, ESAT-6), or lipids, which is then converted into an electrical signal using an electrochemical transducer. A biosensor typically consists of three parts: a recognition element (DNA probe, antibody, or aptamer), an electrochemical transducer (usually modified electrodes), and a signal processor (Khaleque et al., 2024). To ensure high surface area, conductivity, and biological compatibility, biorecognition elements are integrated onto functional electrodes. Upon interaction between immobilized receptors and MTB targets and biological materials such as sputum, some interactions alter the electrical properties of the electrode-solution interface. These alterations can be quantified as changes in current (amperometry), potential (potentiometry), impedance (EIS), or charge- transfer resistance. DNA biosensors specifically designed for MTB sequences are based on hybridization between the probe and target DNA, which, after altering the charge distribution on the electrode surface, can be measured by techniques such as DPV or EIS. For signal amplification, nanomaterials or redox labels including methylene blue, can be used. Similarly, protein biosensors are based on immobilized antibodies or aptamers specific for MTB antigens such as ESAT-6 or CFP-10, which, after modification of the electrode surface, result in detectable electrochemical signals. The methods can be both labeled and label-free types, namely enzyme or redox labels. Moreover, rapid PoC analysis is enabled through portable potentiostats. Tensitivity and specificity are improved through microfluidics and CRISPR modules. The field-portable tool that is cost-effective in detecting MTB in resource-limited settings is based on the field of nanotechnology, molecular biology, and electrochemistry.

5.2.2 Advantages and limitations

Because of the various benefits, electrochemical biosensors have widely been used in detecting MTB. It has a high level of sensitivity, ranging from femtomolar to attomolar levels when used with nanomaterials. When planning to have a proof-of-concept in a resource-constrained environment, they are considered the best choice, as they are miniaturized. Other advantages include quick detection capability, lower sample handling, cost-effectiveness, and microfluidics, enabling the analysis to be performed in real time or with multiplexing. However, electrochemical biosensors have a number of disadvantages in spite of the above-stated advantages. Reproducibility and reliability of sensor responses are hard to achieve, particularly when dealing with complex biological fluids such as sputum or blood, which are prone to fouling and interference. Consistent biomolecule immobilization is difficult to maintain, and drift is common. High specificity is particularly difficult to achieve in regions with high prevalence of coinfections or latent TB patients. Moreover, the short shelf life of biological components such as enzymes and aptamers, and the complexity and expense of amplification strategies, make them impractical.

5.3 Optical biosensors for *Mycobacterium Tuberculosis* detection

Optical biosensors are highly flexible analytical instruments that are capable of detecting a very broad range of biological analytes in real time, without the need for labeling, and with a high degree of specificity, including antigens, antibodies, enzymes, nucleic acids, receptors, and infections. These biosensors operate by translating the reaction between a biological recognition component and a target analyte into a visible signal (Akgönüllü and Denizli, 2022). Key characteristics include high sensitivity and specificity, low sample volume, real-time analysis, label-free analysis, and multiplex analysis, which is the ability to analyze multiple analytes simultaneously using different wavelengths of light. SPR is a general optical phenomenon that can detect changes in the refractive index near a metal surface caused by biomolecular binding, thus making it very suitable for real-time analysis of interactions such as antibody-antigen or DNA hybridization (Nguyen et al., 2015). They are further divided into two types: SPR-based immunosensors and genosensors. SPR-based immunosensors sense the immunocomplex formed because of the reaction between the surface-immobilized antibody and the antigen. SPR-based genosensors identify the variation in the relative refractive index because of the hybridization of the single-stranded probe DNA (ssDNA) and its complementary target (Chaturvedi et al., 2024). Likewise, surface-enhanced Raman scattering (SERS) takes advantage of plasmonic nanostructures to enhance weak Raman scattering signals of target molecules, achieving ultrasensitive, label-free, and multiplex analysis, even in biological samples (Rojas Martínez et al., 2024). FRET tracks the transfer of energy between two closely located fluorophores, enabling scientists to study protein interactions and enzyme activity (Soleja and Mohsin, 2024). FRET-based biosensors are nonradiative systems that transfer electrical excitation energy from a donor fluorophore to a closely located acceptor fluorophore via dipole-dipole coupling. It allows quantitative analysis of biological molecules such as antigens, antibodies, nucleic acids, and enzymes. The biosensor is capable of detecting conformational changes in DNA-protein and protein-protein interactions.

5.3.1 Working principle and mechanism

Optical biosensors for the detection of MTB are able to provide specific biomolecular interactions that can be translated into optical signals, which can be employed for the detection of MTB biomarkers. Biosensors make use of biological recognition elements such as antibodies, nucleic acid probes, or aptamers to selectively capture MTB-specific targets such as DNA sequences, proteins, or even whole bacterial cells. The interaction is then translated into an optical signal that can be detected using platforms such as SPR, fluorescence, SERS, or colorimetric assays. SPR is a widely used optical biosensing technique that operates on the principle of refractive index changes near the surface of the sensor chip when analytes bind to immobilized receptors (Cho et al., 2025), as depicted in Fig. 5.3. This label-free, real-time analytical technique is highly specific and sensitive to the femtomolar level. Fluorescence-based biosensors, on the other hand, use fluorescent probes, quantum dots, or upconversion nanoparticles conjugated with MTB-specific probes (Zhang et al., 2022). Fluorescence intensity or lifetime changes during hybridization or antigen-antibody recognition, sometimes aided by FRET, enable highly sensitive detection of MTB nucleic acids or proteins (Zhou et al., 2023). SERS-based biosensors utilize plasmonic nanostructures to enhance Raman scattering

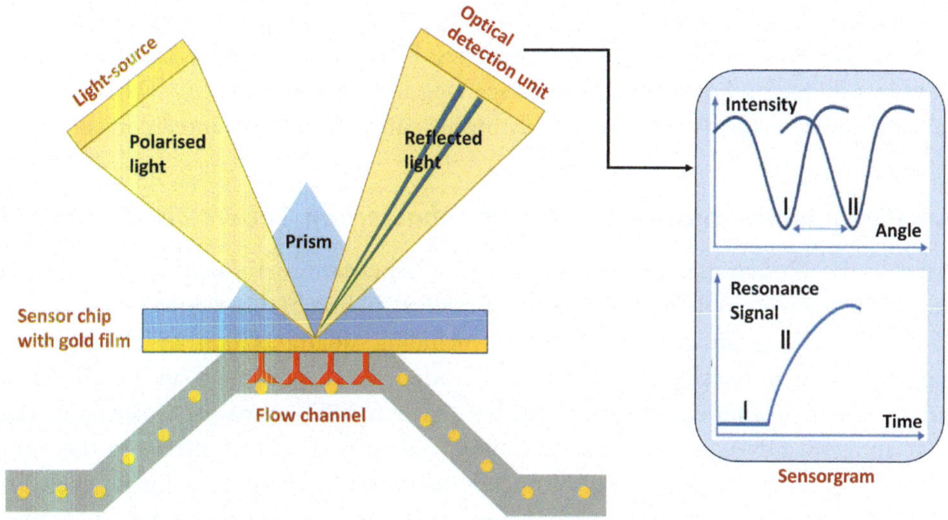

Figure 5.3 Systematic arrangement of biosensors using surface plasmon resonance.

signals from MTB biomarkers, which enables ultrasensitive, multiplexed, and label-free detection of MTB even in complex clinical samples (Botta et al., 2018). On the other hand, colorimetric biosensors provide a simple and visible result based on nanozymes or gold nanoparticle aggregation, making them highly suitable for PoC testing that is fast and inexpensive. The important concept here is the optical property change that is directly proportional to the MTB concentration, which is based on biorecognition events. There have been major advances in the sensitivity, stability, and multiplexing ability of optical biosensors based on advances in nanotechnology, surface science, and photonics, which make them a promising tool for accurate TB diagnosis.

5.3.2 Advantages and limitations

Optical biosensors allow for quick, real-time, and very sensitive detection, even at the femtomolar or single-molecule level, with label-free analysis using SPR and SERS. Fluorescence systems enable multiplexing, whilst colorimetric biosensors provide simple, low-cost detection ideal for resource-constrained environments. Plasmonic nanoparticles, quantum dots, and nanozymes improve performance by allowing the identification of proteins, DNA, and whole MTB cells. Despite their adaptability, several significant obstacles remain to their wider clinical acceptance. Many optical biosensors require carefully controlled laboratory conditions and precise instrumentation, such as lasers, and spectrometers, or microfluidic systems, which are not always possible in resource-limited settings. Sample preparation is also a concern since complex biological fluids may cause nonspecific binding, reducing the test's accuracy.

Colorimetric biosensors are easier to read, but they are frequently less sensitive than SPR, fluorescence, or SERS-based approaches. Fluorescence systems are susceptible to photobleaching, background autofluorescence, and probe instability, whereas SERS devices require properly constructed plasmonic substrates to guarantee consistency.

5.4 Piezoelectric biosensors for *Mycobacterium Tuberculosis* detection

Piezoelectric biosensors represent a category of label-free, extremely sensitive analysis equipment which employs the piezoelectric effect, whereby mechanical stress induces an electric charge in certain materials, including quartz (Chen et al., 2024). In applications to biosensing, these sensors respond to changes in mass on the sensor surface. Upon attachment of a target analyte, such as MTB or its particular biomarkers (e.g., antigens, DNA, or proteins), to a biorecognition element tethered to the crystal surface (e.g., antibodies, aptamers, or nucleic acid probes), the imposed mass results in a measurable change in the resonance frequency of the quartz crystal. The linear relationship between the change in frequency and the mass of the bound target allows quantification. There are three types of piezoelectric biosensors: QCM, multichannel series piezoelectric quartz crystal (MSPQC) sensors, and acoustic wave sensors (Srivastava et al., 2016). The QCM sensor can be employed in a vacuum state, gas state, or liquid state and utilizes the principle of the frequency shift of a quartz crystal (Alassi et al., 2017). The density, permittivity, conductivity, and viscosity of the liquid environment surrounding the sensor are the principles that make it work. On the other hand, the MSPQC sensor can work in a liquid environment and is capable of detecting very small changes in conductivity (Zhang et al., 2019). The sensor functions based on the principle of the change in the electrical properties of the medium of detection because of the conversion of weakly charged or uncharged substrates to highly charged end products as a result of microorganism metabolism (Joshi et al., 2022). The acoustic wave biosensor works by utilizing the detection of the propagation of acoustic waves near the surface of a piezoelectric crystal. The biosensor uses nanomaterials and microfabrication technology in order to increase its sensitivity and specificity in the detection of biological analytes (Zhang et al., 2024).

5.4.1 Working principle and mechanism

Piezoelectric biosensors, particularly QCM biosensors, have been considered for MTB detection owing to their label-free, real-time, and highly sensitive characteristics. The mechanism of the biosensor is based on the piezoelectric effect, whereby an alternating voltage causes the quartz crystal to vibrate at a specific resonance frequency (Fig. 5.4). With the binding of biomolecules to the functionalized surface of the biosensor, a

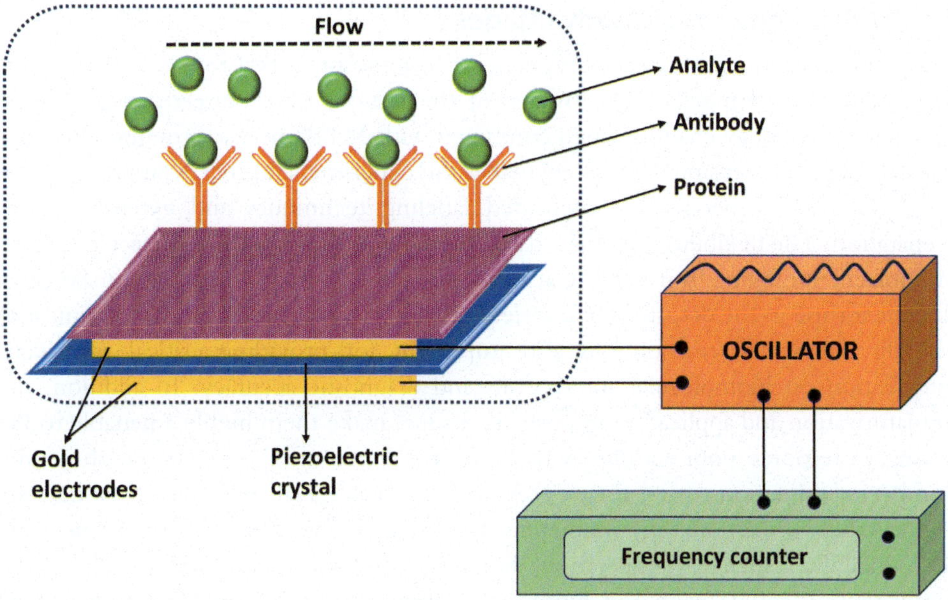

Figure 5.4 Schematic representation of a typical piezoelectric biosensor.

corresponding change in resonance frequency occurs, which is directly proportional to the bound mass, as described by the Sauerbrey equation Skládal (2024):

$$\Delta F = -2.3 \times 10^{-6}\, F^2 \Delta M / A \tag{5.1}$$

where ΔF=frequency change, F=resonant frequency, ΔM=change in deposited mass, and A=area of electrode.

In practice, the surface of the QCM is modified with MTB-specific antibodies or DNA probes that are complementary to the genomic DNA of MTB, allowing for the selective binding of MTB biomarkers. These binding interactions reduce the frequency of oscillations, which is calculated with respect to its concentration. In order to enable for better specificity in diagnoses, modern QCM biosensors have been developed to assess energy dissipation, which is connected with the viscoelastic properties of the bound layer (Johannsmann, 2006). Integration with microfluidics improves sample handling and analysis time for multiplex analysis. Moreover, surface modification using nanomaterials improves mass sensitivity, allowing detection at the femtomolar level for MTB antigens or DNA. While there are limitations in accurately controlling surface chemistry and adsorption, piezoelectric biosensors can still be considered an alternative for rapid, accurate, and affordable TB testing, especially in resource-limited settings.

5.4.2 Advantages and disadvantages

The most important advantage of piezoelectric biosensors is that they are very sensitive to minute mass changes on the surface of the sensor, which directly correspond to observable frequency changes when combined with MTB-specific proteins. This mass-sensing, label-free sensing technology makes the experimental process simpler and faster since there is no need for complicated labeling techniques and extensive sample preparation. The flexibility of QCM-based piezoelectric biosensors to detect a variety of biomolecules, ranging from nucleic acids to proteins and even whole cells, makes them suitable for use in the detection of a variety of MTB indicators. Their ability to monitor binding kinetics in real time is also important for providing critical quantitative information on biomolecular interactions and diagnostic accuracy. In addition, their miniaturization and application in portable devices make them highly amenable to PoC testing in resource-poor settings with high prevalence of TB. However, there are a number of limitations that exist to prevent their widespread use. These include their vulnerability to several environmental factors that may affect their stability and reproducibility, such as temperature, humidity, and vibration. Additionally, surface modifications could make the biosensors complex, but nonspecific binding could influence their specificity. The extensive use under environments of limited resources could also be a challenge in terms of throughput, cost of equipment, and the knowledge technology requires.

5.5 Magnetic biosensors for *Mycobacterium Tuberculosis* detection

Magnetic biosensors have been identified as a promising alternative for MTB detection because of their speed, sensitivity, and ability to conduct label-based detection. Magnetic biosensors utilize the unique magnetic properties of nanomaterials, particularly MNPs, which can be designed for specific detection of MTB biomarkers, such as DNA, proteins, or antigens. The capability of magnetic biosensors to work well in turbid biological fluids, such as sputum or blood, makes them an appropriate technology for diagnostic use (Torres-Chavolla and Alocilja, 2011; León-Janampa et al., 2022). In the magnetic biosensor, the biological recognition component is immobilized on the surface of the MNPs as a sensor and subsequently combined with the sample to sense the analytes in the presence of a magnetic field. These are quite stable at room temperature with enhanced integration flexibility, low background noise, and high sensitivity relative to fluorescent-based biosensors. These biosensors are sufficient enough to capture a clean sample from biological fluids, e.g., blood, urine, and sputum, by magnetic separation. Magnetic biosensors utilize the application of several well-known technologies for the diagnosis of TB, such as magnetoresistive biosensors and magnetic relaxation switch assays, for the detection of the interaction between the biological recognition element and the target analyte (Nabaei et al., 2018).

5.5.1 Working principle and mechanism

The general principle of magnetic biosensors is based on the utilization of MNPs coated with biomolecules such as aptamers, antibodies, or oligonucleotide probes. These probes specifically recognize and bind to MTB biomarkers, including nucleic acid sequences and antigens. Subsequent to the binding of the target biomarker, the MNPs produce complexes detectable through magnetic sensing techniques. In that sense, one of the fundamental principles underlying this process is MRS. During this process, MNPs aggregate or disperse in solution because of target analyte binding. The change in the spin-spin relaxation time of water protons in the surroundings can be detected by means of NMR or magnetic resonance relaxometry because of this (Chen et al., 2015). The change in the value of the relaxation time is proportional to the concentration of MTB biomarkers, thus enabling quantitative detection. Magnetoresistive detection is another significant aspect (Barroso et al., 2018). This method involves the prefunctionalization of the sensor surfaces with MTB biomarker-based probes. When the test substance containing the MTB biomarkers is given, these biomarkers bind to the probes. Later, MNP-conjugated secondary probes bind to the bound target, bringing MNPs close to the sensor surface. The generated local magnetic field affects the electrical resistance of the magnetoresistive sensor, providing a measurable signal. These two techniques, relaxation-based detection and resistance-based detection, form the basis of the working principle of magnetic biosensors. Magnetic biosensors offer a highly sensitive approach for the detection of MTB biomarkers using a combination of biomolecular recognition and magnetic transduction.

Moreover, the ME biosensors work on the basis that when the magnetostrictive strip or sheet is exposed to an alternating magnetic field, it resonates at a particular frequency. The thin, ribbon-like ferromagnetic strip acts as the transducer. The ME strip undergoes elastic oscillations within its basal plane when it is exposed to a magnetic field. A pickup or collection coil can remotely sense the secondary electromagnetic wave generated by the elastic oscillations. The dimensions and properties of the strip (density, Poisson's ratio, and modulus of elasticity) determine the resonant frequency. It can be stated numerically as (Stoyanov and Grimes, 2000):

$$f = \sqrt{}(E/\rho(1 - e^2)2L)$$

where E is the elastic modulus, ρ is the material density, σ is Poisson's ratio, and L is the strip length.

The strip has been designed for biosensing with the use of capture biomolecules, such as antibodies or DNA probes, which selectively bind to MTB biomarkers

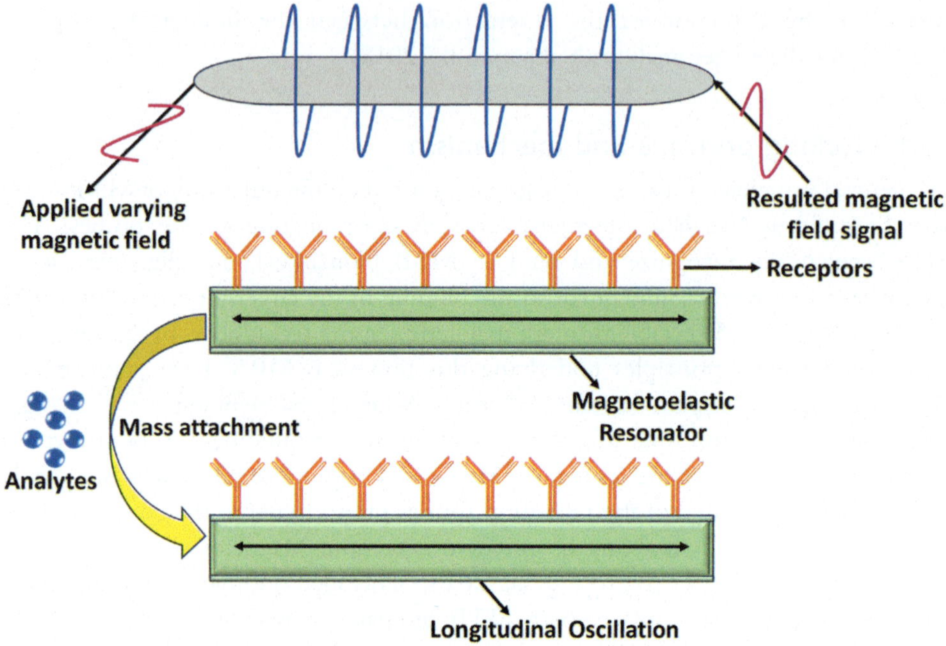

Figure 5.5 Schematic illustrating a magnetoelastic biosensor in which the resonance of the strip shifts by antigen-antibody binding.

(proteins, DNA, or antigens) in a unique manner. The mass of the strip will increase as target molecules bind to it. The following gives this mass-dependent frequency shift (Cai et al., 2000):

$$\Delta f = -f/2\,(\Delta m/M)$$

where f is the initial resonance frequency, M is the initial mass of the strip, Δm is the added mass because of biomolecular binding, and Δf is the resulting frequency shift.

Therefore, the detection of MTB is done by monitoring the resonance frequency change of the ME strip after the binding of the analyte. Fig. 5.5 illustrates the schematic representation of a magnetoelastic biosensor, showing the effect of antigen binding to immobilize antibodies on the resonance of the magnetoelastic strip.

5.5.2 Advantages and limitations

Magnetic biosensors have specific advantages in MTB detection, making them the most suitable for rapid, sensitive, and specific detection. The ability of magnetic biosensors to separate and concentrate biomarkers using magnetic manipulation increases sensitivity and minimizes background noise in complex biological samples. MNPs functionalized with nucleic acid probes or MTB-specific antibodies can specifically bind to MTB antigens or DNA, providing detection sensitivity comparable to or superior to conventional PCR but with less complex equipment and shorter experimental times. Magnetic biosensors are not influenced by optical interference, as in fluorescence or colorimetric sensors, ensuring accurate results even in turbid or colored samples (Pashchenko et al., 2018). They also work well with portable, automated PoC systems using magnetic relaxation assays or magnetoresistive sensors. However, there are still some problems that have been hindering their widespread use. Since aggregation or loss of magnetism will affect sensitivity, MNPs are difficult to functionalize and stabilize. Otherwise, nonspecific binding will affect specificity. Moreover, the use of highly advanced sensors or external magnetic fields will add to the complexity and cost of processing. Finally, low concentrations of MTB may require additional amplification steps, which will make the process of the test more difficult.

5.6 Microfluidic and lab-on-a-chip biosensors for *Mycobacterium Tuberculosis* detection

Other possible alternatives for the detection of MTB include microfluidic and lab-on-a-chip (LOC) biosensors. Biosensors have the ability to analyze samples, identify biomarkers, and interpret results on a single chip, thus cutting down the time and reagents needed for analysis. Biosensors are ideal for parallel analysis because of their accuracy in analyzing small amounts of fluids (Nikoleli et al., 2018). Functionalization with MTB-specific antibodies, nucleic acid probes, or aptamers enables the selective capture and analysis of MTB antigens or DNA (Bernacka-Wojcik et al., 2013; Molloy et al., 2021). Additionally, they are making an appeal for PoC tests in settings that are resource-constrained, in which conventional methods, such as PCR, are less feasible because of their portability potential. It is interesting to notice that through a closed system, microfluidic devices offer reproducibility as well as the prevention of potential contaminations. In order to handle fluids or gases in very small quantities (10^{-9} L to 10^{-18} L), microfluidic devices use microchannels with cross-sectional size measures between 10 μm and 100 μm (Jagannath et al., 2022). In addition, automation, sensitivity, efficiency, and wide analytical capabilities are among improvements in LOC systems based on microfluidics that have contributed to a significant increase in the use of this system in biological applications and diagnostics.

5.6.1 Working mechanism and miniaturization benefits

Microfluidics and LOCs have fully revolutionized how TB diagnosis is being carried out through various complicated laboratory analyses put together into compact, portable devices. These technologies work by accurately managing fluid volumes in microchannels made on silicon, glass, or polymer substrates (Molloy et al., 2021; Mark et al., 2010). In addition, critical processes such as sampling, biomarker extraction, amplification, and detection can be performed more effectively using these technologies. For example, essential processes such as sampling, biomarker extraction, amplification, and detection can be performed more effectively using these technologies. They involve nucleic acid amplification, which can be done by using either PCR or isothermal techniques, coupled with immunoassays (Molloy et al., 2021), and biosensing techniques such as electrochemical, optical, or magnetic approaches for the detection of MTB on a single microchip (Zeng et al., 2025). Favorable characteristics for molecule interactions, such as laminar flow, a high surface-volume ratio, as well as efficient heat and mass transfer, are also microenvironment benefits for a faster rate of reaction with increased sensitivity. Another advantage resulting from miniaturization is that considerable volume reduction is seen in the volume of samples and reagents. This is essential for diagnosing a disease using fewer biological samples, as seen in pediatric or sputum-scarce conditions (Sin et al., 2014). From the engineering perspective, they can multiplex, which means that it can be possible to carry out the testing of multiple pathogens or markers simultaneously in one test, including coinfections. Advances in the technology of smartphone devices help to improve data analysis, portability, and use. Thus, the engineering potential of microfluidics in reducing reagents, speeding up processing, increasing sensitivity, and making it more portable and automated is revolutionizing the improvement of MTB and TB testing methods.

5.6.2 Automation and point-of-care testing advantages

MTB detection has evolved into an efficient, accurate, noncentralized procedure with minimal end-user interaction through automation as well as proof-of-concept testing using microfluidic and LOC technologies. Combined microfluidic technologies minimize detection time and reagents while escalating sensitivity and specificity through the simultaneous integration of several micro-forward analysis processes such as MTB preparation, markers, as well as analysis into a highly compact package. It is vital that MTB diagnosis is accurate to allow for successful treatment of MTB in environments with minimal resources, possibly through automation to reduce inaccuracies. It is also vital that there is the ability to perform simultaneous tests using microfluidic technologies, such as MTB as well as resistance markers. Given the power consumption required by these devices, which is very low, it would be possible to use LOC devices in remote areas without access to lab facilities, which will therefore greatly contribute to the better decision-making process needed to limit the transmission of this disease. Moreover, telemedicine and real-time epidemiological surveillance will also result from the wireless data transmission and results that can be viewed on a smartphone. MTB detection will be fast and sensitive because of the automated nucleic acid amplification techniques, including loop-mediated isothermal amplification

(LAMP) (Mori and Notomi, 2009; Zeng et al., 2025). Generally speaking, the benefits of the automation and PoC of microfluidic and LOC technologies not only provide access to improved MTB diagnostics but also offer a solution that is essential and cost-effective in the fight against TB throughout the globe.

5.6.3 Using biosensors with high-throughput *Mycobacterium Tuberculosis* identification techniques

The detection of effective MTB has seen a new paradigm with the advent of the amalgamation of both microfluidic technology and LOC with biosensors. When working with nanoliter to microliter numbers, the microfluidic technique ensures target enrichment, quick reaction speeds, and efficient analysis. When combined with electrochemical, optical, magnetic, and piezoelectric biosensors, microfluidic technology can automate traditional TB analysis techniques, including the simultaneous examination of several samples. The most significant advantage of microfluidic devices is the high specificity and low reagent requirements of multiplex analysis of various MTB targets, such as nucleic acids, antigens, and metabolites. LOC devices also include sample lysis, nucleic acid amplification (LAMP/RPA), and real-time analysis, which allows the time of analysis to be reduced from hours or days to less than 1 hour (Li et al., 2023). The ability to perform high-throughput analysis is enabled by multiplexed sensor arrays or microchannel designs that are in parallel, which enables the analysis of multiple patients or biomarkers simultaneously. Additionally, real-time analysis and automated interpretation of results are enabled by the integration of LOC biosensor technology with smartphone imaging and AI analysis, which enhances accessibility. A comparison of different biosensors available for MTB detection is presented in Table 5.1.

5.7 Challenges and future prospects

5.7.1 Biosensor stability and reproducibility issues

The key limitations of biosensors are the stability and reproducibility of the biosensors, making them unfit for practical applications in hospitals and industries. Environmental factors such as temperature, pH, and humidity can lead to the degradation and denaturation of biological recognition molecules such as enzymes, antibodies, and aptamers. Drift and shortened shelf life are the consequences of this. Sensitivity and repeatability are also impaired by surface fouling and nonspecific binding because of complex biological molecules. Variability between batches is also introduced by variability in the synthesis of nanomaterials and bioconjugation methods. New synthetic receptors such as molecularly imprinted polymers (MIPs) and new immobilization methods such as covalent binding are being investigated to overcome these limitations. They will be more practical for real-world applications because of advances in self-calibrating algorithms and antifouling surfaces. When

Table 5.1 Comparison of the advantages and disadvantages of different biosensors in the detection of *Mycobacterium Tuberculosis.*

Biosensor type	Advantages	Disadvantages	Current research trends	References
Electrochemical biosensors	Can detect both low and high concentrations of analytes Ion detection possible even at zero current High precision	Highly sensitive to temperature and environmental variations Limited to aqueous media	Development of portable, flexible, and cost-effective designs for PoC testing	Ronkainen et al. (2010), Wu et al. (2023), Hammond et al. (2016)
Optical biosensors	High sensitivity and rapid response Real time, reusable measurement Requires minimal sample	Immobilization issues Expensive instruments and chips required	Miniaturization efforts to improve portability and affordability	Damborský et al. (2016), Chen and Wang (2020), Uniyal et al. (2023)
Piezoelectric biosensors	Lightweight and easy to use High sensitivity and label-free detection Can operate under high pressure and temperature	Brittleness of materials Sensitive to moisture strongly dependent on sample preparation	Development of composite-based piezoelectric materials to enhance durability and robustness	Skládal (2016), Pohanka (2017), Swain et al. (2023)
Magnetic biosensors	Long-term probe stability in culture Detects low protein concentrations with high sensitivity Magnetic field provides controlled detection environment	Particle aggregation may cause false positives Requires high-frequency operation	Superparamagnetic beads are being developed to prevent aggregation	Xianyu et al. (2018), Mitu et al. (2023), Chen et al. (2017)

Microfluidic and lab-on-a-chip biosensors	Require very small sample and reagent volumes Enable rapid and multiplexed analysis Suitable for integration with portable diagnostic devices High potential for PoC tuberculosis (TB) testing	Fabrication complexity and cost Risk of channel clogging and biofouling Requires precise fluid control systems	Integration with electrochemical/optical sensors for highly sensitive, portable, and low-cost TB diagnostics	Luka et al. (2015), Jamshaid et al. (2016), Pires et al. (2014)

paired with machine learning-assisted signal processing and biointerfaces, they will offer stable and robust biosensors for PoC and environmental monitoring.

5.7.2 Cost-effectiveness and scalability for mass deployment

Scalability and cost-effectiveness remain significant challenges in the use of advanced biosensors for TB and other infectious diseases. Although the sensitivity and specificity of electrochemical, optical, and microfluidic biosensors are very high, their scalability is challenged by the complexity of processing, the cost of nanomaterials, and the need for advanced equipment. In resource-limited settings and areas with a high prevalence of TB, the use of nanostructures and surface modifications makes the technology very expensive and unaffordable. Scalability is a problem and regulatory approval is complicated by the absence of clearly defined processing and processing unpredictability.

5.7.3 Clinical validation and regulatory challenges

Poor repeatability, insufficient validation across varied populations, and the expensive expenses and time needs of multicenter studies are the main obstacles to the clinical translation of TB biosensors. Approval is further delayed by stringent regulatory criteria (FDA/EMA), difficulties with nanomaterials and microfluidics, and the lack of standardized validation procedures. Standardized protocols, AI-powered validation systems, and international cooperation are necessary for faster adoption.

5.8 Conclusion

In terms of biosensing technologies used for MTB detection, it has improved significantly, providing sensitive, specific, and rapid alternatives to conventional diagnosis of TB, such as optical, electrochemical, magnetic, piezoelectric, or microfluidic biosensors. This technology can provide fast, specific, and cost-effective tools for TB diagnosis, especially in resource-poor, TB-endemic countries. However, it is difficult to translate these technologies into clinically approved products because of clinical validation on a large scale, regulatory approval, quality and safety, and finally, lack of standard validation procedures, particularly for nanobased technologies used to develop biosensing systems. Clinicians and researchers must work as a team to develop effective biosensing technologies for TB diagnosis.

References

Akgönüllü S, Denizli A. Recent advances in optical biosensing approaches for biomarkers detection. Biosensors and Bioelectronics: X 2022;12:100269. Available from https://doi.org/10.1016/j.biosx.2022.100269.

Alassi A, Benammar M, Brett D. Quartz crystal microbalance electronic interfacing systems: a review. Sensors (Basel) 2017;17(12):2799. Available from https://doi.org/10.3390/s17122799.

Bai L, Chen Y, Liu X, Zhou J, Cao J, Hou L, et al. Ultrasensitive electrochemical detection of Mycobacterium tuberculosis IS6110 fragment using gold nanoparticles decorated fullerene nanoparticles/nitrogen-doped graphene nanosheet as signal tags. Analytica Chimica Acta 2019;1080:75–83.

Barroso TG, Martins RC, Fernandes E, Cardoso S, Rivas J, Freitas PP. Detection of BCG bacteria using a magnetoresistive biosensor: a step towards a fully electronic platform for tuberculosis point-of-care detection. Biosensors and Bioelectronics 2018;100:259–65.

Bernacka-Wojcik I, Lopes P, Catarina Vaz A, Veigas B, Jerzy Wojcik P, Simões P, et al. Bio-microfluidic platform for gold nanoprobe based DNA detection—application to Mycobacterium tuberculosis. Biosensors and Bioelectronics 2013;48:87–93. Available from https://doi.org/10.1016/j.bios.2013.03.079.

Botta R, Chindaudom P, Eiamchai P, Horprathum M, Limwichean S, Chananonnawathorn C, et al. Tuberculosis determination using SERS and chemometric methods. Tuberculosis 2018;108:195–200.

Cai QY, Cammers-Goodwin A, Grimes CA. A wireless, remote query magnetoelastic CO2 sensor. Journal of Environmental Monitoring 2000;2(6):556–60.

Chaturvedi M, Patel M, Tiwari A, Dwivedi N, Mondal D, Srivastava AK, et al. An insight to the recent advancements in detection of Mycobacterium tuberculosis using biosensors: a systematic review. Progress in Biophysics and Molecular Biology 2024;186:14–27.

Chen C, Wang J. Optical biosensors: an exhaustive and comprehensive review. Analyst 2020;145(5):1605–28.

Chen Y, Xianyu Y, Wang Y, Zhang X, Cha R, Sun J, et al. One-step detection of pathogens and viruses: combining magnetic relaxation switching and magnetic separation. ACS nano 2015;9(3):3184–91.

Chen Y, Zhang X, Lu C. Flexible piezoelectric materials and strain sensors for wearable electronics and artificial intelligence applications. Chemical Science 2024;15(40):16436–66. Available from https://doi.org/10.1039/d4sc05166a.

Chen Y-T, Kolhatkar AG, Zenasni O, Xu S, Lee TR. Biosensing using magnetic particle detection techniques. Sensors 2017;17(10):2300.

Cho SH, Choi S, Suh JM, Jang HW. Advancements in surface plasmon resonance sensors for real-time detection of chemical analytes: sensing materials and applications. Journal of Materials Chemistry C 2025;13(13):6484–507.

Costa MP, Andrade CA, Montenegro RA, Melo FL, Oliveira MD. Self-assembled monolayers of mercaptobenzoic acid and magnetite nanoparticles as an efficient support for development of tuberculosis genosensor. Journal of Colloid and Interface Science 2014;433:141–8.

Damborský P, Švitel J, Katrlík J. Optical biosensors. Essays in Biochemistry 2016;60(1):91–100.

Diouani MF, Ouerghi O, Refai A, Belgacem K, Tlili C, Laouini D, et al. Detection of ESAT-6 by a label free miniature immuno-electrochemical biosensor as a diagnostic tool for tuberculosis. Materials Science and Engineering: C 2017;74:465–70.

Generoso M, De Rosa M, De Rosa R, De Magistris L, Secondulfo M, Fiandra R, et al. Cellobiose and lactulose coupled with mannitol and determined using ion-exchange chromatography with pulsed amperometric detection, are reliable probes for investigation of intestinal permeability. Journal of Chromatography. B, Analytical Technologies in the Biomedical and Life Sciences 2003;783(2):349–57. Available from https://doi.org/10.1016/s1570-0232(02)00766-3.

Gordillo-Marroquín C, Gómez-Velasco A, Sánchez-Pérez HJ, Pryg K, Shinners J, Murray N, et al. Magnetic nanoparticle-based biosensing assay quantitatively enhances acid-fast bacilli count in paucibacillary pulmonary tuberculosis. Biosensors 2018;8(4):128.

La Grasta A, De Carlo M, Di Nisio A, Dell'Olio F, Passaro VMN. Potentiometric chloride Ion biosensor for cystic fibrosis diagnosis and management: modeling and design. Sensors 2023;23(5):2491.

Hammond JL, Formisano N, Estrela P, Carrara S, Tkac J. Electrochemical biosensors and nanobiosensors. Essays in Biochemistry 2016;60(1):69–80.

Jagannath A, Cong H, Hassan J, Gonzalez G, Gilchrist MD, Zhang N. Pathogen detection on microfluidic platforms: recent advances, challenges, and prospects. Biosensors and Bioelectronics: X 2022;10:100134. Available from https://doi.org/10.1016/j.biosx.2022.100134.

Jamshaid T, Neto ETT, Eissa MM, Zine N, Kunita MH, El-Salhi AE, et al. Magnetic particles: from preparation to lab-on-a-chip, biosensors, microsystems and microfluidics applications. TrAC Trends in Analytical Chemistry 2016;79:344–62.

Johannsmann D. Studies of viscoelasticity with the QCM. In: Steinem C, Janshoff A, editors. Piezoelectric Sensors. Springer Series on Chemical Sensors and Biosensors, vol 5. Berlin, Heidelberg: Springer; 2006. p. 49–109. Available from https://doi.org/10.1007/5346_024.

Joshi H, Kandari D, Maitra SS, Bhatnagar R. Biosensors for the detection of Mycobacterium tuberculosis: a comprehensive overview. Critical Reviews in Microbiology 2022;48(6):784–812.

Kaewphinit T, Santiwatanakul S, Chansiri K. Gold nanoparticle amplification combined with quartz crystal microbalance DNA based biosensor for detection of Mycobacterium tuberculosis. Sensors & Transducers 2012;146(11):156.

Khaleque MA, Hossain SI, Ali MR, Aly MAS, Abuelmakarem HS, Al Mamun MS, et al. Bioreceptor modified electrochemical biosensors for the detection of life threating pathogenic bacteria: a review. RSC Advances 2024;14(39):28487–515.

Kooti S, Kadivarian S, Abiri R, Mohajeri P, Atashi S, Ahmadpor H, et al. Modified gold nanoparticle colorimetric probe-based biosensor for direct and rapid detection of Mycobacterium tuberculosis in sputum specimens. World Journal of Microbiology and Biotechnology 2023;39(5):118.

León-Janampa N, Shinkaruk S, Gilman RH, Kirwan DE, Fouquet E, Szlosek M, et al. Biorecognition and detection of antigens from Mycobacterium tuberculosis using a sandwich ELISA associated with magnetic nanoparticles. Journal of Pharmaceutical and Biomedical Analysis 2022;215:114749.

Li J, Hu K, Zhang Z, Teng X, Zhang X. Click DNA cycling in combination with gold nanoparticles loaded with quadruplex DNA motifs enable sensitive electrochemical quantitation of the tuberculosis-associated biomarker CFP-10 in sputum. Microchimica Acta 2019;186(9):662.

Li L, Yuan Y, Chen Y, Zhang P, Bai Y, Bai L. Aptamer based voltammetric biosensor for Mycobacterium tuberculosis antigen ESAT-6 using a nanohybrid material composed of reduced graphene oxide and a metal-organic framework. Microchimica Acta 2018;185(8):379.

Li Z, Xu X, Wang D, Jiang X. Recent advancements in nucleic acid detection with microfluidic chip for molecular diagnostics. TrAC Trends in Analytical Chemistry 2023;158:116871.

Liang L, Chen M, Tong Y, Tan W, Chen Z. Detection of Mycobacterium tuberculosis IS6110 gene fragment by fluorescent biosensor based on FRET between two-dimensional metal-organic framework and quantum dots-labeled DNA probe. Analytica Chimica Acta 2021;1186:339090.

Lino C, Barrias S, Chaves R, Adega F, Martins-Lopes P, Fernandes JR. Biosensors as diagnostic tools in clinical applications. Biochimica et Biophysica Acta (BBA) - Reviews on Cancer 2022;1877(3):188726. Available from https://doi.org/10.1016/j.bbcan.2022.188726.

Luka G, Ahmadi A, Najjaran H, Alocilja E, DeRosa M, Wolthers K, et al. Microfluidics integrated biosensors: a leading technology towards lab-on-a-chip and sensing applications. Sensors 2015;15(12):30011–31.

Manoharan Nair Sudha Kumari S, Thankappan Suryabai X. Sensing the future—frontiers in biosensors: exploring classifications, principles, and recent advances. ACS Omega 2024;9(50):48918–87. Available from https://doi.org/10.1021/acsomega.4c07991.

Maphanga C, Manoto S, Ombinda-Lemboumba S, Ismail Y, Mthunzi-Kufa P. Localized surface plasmon resonance biosensing of Mycobacterium tuberculosis biomarker for TB diagnosis. Sensing and Bio-Sensing Research; a Journal of Science and its Applications 2023;39:100545.

Mark D, Haeberle S, Roth G, Von Stetten F, Zengerle R. Microfluidic lab-on-a-chip platforms: requirements, characteristics and applications. Microfluidics based microsystems: fundamentals and applications 2010:305–76.

Mitu SA, Ahmed K, Bui FM. Magnetic biosensors. Biosensors Nanotechnology 2023:201–13.

Molloy A, Harrison J, McGrath JS, Owen Z, Smith C, Liu X, et al. Microfluidics as a Novel Technique for Tuberculosis: From Diagnostics to Drug Discovery. Microorganisms 2021;9(11). Available from https://doi.org/10.3390/microorganisms9112330.

Montoya A, March C, Montagut YJ, Moreno MJ, Manclus JJ, Arnau A, et al. A high fundamental frequency (HFF)-based QCM Immunosensor for tuberculosis detection. Current Topics in Medicinal Chemistry 2017;17(14):1623–30.

Mori Y, Notomi T. Loop-mediated isothermal amplification (LAMP): a rapid, accurate, and cost-effective diagnostic method for infectious diseases. Journal of Infection and Chemotherapy Official Journal of the Japan Society of Chemotherapy 2009;15(2):62–9. Available from https://doi.org/10.1007/s10156-009-0669-9.

Nabaei V, Chandrawati R, Heidari H. Magnetic biosensors: modelling and simulation. Biosensors and Bioelectronics 2018;103:69–86.

Nguyen HH, Lee SH, Lee UJ, Fermin CD, Kim M. Immobilized enzymes in biosensor applications. Materials 2019;12(1):121.

Nguyen HH, Park J, Kang S, Kim M. Surface plasmon resonance: a versatile technique for biosensor applications. Sensors (Basel) 2015;15(5):10481–510. Available from https://doi.org/10.3390/s150510481.

Nikoleli G-P, Siontorou CG, Nikolelis DP, Bratakou S, Karapetis S, Tzamtzis N. Biosensors based on microfluidic devices lab-on-a-chip and microfluidic technology. Nanotechnology and Biosensors 2018:375–94.

Pashchenko O, Shelby T, Banerjee T, Santra S. A comparison of optical, electrochemical, magnetic, and colorimetric point-of-care biosensors for infectious disease diagnosis. ACS Infectious Diseases 2018;4(8):1162–78. Available from https://doi.org/10.1021/acsinfecdis.8b00023.

Pires NMM, Dong T, Hanke U, Hoivik N. Recent developments in optical detection technologies in lab-on-a-chip devices for biosensing applications. Sensors 2014;14(8):15458–79.

Pohanka M. The piezoelectric biosensors: principles and applications, a review. International Journal of Electrochemical Science 2017;12(1):496–506.

Ribeiro LV, Cancino-Bernardi J, Razzino CDA, Machado TR, Tuesta MA, Zucolotto V. Exploring electrochemical impedance spectroscopy for the early diagnosis of Mycobacterium tuberculosis using CFP10: ESAT6 protein detection. Frontiers in Sensors 2024;5:1512936.

Rojas Martínez V, Lee E, Oh J-W. Exploring plasmonic standalone surface-enhanced raman scattering nanoprobes for multifaceted applications in biomedical, food, and environmental fields. Nanomaterials 2024;14(22):1839.

Ronkainen NJ, Halsall HB, Heineman WR. Electrochemical biosensors. Chemical Society Reviews 2010;39(5):1747–63.

Sandoval Bojórquez DI, Janićijević Ž, Palestina Romero B, Oliveros Mata ES, Laube M, Feldmann A, et al. Impedimetric nanobiosensor for the detection of SARS-CoV-2 antigens and antibodies. ACS Sensors 2023;8(2):576–86. Available from https://doi.org/10.1021/acssensors.2c01686.

Shanbhag MM, Manasa G, Mascarenhas RJ, Mondal K, Shetti NP. Fundamentals of bio-electrochemical sensing. Chemical Engineering Journal Advances 2023;16:100516. Available from https://doi.org/10.1016/j.ceja.2023.100516.

Sin ML, Mach KE, Wong PK, Liao JC. Advances and challenges in biosensor-based diagnosis of infectious diseases. Expert Review of Molecular Diagnostics 2014;14(2):225–44. Available from https://doi.org/10.1586/14737159.2014.888313.

Singh A, Sharma A, Ahmed A, Sundramoorthy AK, Furukawa H, Arya S, et al. Recent Advances in Electrochemical Biosensors: Applications, Challenges, and Future Scope. Biosensors 2021;11(9):336.

Skládal P. Piezoelectric biosensors. TrAC Trends in Analytical Chemistry 2016;79:127–33.

Skládal P. Piezoelectric biosensors: shedding light on principles and applications. Mikrochimica Acta 2024;191(4):184. Available from https://doi.org/10.1007/s00604-024-06257-9.

Soleja N, Mohsin M. Exploring the landscape of FRET-based molecular sensors: design strategies and recent advances in emerging applications. Biotechnology Advances 2024;77:108466. Available from https://doi.org/10.1016/j.biotechadv.2024.108466.

Srivastava SK, Van Rijn CJ, Jongsma MA. Biosensor-based detection of tuberculosis. RSC advances 2016;6(22):17759–71.

Stoyanov PG, Grimes CA. A remote query magnetostrictive viscosity sensor. Sensors and Actuators A: Physical 2000;80(1):8–14.

Swain J, Swain S, Singh D, Jena A, Samantaray R, Sahu R. Role of piezoelectric biosensors. Point-of-Care Biosensors for Infectious Diseases 2023:129–45.

Torres-Chavolla E, Alocilja EC. Nanoparticle based DNA biosensor for tuberculosis detection using thermophilic helicase-dependent isothermal amplification. Biosensors and Bioelectronics 2011;26(11):4614–8.

Uniyal A, Srivastava G, Pal A, Taya S, Muduli A. Recent advances in optical biosensors for sensing applications: a review. Plasmonics 2023;18(2):735–50.

Wu J, Liu H, Chen W, Ma B, Ju H. Device integration of electrochemical biosensors. Nature Reviews Bioengineering 2023;1(5):346–60.

Xianyu Y, Wang Q, Chen Y. Magnetic particles-enabled biosensors for point-of-care testing. TrAC Trends in Analytical Chemistry 2018;106:213–24.

Yuanfeng P, Ruiyi L, Qingqing X, Xiaofen C, Yongqiang Y, Zaijun L. Electrochemical detection of Mycobacterium tuberculosis IS6110 gene fragments based on the gold nanocrystals with uniform morphology and highly exposed high-index facets and target DNA-induced recycling amplification. Sensors and Actuators B: Chemical 2020;314:128061.

Zeng M, Wang X, Tan Z, Guo W, Deng Y, Li S, et al. A novel rapid detection method for Mycobacterium tuberculosis based on scattering-light turbidity using loop-mediated isothermal amplification. Biosensors 2025;15(3):162.

Zhang H, Chen P, Yang L, Wang H, Zhu Z. Fabrication of surface acoustic wave biosensors using nanomaterials for biological monitoring. Nanomanufacturing 2024;4(3):159–72.

Zhang J, Huang J, He F. The construction of Mycobacterium tuberculosis 16S rDNA MSPQC sensor based on exonuclease III-assisted cyclic signal amplification. Biosensors and Bioelectronics 2019;138:111322.

Zhang Y, Tang H, Chen W, Zhang J. Nanomaterials used in fluorescence polarization based biosensors. International Journal of Molecular Sciences (IJMS) 2022;23(15):8625. Available from https://doi.org/10.3390/ijms23158625.

Zhou M, Li X, Wen H, Huang B, Ren J, Zhang J. The construction of CRISPR/Cas9-mediated FRET 16S rDNA sensor for detection of Mycobacterium tuberculosis. Analyst 2023;148(10):2308–15.

CHAPTER 6

Electrochemical detection of *Mycobacterium tuberculosis*

Priyanshi Gupta, Vanshika Kumari, Shikha Teotia, and Maumita Das Mukherjee
Amity Institute of Applied Sciences, Amity University, Uttar Pradesh, India

6.1 Introduction

Electrochemical biosensors are widely utilized because of features such as rapid response times, ability to detect multiple analytes and less costly.

The exceptional sensitivity and potential to have miniaturization ability make them well suited for portable and POC devices (Das et al., 2012). Furthermore, their compatibility with real-time monitoring and signal-processing methods adds to their practicality. These advantages collectively position electrochemical sensors as a preferred choice for precise, rapid, and accessible detection in both clinical and environmental settings (Mukherjee et al., 2015; Kim et al., 2021). A schematic comparison of various biosensors, highlighting the significant role of electrochemical biosensors in the detection of MTB, is presented in Fig. 6.1.

Electrochemical biosensors are diagnostic tool which can detect analyte using an electrochemical transduction. These sensors translate various kind of biological interactions—such as binding of a target analyte into a measurable signals such as electrical signals, including current, potential, electrochemical impedance, or electroluminescence, as illustrated in Fig. 6.2. The magnitude of the signal is generally proportional to concentration of the target analyte, enabling accurate quantitative analysis (Singh et al., 2021). An electrochemical biosensor comprises of an electrochemical cell, which have two or three electrodes: a working electrode (WE), a reference electrode (RE) and a counter or auxiliary electrode (CE/AE) (Das et al., 2012; Jjagwea et al., 2024; Grieshaber et al., 2008). All electrodes are generally immersed in a common electrolyte solution. Notably, in potentiometric and conductometric systems, the counter electrode is not required (Shanbhag et al., 2023). The reference electrode—commonly composed of Ag/AgCl or a standard calomel electrode (SCE)—is positioned at a distance from the reaction interface to maintain a stable and well-defined potential. The working electrode, which acts as the primary site for the redox reaction, is typically fabricated from materials such as carbon paste (CPE), indium tin-oxide (ITO), glassy carbon (GCE), or gold (AuE) (Das et al., 2012; Jjagwea et al., 2024; Grieshaber et al., 2008). The binding events between the biological

Advances in Biosensor Technology for Mycobacterium Tuberculosis Detection
DOI: https://doi.org/10.1016/B978-0-443-44589-7.00005-4

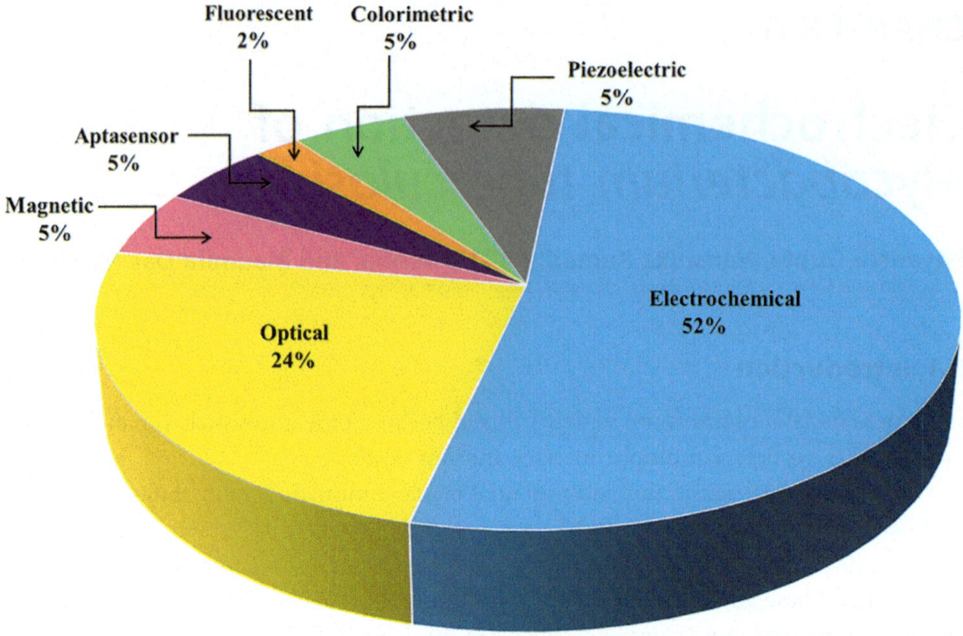

Figure 6.1 Comparison of biosensors, emphasizing the prominent role of electrochemical biosensors in detecting *Mycobacterium tuberculosis* (MTB). Comparison of biosensors, emphasizing the prominent role of electrochemical biosensors in detecting MTB.

element and the target analyte are recognized by a WE, while the counter electrode completes the electrical circuit by sustaining current flow through the electrolytic solution to the WE (Grieshaber et al., 2008). In standard three-electrode systems, a potential is given between WE and RE, while produced current is quantified between the working and counter electrodes. Electrode surfaces are often modified using different nanomaterials such as Nano gold, CNT, nano-structured polymers, 2-D nanomaterials like Graphene, which introduce functional groups which will help in immobilizing biorecognition elements (Das et al., 2012). Detection of analyte using such a three electrode-system, provides high precision and sensitivity in detecting biochemical interactions, making electrochemical biosensors an attractive platform for the rapid and reliable detection of MTB. Based on the type of biorecognition element, electrochemical biosensors can be broadly classified into two main categories: biocatalytic sensors and affinity-based sensors. Biocatalytic sensors utilize enzymes as the biorecognition element, influencing their catalytic activity to drive specific biochemical reactions with the target analyte, which are subsequently transduced into measurable signals. In contrast, affinity-based sensors rely on selective binding interactions between the target analyte and biomolecules such as antibodies, aptamers, nucleic acids, or receptor proteins. Although a large variety of biorecognition elements

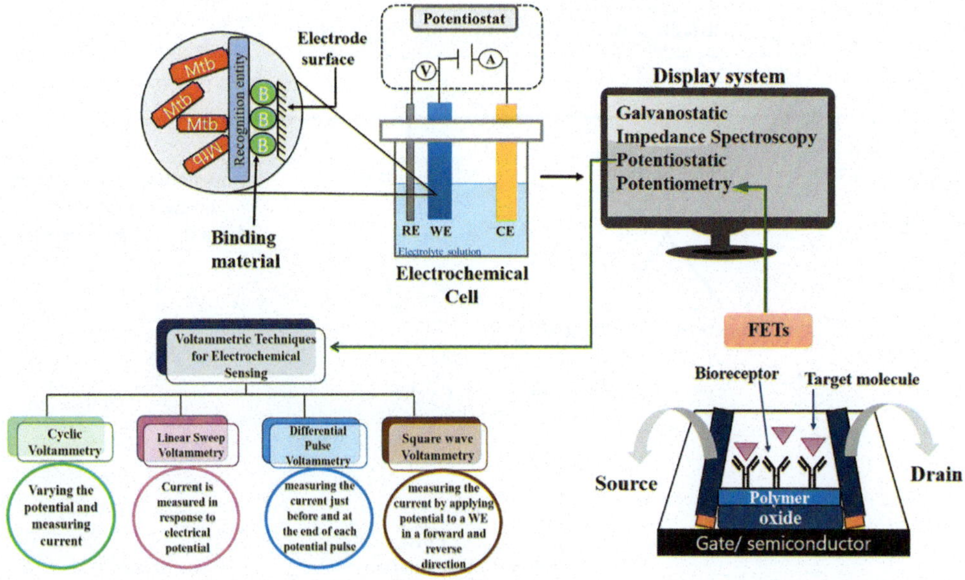

Figure 6.2 An electrochemical biosensing platform for analyte detection, utilizing a combination of different analytical approaches. An electrochemical biosensing platform for analyte detection, utilizing a combination of different analytical approaches.

are employed in biosensor design, still electrochemical biosensors most commonly operates on enzymes. This preference arises from enzymes' dual advantages, their high specificity for target substrates and their intrinsic biocatalytic activity, which enables the generation of detectable electrochemical signals through substrate conversion. These properties make enzyme-based electrochemical biosensors particularly effective for sensitive and selective analyte detection.

6.2 Different techniques of electrochemical biosensors

An electrochemical biosensor interprets the biochemical interaction between a biomolecule and an analyte into an electrical signal, and the signal vary proportionally to the analyte concentration. The modified electrode is composed of different types of materials especially nanomaterials such as gold, silver, carbon, or platinum, which detect the analyte by changing different electrochemical parameters. Depending on the specific detection principle employed, electrochemical biosensors are classified as Potentiometry, Potentiostatic, Galvanostatic, and Impedance Spectroscopy as shown in Fig. 6.3.

1. Potentiometric
2. Field-effect transistors

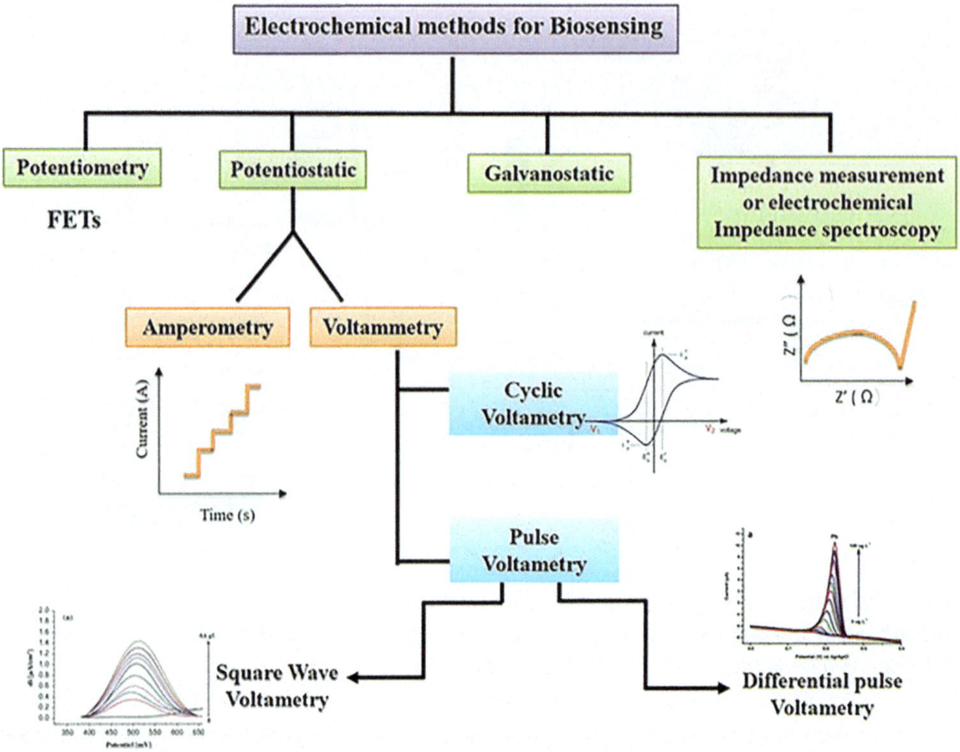

Figure 6.3 Flow chart showing different methods of electrochemical biosensing.

Field-effect transistor (FET) biosensors, also known as semiconductor biosensors are ultrasensitive, label-free, and miniaturized devices that detect biological analytes by converting molecular binding events into electrical signals. These sensors typically comprise a semiconductor channel (e.g., graphene, silicon) connected to two electrodes (source and drain) (Reddy et al., 2012). Where the gate electrode is replaced by a biorecognition layer; binding charges alter the channel's surface potential and conductance, enabling real-time, low-cost, and rapid point-of-care diagnostics. During the biorecognition process, with the interaction of a charged analyte, electrostatic changes occurs, known as the gating effect, alter the electrical properties of the channel. These changes are detected as variations in the drain-source current or channel conductance. FET biosensors offer distinct advantages, including high sensitivity, rapid response, and label-free detection, making them highly effective for real-time, and precise biomolecular sensing (Hou et al., 2016). Wang et al. (2025) developed a graphene field-effect transistor (GFET) biosensor integrated with the CRISPR/Cas12a system for detecting Mycobacterium tuberculosis DNA. The CRISPR/Cas12a enzyme provides highly specific recognition and cleavage of target DNA, while the GFET platform translates these molecular interactions into measurable electrical signals with inherent amplification. This combination enables rapid, ultrasensitive detection without the need for nucleic acid preamplification, achieving a remarkably low detection limit of 2.42×10^{-18} M (Wang et al., 2025).

6.2.1 Potentiostatic

6.2.1.1 Amperometry

Amperometric determination is widely used because of its high sensitivity and accuracy in analyte detection. In this electroanalytical techniques, we measure the electric current produced due to oxidation or reduction of an electroactive species present on the electrode and which is held at a constant potential over time. The setup involves an electrochemical cell with two or three electrodes placed in a suitable electrolyte solution. The working electrode is where the target reaction takes place, while the reference electrode (like Ag/AgCl or Hg/Hg_2Cl_2) maintains a stable voltage to ensure accurate measurements. A third electrode, called the auxiliary or counter electrode, is made from inert materials such as platinum or graphite, helps complete the circuit. To keep the ionic strength stable and to reduce resistance in the solution, a supporting electrolyte is added, which also prevents unwanted ion movement (Mukherjee et al., 2015; Manjushree et al., 2024). The performance of a sensor widely depends on electrode material. As a result, creating and maintaining electrodes has taken a lot of work (Kangmennaa et al., 2024). Heyrovsky's development of the dropping mercury electrode, which earned him the Nobel Prize in 1922, laid the foundation for classical electrochemical analysis. However, in recent years, solid electrodes composed of noble metals and various carbon-based materials have emerged as the preferred choice for sensor applications because of their enhanced stability, sensitivity, and versatility (Manjushree et al., 2024). Mercury was long favored as an electrode material because of its wide cathodic potential window, excellent reproducibility, and easily renewable surface. Among mercury-based electrodes, the hanging mercury drop electrode (HMDE) and mercury film electrode (MFE) were most employed in stripping analysis (Keawkim et al., 2013). However, the main limitations of this approach are the limited anodic potential range of mercury electrodes and the toxicity of mercury. As a result, metal electrodes are now more commonly used as alternatives.

6.2.1.2 Voltammetry

Voltammetry is an electroanalytical technique that measures the current response resulting from the application of a time-varying potential, enabling the quantitative determination of analytes (Kurbanoglu et al., 2017). The technique can be further classified based on the material of the working electrode and the nature of the applied potential, giving rise to methods such as cyclic voltammetry (CV), linear sweep voltammetry, differential pulse voltammetry, and square-wave voltammetry, among others (Chaudhary et al., 2023).

6.2.2 Cyclic voltammetry

CV is a key electrochemical technique used to study redox reactions and electrode processes. It measures the current at the working electrode while scanning the potential between two limits in a back-and-forth manner. A cyclic graph obtained in this process gives the information about redox potentials, reaction kinetics, electron-transfer mechanism and so on (Dzulkurnain et al., 2021). CV uses a three-electrode setup: working, counter, and reference electrodes. It is very useful techniques for redox-active species, sensors application, and catalysis (Alyamni et al., 2024).

6.2.3 Linear sweep voltammetry

Linear Sweep Voltammetry (LSV) is an electrochemical method used to investigate redox behavior by applying a linearly varying potential of the working electrode and recording resulting current. At a constant rate, the potential is swept in one direction—either anodic or cathodic. Unlike CV, LSV does not consider the reverse scan, making it ideal for studying irreversible reactions or systems with slow electron-transfer kinetics (Olabintan et al., 2024). LSV is widely used for understanding the behavior of electrode materials, analyzing its mechanisms of reaction and evaluating electrocatalytic activity. It helps determining different electrode parameters such as peak potential, onset potential, and electron-transfer characteristics (Olabintan et al., 2024). LSV is extensively been used in the field of food analysis, sensing, environmental remediation and many more because of its simplicity, sensitivity, and ability to probe a wide range of chemical and biological systems (Forster and Walsh, 2005).

6.2.4 Differential pulse voltammetry

By reducing background (non-faradaic) currents, differential pulse voltammetry (DPV) a highly sensitive electroanalytical method improves signal detection. It entails applying a working electrode with a linearly increasing base potential and a sequence of voltage pulses on top of it (Laborda et al., 2014). Before and after each pulse, the current is measured twice, and the difference is plotted against the applied potential. By suppressing capacitive currents and enhancing the faradaic signal, this differential technique enables accurate detection of low analyte concentrations. DPV produces distinct, sharp peaks that are particularly useful for separating compounds with similar redox potentials (Uslu and Ozkan, 2011). It is ideal for trace-level analysis of biomolecules, environmental toxins, heavy metals, and pharmaceutical residues because of its high sensitivity and resolution. The method is widely used in the development of electrochemical biosensors and continues to advance through surface modifications and nanomaterial enhancements, with improvements made in selectivity, reproducibility, and detection limits in complex samples. DPV is regarded as a key tool in both research and applied analytical chemistry (Abdullahi et al., 2023). An electrochemical aptasensor was established by Thakur et al. (2017c) for the detection of the MTB-specific antigen MPT64. A graphene-iron oxide-chitosan (CHIT-IO-GR) nanocomposite film was deposited onto a fluorine-doped tin oxide (FTO) electrode to construct the sensor, providing a conductive and biocompatible surface. A biotinylated DNA aptamer specific to MPT64 was immobilized via strong streptavidin-biotin binding. High sensitivity was shown by the resulting aptasensor, achieving a detection limit of 0.9 fg/mL at an operating potential of 0.44 V within 20 minutes. Good reusability and reliability in real-sample conditions were also demonstrated, with recovery rates recorded between 83–95% in spike-in studies (Thakur et al., 2017c).

6.2.5 Square-wave voltammetry

Square-wave voltammetry (SWV) is a fast and sensitive electrochemical method that measures the difference between forward and reverse currents using square-shaped potential pulses. This method improves faradaic signals by minimizing the differential measurement. SWV offers significantly higher peak currents—up to four times higher than differential pulse voltammetry (DPV)—as well as faster data acquisition usually producing a complete voltammogram in a matter of seconds. The devices exceptional signal-to-noise ratio and high scan rate (~500 mV/s) make it perfect for identifying trace analytes in complex samples (Levey and Macpherson, 2024). SWV is frequently used in food safety, environmental analysis and biomedical diagnostics. It is a useful tool in sensor development including those that target metal ions, biomarkers, and pollutants, because of its capacity to differentiate closely related compounds and identify low analyte concentrations. By adding nanomaterials and altering the surface, SWV performance can be further improved bolstering its expanding use in high-throughput analytical applications (Cobb and Macpherson, 2019). In order to detect the TB virulence factor ESAT-6 Li et al. created an ultrasensitive electrochemical aptasensor. (2018). The sensor uses a glassy carbon electrode modified by reduced graphene oxide-metal-organic framework composite (MOF-rGO) to enhance electron-transfer efficiency and facilitate stable immobilization of the redox probe toluidine blue. Platinum-gold core-shell nanoparticles (Pt@Au) were used to anchor thiolated ESAT-6-specific aptamers in order to boost the electrochemical signal and increase sensitivity. Consequently, the aptasensor demonstrated a very low limit of detection and a wide linear detection range (Li et al., 2018). Galvanostatic electrochemical methods entail applying a steady current to an electrochemical system and tracking the voltage change that results over time. Investigating electrode materials corrosion behavior and charge/discharge processes in batteries and supercapacitors are all common uses for this technique. Unlike potentiostatic methods, which regulate the voltage, galvanostatic methods provide insights into reaction kinetics and the stability of electroactive materials under constant-load conditions. The study of nanostructured electrodes and biosensors has benefited from recent developments in galvanostatic methods, which allow better simulation of real-world operating conditions and precise control over redox reactions (Wang et al., 2021).

6.2.5.1 Impedance spectroscopy

A potent and nondestructive method for examining the electrical properties of electrode systems in both the bulk phase and at the electrode-electrolyte interface is electrochemical impedance spectroscopy (EIS) (Maalouf et al., 2007). It makes it possible to extract important electrochemical process parameters such as diffusion coefficients double-layer capacitance and charge-transfer resistance. EIS provides a distinct electrical fingerprint of the interface by sensitively detecting changes in interfacial properties that arise from particular biorecognition events on modified electrode surfaces. Multiple elements are usually involved in electrochemical reactions at the electrode, such as mass transport from the bulk solution to the surface charge, transfer across the interface, solution resistance, and adsorption of electroactive species (Kim et al., 2022). Nyquist or Bode plots are frequently used to visualize the impedance response, which is measured as a function of frequency. Frequently shown is a Nyquist plot which has a semicircular area and a linear tail. The linear region at lower frequencies indicates diffusion-controlled behavior, whereas the semicircle seen at higher frequencies represents the kinetics of the charge-transfer process (Elshafey et al., 2013). Rapid electron-transfer systems may only exhibit the linear diffusion component and lack the semicircle. On the other hand, slow electron-transfer produces a noticeable semicircle whose diameter is directly proportional to the charge-transfer resistance (Wang, 2006). Understanding and managing interfacial phenomena are crucial in battery diagnostics, corrosion research, and biosensor development all of which benefit greatly from EIS. EISs capacity to precisely analyze intricate electrochemical processes has been further improved by developments in data modeling and equivalent circuit analysis. Xia et al. (2023) created a label-free electrochemical impedance sensor to identify the ESAT-6 tuberculosis biomarker. By attaching monoclonal antibodies to nitrogen-doped graphene oxide, a stable and conductive sensing interface was produced. Monitoring impedance changes brought on by ESAT-6s specific binding allowed for detection over a broad range (0. 01–100 ng/mL) with a very low detection limit (7. 0 pg/mL). The sensors potential for quick and accurate TB diagnostics was highlighted by its superior sensitivity, reproducibility, stability, and specificity when compared to traditional techniques (Xia et al., 2023).

6.3 Electrochemical based detection of *mycobacterium tuberculosis*

6.3.1 Immunosensors

6.3.1.1 Amperometric based immunosensors

Changes in current resulting from the formation of immunocomplexes between antigens and antibodies on the working electrode (WE) surface are measured by amperometric immunosensors. The design of amperometric immunosensors combines the high specificity of conventional immunoassays with the enhanced sensitivity and low detection limits that are achievable through modern electrochemical techniques. Electrode selection and the choice of immobilization reagents are two critical factors in the fabrication of electrochemical immunosensors. Various electrode materials have been employed in electrochemical immunosensor design, including carbon paste electrodes (CPE), glassy carbon electrodes (GCE), gold electrodes, and screen-printed carbon electrodes (SPCE). Among these, SPCEs have gained significant attention because of their efficient electrochemical transduction, low cost, miniaturized format, ease of handling, and disposability (Yuan et al., 2010). A study presented by Hiraiwa et al. (2015) provides a cost-effective, rapid method for electrically detecting MTB using a microtip sensor. Signal amplification is accomplished through a duplicated coffee-ring effect, where capillary forces concentrate the bacteria and fluorescent antibodies at the edge of the microtip as it is withdrawn from the sample. This enhanced the formation of immunocomplexes, boosting both fluorescence and electrical signals. The microtip is coated with capture antibodies and designed to work with complex samples such as sputum. The detection limit of 100 CFU mL^{-1} which is comparable with previously mentioned methods (Hiraiwa et al., 2015). Diouani et al. (2016) reported a promising diagnostic tool for the rapid and accurate detection of tuberculosis based on measurement of the antigenic target 6-kDa early secreted antigenic target protein (ESAT-6). ESAT-6 is one of the earliest proteins secreted by MTB, this immunosensor enables TB diagnosis within 2–3 days of bacterial culture—much faster than the traditional 3–4 weeks. This rapid detection could reduce unnecessary and potentially harmful antibiotic treatments often given to patients while waiting for confirmation. The interaction with the ESAT-6 antigen was measured using square-wave voltammetry (SWV) with a ferricyanide redox probe demonstrated high sensitivity, with a detection limit of 7 ng/mL (Diouani et al., 2016). On the other hand, a disposable sensor introduced by Chutichetpong et al. (2018) offers an easy and sensitive electro-chemical sandwich immunoassay for rapid drug susceptibility testing (DST) of tuberculosis. The test detects the MPT64 protein, a marker of MTB growth, to differentiate between drug-sensitive (DS-TB) and multidrug-resistant TB (MDR-TB) within the time period of 3 days and 4–6 days (Chutichetpong et al., 2018). Another sandwich-type immunoassay is syntheised by Tufa et al. (2018) comprising a core-shell Fe_3O_4/Ag/GQDs nanotriplex for detecting MTB via the CFP-10 antigen.

The behavior of electron-transfer (ET) on the nanocomposite-modified electrode surface was studied using the heterogeneous rate constant (k_0) through the Butler-Volmer (BT) model. Both voltammogram and Nyquist plots confirmed that lower ET rates lead to increased overpotentials and higher electron-transfer resistance (Re). The calculated electrochemical active surface area of the modified electrode was approximately 0.33 ± 0.005 cm^2. Furthermore, effective recovery rates in human urine samples spiked with CFP-10 have been demonstrated, verifying possible appropriateness for point-of-care and clinical diagnostics in the real world (Tufa et al., 2018). Mohd Bakhori et al. developed a modified electrode for TB detection using functionalized silica nanoparticles (SiNPs) and a quantum. (CdSe/ZnS) on an SPCE. using an indirect enzyme-linked immunosorbent assay (ELISA) in 2019. The enzyme catalase is linked to the electrode through the binding of antigen to antibody to detect CFP10–. differential pulse voltammetry (DPV) of the ESAT6 antigen. Excellent electrochemical behavior and enhanced electron-transfer were shown by CV characterization, with a 4.14-fold increase in active surface area compared to bare SPCE. High selectivity for CFP10–ESAT6 and a linear detection response (R^2=0.9937) within a range of 40 to 100 ng/mL, with a limit of detection as low as 1.5 × 10^-10 g/mL, were demonstrated by the method. The method showed excellent reproducibility, with a relative standard deviation of 1.45% (Mohd Bakhori et al., 2020). A recent study presented by Braz et al. (2025) developed a low-cost, rapid immunosensor using a lab-made SPE with the peptide TB 68-G as the recognition site. The peptide, designed for both graphene affinity and interaction with anti-M. tuberculosis antibodies, enabled effective electrode modification. The label-free electrochemical sensor achieved a detection limit of 192 ng/mL with an R^2 of 0.98 (Braz et al., 2025).

6.3.1.2 Impedimetric based immunosensors

Impedimetric immunosensors operate by measuring fluctuations in impedance resulting from the formation of antigen-antibody immunocomplexes on the electrode surface. Kim et al. (2014) developed an anodized based capacitive sensor for the detection of interferron-gamma (IFN-γ), a biomarker for latent tuberculosis infection (LTBI). AN anti-IFN-γ body immobilized on the surface of an anodized aluminum oxide (AAO) acting as a bioreceptor measures the capacitance change that detects IFN-γ producing a detection range of ~0.1 pg/mL to 10,000 pg/mL (Kim et al., 2014). Another study presented by Yang et al. (2015) developed a label-free electrochemical impedance immunosensor for the sensitive detection of bovine interferon gamma (BoIFN-γ), important for early diagnosis of MTB infections. TiO2 nanorods, synthesized by a hydrothermal method, were used to immobilize BoIFN-γ antibodies on a glassy carbon electrode. Binding between BoIFN-γ and the antibody increased impedance, enabling detection over a wide range of 0.0001–0.1 ng/mL with a low limit of 0.1 pg/mL, outperforming traditional ELISA methods (Yang et al., 2015). An impedimetric biosensor for detecting the 16 kDa heat shock protein (HSP) from MTB, a useful marker for tuberculosis diagnosis, is developed by Gopinath et al. (2016). A nanogapped dielectric surface, made of silver coated with zinc oxide and gold was used to immobilize specific antibodies. Sensitivity was enhanced by gold nanoparticle conjugation with the antigen, enabling detection as low as 100 fM across a range from 100 fM to 1 nM (Gopinath et al., 2016). A noninvasive impedance analysis method to monitor the fabrication of electroimmunosensors for detecting MTB was presented by Sepulveda et al. (2017). Gold nanolayers were deposited on copper to make the sensors, the surface was then modified with a self-assembled monolayer, and polyclonal antibodies were attached. Each fabrication step was tracked by impedance measurements, and strong responses to the ESAT-6 antigen were shown, confirming the method's effectiveness for both sensor assembly monitoring and antigen detection (Sepulveda et al., 2017). Xia et al. have recently developed an ultrasensitive label-free electrochemical impedance immunosensor that targets the ESAT-6 protein to detect MTB. in 2023. The recombinant His-CFP10-ESAT-6 protein was used to generate stable monoclonal antibodies against ESAT-6. An electrode modified with nitrogen-doped graphene oxide (N-GO) was used to immobilize the best antibody (4D7). The sensor detected ESAT-6 over a wide range of 0.01–100 ng/mL with a low detection limit 7.0 pg/mL, showing high sensitivity, specificity, reproducibility, and stability compared to traditional methods (Xia et al., 2023).

6.3.2 Genosensors

6.3.2.1 Amperometric based genosensors

Amperometric genosensors operate on the principle of DNA hybridization, where a single-stranded DNA (ssDNA) probe immobilized on an electrode surface selectively binds to a complementary TB-specific target sequence. Upon hybridization, the system generates an electrical signal through redox reactions involving enzymatic labels or redox-active indicators, which is then measured as a current at a fixed potential. The incorporation of nanomaterials such as gold nanoparticles, graphene oxide, and carbon nanotubes has significantly enhanced the sensitivity and electron-transfer capabilities of these sensors. The Barreda-Gracia group. 2014 used a combination of electrochemical detection and isothermal DNA amplification to identify DNA sequences with high sensitivity and specificity. A target MTB sequence is amplified at 65°C and captured on magnetic beads using asymmetric helicase-dependent amplification (HDA). As few as 15 gene copies (0.5 aM) can be detected in less than 4 hours thanks to a hybridization step that improves specificity. The method is a flexible tool for molecular diagnostics since it can be modified to detect other DNA targets and is effective on clinical samples such as sputum urine and pleural fluid (García et al., 2015). Zribi et al. (2016) created an electrochemical biosensor based on microfluidics that can detect DNA directly, label-free, and without amplification. The sensor uses ferrocene as a redox-active marker and incorporates multiwalled carbon nanotubes (MWCNTs) to improve electron-transfer efficiency. With a wide dynamic range from 0.1 fM to 1 pM this system showed high sensitivity, allowing the detection of the rpoB gene of MTB at femtomolar concentrations (Zribi et al., 2016). Khoder et al. present a different study that uses ferrocene as a redox-active marker. 2019 for MTB detection. The application of aligned polypyrrole nanowires (nw-PPy) which are produced without a template, as DNA detection transducers was explained. Polyamidoamine dendrimers (PAMAM) and ethylenediamine (EDA) which contain amines are used to further functionalize the nw-PPy surface. Their electrochemical behavior is monitored by redox molecules such as ferrocene. This method uses single nucleotide polymorphism (SNP) analysis to identify drug-resistant strains and detect MTB (Khoder and Youssouf, 2019). Jaroenrama et al. (2020) introduced screen-printed graphene electrodes (SPGE) for label-free electrochemical detection and loop-mediated isothermal amplification (LAMP) for DNA amplification. Integrated with a portable, in-house potentiostat, the device provides a rapid, qualitative positive/negative result displayed on an LCD screen. It detects as little as 1 pg of DNA (about 40 MTB genomes) with 100% accuracy across 104 clinical sputum samples. Cost-effective (~$2.5/ test), fast (< 65 minutes), and instrument-free, this method is well-suited for TB detection in both low- and high-resource settings (Jaroenrama et al., 2020). A highly repetitive mobile genetic marker widely used for identifying members of the MTB complex is IS6110. Due to its high copy number (1–20 copies per cell) and strong presence in the genome, IS6110 serves as a gold standard biomarker for TB detection. The sensor's performance developed by Javed et al. (2021) was evaluated using CV and DPV. The goal is to develop a reliable, accurate, and portable diagnostic tool for identifying active TB cases, especially those that are often missed, to improve timely treatment and disease control exhibiting a detection limit of 3.40pM (Javed et al., 2021). Chaturvedi et al. (2023) presented a highly sensitive and selective electrochemical biosensor for detecting MTB, based on a ternary nanocomposite of reduced graphene oxide, polydopamine, and gold nanoparticles (rGO-PDA-AuNP). The MTB-specific probe DNA is immobilized on a glassy carbon electrode via avidin-biotin coupling. Structural and surface characterizations were confirmed using UV-Vis, Raman, XRD, TEM, FESEM, and FT-IR. Electrochemical analysis using CV and LSV showed the sensor detect target DNA down to 0.1 × 10-7 mM with fast response (5 seconds) and high sensitivity. This platform offers an approach for early-stage MTB diagnosis and can be extended to detect other pathogens and biomarkers (Chaturvedi et al., 2023).

6.3.2.2 Impedimetric based Genosensors

Impedimetric genosensors detect changes in impedance as target complementary single-stranded DNA (ssDNA) hybridizes with the ssDNA probe on the electrode surface. The sensitivity of these sensors to target DNA (tDNA) is assessed by observing variations in the electron transfer resistance. The tDNA concentration is measured by these biosensors using a linear relationship between the tDNA concentration and the relative electron transfer resistance. Costa and associates. A genosensor for the electrochemical detection of MTB genomic DNA was presented in 2014. The biosensor immobilizes a DNA probe using SAMs of magnetite nanoparticles (Fe_3O_4 NPs) and mercaptobenzoic acid (MBA) on a gold electrode. CV electrochemical impedance spectroscopy (EIS) and a redox pair were used to assess the sensors interaction with the DNA. With a detection limit of 6 ng/μL the sensor can identify TB DNA at low concentrations (Costa et al., 2014). A highly sensitive and specific diagnostic system is required because conventional diagnostic techniques can be unreliable. Thus, Perumal et al. (2018) offers a method that combines impedimetric sensing with nanostructured composite materials. Three-dimensional graphene was used to create gold (Au) nanorods with Au nanoparticles forming rod-like structures on the graphene surface. With a detection range of 10 fM to 0. 1 μM this composite allowed for efficient bio-capturing and impedimetric sensing of tuberculosis DNA. High-performance biosensing for tuberculosis detection is provided by the 3D graphene/Au nanocomposite (Perumal et al., 2018).

6.3.3 Aptasensor

6.3.3.1 Amperometric based Aptasensor

An ssDNA probe known as an aptasensor can bind to a wide range of substances such as proteins peptides amino acids cytokines medications and even entire cells (Tombelli et al., 2007). The Systematic Evolution of Ligands by Exponential Enrichment (SELEX) method is used in vitro to create these probes. Aptasensors work especially well for identifying toxic or non-immunogenic substances (Jayasena, 1999). In addition to their high specificity chemical stability reusability reproducibility and sensitivity aptasensors offer superior binding affinity over traditional immunosensors because of their capacity to undergo structural changes upon target interaction (Song et al., 2008). The electrochemical aptasensors that are produced when coupled with electrochemical transducers have improved sensitivity reduced cost and quicker detection. As a result they are increasingly being used in a variety of fields such as bioanalysis environmental monitoring and clinical diagnostics (Joshi et al., 2022). Interferon-γ (IFN-γ) is a biomarker used to diagnose tuberculosis a worldwide health concern. Liu and others. Created in 2015 this innovative electrochemical aptasensor uses TdT-mediated signal amplification and exonuclease-assisted target recycling to detect IFN-γ. The sensor is based on a graphene-gold nanoparticle modified electrode and produces a powerful electrochemical signal by binding to amplified DNA with $[Ru(NH_3)_6]^{3+}$. For clinical applications it provides stability specificity and high sensitivity with a detection limit of 0. 003 ng/mL (Liu et al., 2015). Bai et al. present an ultrasensitive electrochemical aptasensor to identify the MTB antigen MPT64. (2017). Novel polyaniline (C_{60}-PAn) nanohybrids doped with fullerene (C_{60}) are employed as electrocatalysts and redox probes for signal amplification. In order to produce a detectable electrochemical signal for a sandwich assay a signal aptamer is attached to a gold nanoparticle-decorated $C\beta^1$-PAn. With a limit of detection of 20 fg/mL and a broad detection range (0. 02–1000 pg/mL) the sensor exhibits high sensitivity specificity and potential for serum testing (Bai et al., 2017). Thakur et al. develop another study based on the detection of MPT64. (2017). Poly(34-ethylenedioxythiophene) (PEDOT) doped with carbon nanotubes (CNTs) was used to create an electrochemical DNA aptasensor for the detection of MTB antigen MPT64. Streptavidin-functionalized carbon nanotubes were used to immobilize the biotinylated aptamer. Fabrication steps were verified by FT-IR FE-SEM EIS and DPV characterization. The aptasensor demonstrated a detection limit of 0. 5 fg/mL in 15 minutes stability for 27 days and reusability for seven cycles using $[Fe(CN)^2]^{3-}/\Psi^-$ as a redox probe (Thakur et al., 2017a). Abnous et al. describe an electrochemical aptasensor for detecting interferon-gamma (IFN-γ) in blood which is essential for diagnosing latent tuberculosis. (2017). The sensor offers rapid detection (50 min) sensitivity and selectivity by combining amperometry with a triple-helix molecular switch (THMS). The THMS breaks down when IFN-γ is present enabling Methylene Blue (MB) to attach to the signal probe and create a weak current. There is a strong current when IFN-γ is absent. With a linear range of 10 to 1500 pg/mL and a detection limit of 3 pg/mL the sensor can recover between 91. 3 and 102. 8 percent of spiked serum samples (Abnous et al., 2017). Li et al. have developed a voltammetric aptasensor for the ultrasensitive detection of ESAT-6 a virulence factor secreted by MTB. in 2018. metal-organic framework-doped reduced graphene oxide (MOF@rGO) on a glassy carbon electrode to improve electron transfer and Toluidine Blue (TB) immobilization. The response is amplified by platinum-gold core-shell nanoparticles. A detection range from 1.0×10^{-4} ng/mL to 2.0×10^{2} ng/mL is shown by the sensor, with a limit of detection of 3.3×10^{-5} ng/mL (Li et al., 2018). Gou et al. present an electrochemical immunoassay for ultrasensitive detection of MTB antigen MPT64 which is essential for early diagnosis. in 2018. Antibodies against MPT64 were immobilized on a gold electrode using Protein G. A nanocomposite ($GO@Fe3O_4@Pt$) was created by combining graphene oxide with magnetite (Fe3O4) and platinum (Pt) nanoparticles for signal amplification. With a limit of detection of 0. 34 fg/mL and a detection range of 5. 0 fg/mL to 1. 0 ng/mL the assay can be finished in 4 hours (Gou et al., 2018). Yunus et al. present a sensitive sandwich biosensor based on aptamers and antibodies for the early detection of MTB CFP10 antigen. using chronoamperometry in 2021. An SPCE was used to immobilize an anti-CFP10 aminated aptamer. Horseradish peroxidase (HRP) and H_2O_2 amperometric responses were used to quantify CFP10 with hydroquinone acting as an electron mediator. A detection limit of 1.22 ng/mL in buffer and 1.05 ng/mL in serum is shown by the sensor, demonstrating high sensitivity and rapid performance for point-of-care tuberculosis detection (Yunus et al., 2021). Similarly, Azmi et al. developed a portable electrochemical aptamer-antibody-based biosensor for the detection of MTB in sputum samples. The biosensor utilizes a CFP10-ESAT6 binding aptamer immobilized on a graphene@polyaniline@modified gold electrode. The aptamer captures the target antigen, enhanced by Fe3O4/Au MNPs-conjugated antibodies. The sensor detects CFP10-ESAT6 in the 5–500 ng/mL range, with a detection limit of 1.5 ng/mL. It demonstrated high specificity, sensitivity (100%), and reproducibility (RSD 2.5%), making it suitable for clinical applications (Azmi et al., 2021). Xie et al. develop a further breakthrough in ESAT6 detection. 2021 using an electroactive toluidine blue (Tb) and nitrogen-doped graphene (NG) decorated bimetallic organic framework (Zr-MOF-on-Ce-MOF). This nanohybrid binds ESAT-6 binding aptamers and has a large surface area and good hydrophilicity. The aptasensor shows a linear range from 100 fg/mL to 10 ng/mL and a detection limit of 12 fg/mL, with high reproducibility and potential for clinical tuberculosis diagnosis (Xie et al., 2021). A sensitive electrochemical aptasensor was developed by Li et al. (2022) for MTB antigen MPT64 detection using a new redox nanoprobe. Anthraquinone-2-carboxylic acid (AQCA) was confined in CMK-3, enhancing its electroactivity and signal amplification. The AQCA/CMK-3 nanohybrid, combined with urchin-like Ce-MOFs loaded with gold nanocrystals, facilitated dense capture probe immobilization. The aptasensor demonstrated a linear range of 100 fg/mL to 10 ng/mL, with a detection limit of 67.6 fg/mL, and effectively detected MPT64 in serum samples (Li et al., 2022).

6.3.3.2 Impedimetric based Aptasensor

In impedimetric aptasensors, target recognition is transduced through changes in charge transfer resistance (R_ct) arising from the formation of the aptamer–target complex on the electrode surface. Ding et al. (2017) reported the first use of an interdigitated electrode (IDE) for label-free detection of interferon-γ (IFN-γ) using immobilized RNA aptamers. The gold IDE, consisting of 500 microbands, employs electrochemical impedance spectroscopy (EIS) as the sensing technique. The interdigitated architecture significantly enhances mass transport, accelerates reaction kinetics, and improves the signal-to-noise ratio, thereby increasing detection sensitivity. To minimize nonspecific adsorption and improve selectivity, a ternary surface monolayer composed of mercaptohexanol (MCH) and hexanedithiol (HDT) was employed. The aptasensor detects IFN-γ in a range of 22.22 pM to 0.22 nM with a detection limit of 20.22 pM in buffer and 11.56 pM in FBS, making it ideal for rapid, label-free in-field diagnostics and multiplexed disease detection (Ding et al., 2017). Thakur et al. (2017b) developed an aptaelectrode for detecting MTB-specific antigen MPT64 using a graphene-modified iron-oxide chitosan hybrid (CHIT@IO@GR) nanocomposite film on fluorine tin oxide (FTO). A biotinylated DNA aptamer targeting MPT64 was immobilized on the electrode using streptavidin-biotin interactions. The aptasensor, with a detection limit of 0.9 fg/mL, operated optimally at 0.44 V and maintained 80% of its activity after 10 uses. Recovery studies showed 83–95% accuracy (Thakur et al., 2017b). Another study based on the detection of MTB secreted protein MPT64 using an IDE combined with EIS is presented by Sypabekova et al. (2019). The assay employs a ssDNA aptamer that specifically recognizes MPT64. The aptamer is immobilized on the IDE using a co-adsorbent method. For serum and sputum clinical samples from TB patients the EIS aptasensor exhibits 100% specificity and high sensitivity detecting MPT64 with a limit of detection of 4. 1 fM. Fast and sensitive TB diagnosis is offered by the aptasensor (Sypabekova et al., 2019).

6.4 Conclusion

TB is still a serious global health issue especially in low- and middle-income nations where timely and accurate diagnosis is scarce. The TST chest radiography and NAATs are examples of conventional diagnostic techniques that have several drawbacks such as false positives high expenses reliance on infrastructure and delayed results. These difficulties make it more difficult to diagnose TB early and treat the illness effectively especially in areas with limited resources where the disease is most prevalent. Biosensor technologies particularly electrochemical biosensors have become viable substitutes in response. POC applications are made possible by the high sensitivity specificity quick detection and miniaturization potential of these devices. Amperometry voltammetry potentiometry and electrochemical impedance spectroscopy are examples of electrochemical methods that offer reliable platforms for TB biomarker detection with small sample volumes and quick turnaround times. This chapter emphasizes the shift from traditional diagnostic methods to biosensor-based platforms with an emphasis on electrochemical systems that can meet the worldwide demand for dependable and easily accessible tuberculosis diagnostics. Electrochemical biosensors have the potential to facilitate decentralized real-time testing which can enhance TB detection and decrease transmission. By incorporating these biosensors into public health systems disease control initiatives may be strengthened and efforts to eradicate tuberculosis may be aided.

References

Abdullahi AB, Ismail S, Alshana U. Journal of Food Composition and Analysis 2023;118.

Abnous K, Danesh NM, Ramezani M, Alibolandi M, Hassanabad KY, Emrani AS, et al. Michrochimica Acta 2017;184.

Alyamni N, Abot JL, Zestos AG. ECS Sensor Plus 2024;3(2).

Azmi UZM, Yusof NA, Abdullah J, Mohammad F, Ahmad SAA, Suraiya S, et al. Nanomaterials 2021:11.

Bai L, Chen Y, Bai Y, Chen Y, Zhou J, Huang A. Biomaterials 2017;133.

Braz BA, Santiani MH, Martins G, Moscardi APZ, Beirão BCB, Soccol CR, et al. Talanta 2025;283.
Chaturvedi M, Patel M, Bisht N, Shruti, Mukherjee MD, Tiwari A, et al. Biosensors 2023;13(3).
Chaudhary M, Kumar A, Devi A, Singh BP, Malhotra BD, Singhal K, et al. Materials Advances 2023;4.
Chutichetpong P, Cheeveewattanagul N, Srilohasin P, Rijiravanich P, Chaiprasert A, Surareungchai W. Analytica Chimica Acta 2018;1025.
Cobb SJ, Macpherson JV. Analytical chemsitry 2019;91(12).
Costa MP, Andrade CAS, Montenegro RA, Melo FL, Oliveira MDL. Journal of Colloid and Interface Science 2014;433.
Das M, Dhand C, Sumana G, Srivastava AK, Nagarajan R, Malhotra BD. Journal of Material Chemistry 2012;22:2727.
Ding S, Mosher C, Lee XY, Das SR, Cargill AA, Tang X, et al. ACS Sensors 2017;2(2).
Diouani MF, Ouerghi O, Refai A, Belgacem K, Tlili C, Laouini D, et al. Materials Science & Engineering C 2016:74.
Dzulkurnain NA, Mokhtar M, Rashid JIA, Knight VF, Wan Yunus WMZ, Ong KK, et al. Polymers 2021;13(16).
Elshafey R, Tavares AC, Siaj M, Zouro M. Biosensors Bioelectronics 2013;50.
Forster RJ, Walsh DA. Encyclopedia of Analytical Science (Second Edition) 2005.
García SB, González-Álvarez MJ, de-los-Santos-Álvarez N, J.Palacios-Gutiérrez J, Miranda-Ordieres AJ, Lobo-Castañón MJ. Biosensors and Bioelectronics 2015:68.
Gopinath SCB, Perumal V, Kumaresan R, Lakshmipriya T, Rajintraprasad H, Rao BS, et al. Microchimica Acta 2016;183.
Gou D, Xie G, Li Y, Zhang X, Chen H. Michrochimica Acta 2018;185:436.
Grieshaber D, MacKenzie R, Vörös J, Reimhult E. Sensors (Basel, Switzerland) 2008;8(3).
Hiraiwa M, Kim JH, Lee HB, Inouel S, Becker AL, Weigel KM, et al. Journal of Micromechanics and Microengineering 2015;25.
Hou S, Zhang A, Su M. Nanomaterials 2016;6.
Jaroenrama W, Kampeeraa J, Arunrut N, Karuwan C, Sappat A, Khumwana P, et al. Journal of Pharmaceutical and Biomedical Analysis 2020;186.
Javed A, Abbas SR, Hashmi MU, Babar NUA, Hussain I. International journal of nanomedicine 2021;16.
Jayasena SD. Clinical Chemistry 1999;45(9).
Jjagwea J, Olupot PW, Kulabako R, Carrara S. Heliyon 2024;10(8).
Joshi H, Kandari D, Maitra SS, Bhatnagar R. Critical Reviews in Microbiology 2022;48(6).
Kangmennaa A, Forkuo RB, Agorku ES. Sensing Technology 2024;2(1).
Keawkim K, Chuanuwatanakul S, Chailapakul O, Motomizu S. Food control 2013;31(1).
Khoder R, Youssouf HK. Materials Science & Engineering C 2019:108.
Kim JH, Chang YW, Bok E, Kim HJ, Lee H, Cho SN, et al. Biosensors and Bioelectronics 2014;51.
Kim J, Park S, Hwang S, Yoon W-S. Journal of Electrochemical Science and Technology 2022;13(1).
Kim JH, Suh YJ, Park D, Yim H, Kim H, Kim HJ, et al. Biomedical Engineering Letters 2021;11.
Kurbanoglu S, Uslu B, Ozkan SA. Nanostructures for oral medicine 2017.
Laborda E, González J, Molina Á. Electrochemistry communications 2014;43.
Levey KJ, Macpherson JV. Analytical Chemistry 2024;96(23).
Liu C, Xiang G, Jiang D, Liu L, Liu F, Luo F, et al. Analyst 2015;140(22).
Li Y, Peng D, Guo S, Yang B, Zhou J, Zhou J, et al. Bioelectrochemistry 2022;147.
Li L, Yuan Y, Chen Y, Zhang P, Bai Y, Bai L. Michrochimica Acta 2018;185:379.
Maalouf R, Chebib H, Saïkali Y, Vittori O, Sigaud M, Jaffrezic-Renault N. Biosensors Bioelectronics 2007;22.
Manjushree SG, Ashoka S, Adarakatti PS. IOP Science 2024.
Mohd Bakhori N, Yusof NA, Abdullah J, Wasoh H, Ab Rahman SK, Abd Rahman SF. Materials 2020;13(1).
Mukherjee MD, Dhand C, Dwivedi N, Singh BP, Sumana G, Agarwal VV, et al. Sensors and Actuators B 2015;210:281–9.
Olabintan AB, Abdullahi AS, Yusuf BO, Ganiyu SA, Saleh TA, Basheer C. Microchemical Journal 2024;204.

Perumal V, Saheed MSM, Mohamed NM, Saheed MSM, Murthe SS, Gopinath SCB, et al. Biosensors and Bioelectronics 2018;116.
Reddy D, Register LF, Carpenter GD, Banerjee SK. Journal of Physics D: Applied Physics 2012;45.
Sepulveda D, Aroca MA, Varela A, Portillo PD, Osma JF. Sensors (Basel, Switzerland) 2017;17(10).
Shanbhag MM, Manasa G, Mascarenhas RJ, Mondal K, Shetti NP. Chemical Engineering Journal Advances 2023;16.
Singh A, Sharma A, Ahmed A, Sundramoorthy AK, Furukawa H, Arya S, et al. Biosensors 2021;11(9).
Song S, Wang L, Li J, Fan C, Zhao J. Trends in Analytical Chemistry 2008;27(2).
Sypabekova M, Dukenbayev K, Tsepke A, Akisheva A, Oralbayev N, Kanayeva D. Scientific Reports 2019;9:16273.
Thakur H, Kaur N, Sabherwal P, Sareen D, Prabhakar N. Michrochimica Acta 2017;184.
Thakur H, Kaur N, Sareen D, Prabhakar N. Talanta 2017;171.
Thakur H, Kaur N, Sareen D, Prabhakar N. Michrochimica Acta 2017;184.
Tombelli S, Minunni M, Mascini M. Biomolecular Engineering 2007;24(2).
Tufa LT, Oh S, Tran VT, Kim J, Jeong KJ, Park TJ, et al. Electrochimica acta 2018;290.
Uslu B, Ozkan SA. Analytical Letters 2011;44(16).
Wang W, Du H, Dai C, Ma H, Luo S, Wang X, et al. Nanoscale 2025;17.
Wang Y, Graham ES, Unsworth CP. Journal of neural engineering 2021;18(4).
J. Wang, New Jersey:John Wiley & Sons, Inc., 2006.
Xia Y, Zhu H, Liu R, Li X, Xie X, Zhao J, et al. Talanta 2023;253.
Xie J, Mu Z, Yan B, Wang J, Zhou J, Bai L. Michrochimica Acta 2021;188:404.
Yang Z, Jian Z, Chen X, Li J, Qin P, Zhao J, et al. Biosensors and Bioelectronics 2015;63.
X. Yuan, C. Song, H. Wang, J. Zhang, London: Springer-Verlag, 2010.
Yunus MH, Yusof NA, Raston NHA, Noor SSM, Sulaiman Y, Abdullah J. IEEE sensors journal 2021;21(20).
Zribi B, Roy E, Pallandre A, Chebil S, Koubaa M, Mejri N, et al. Biomicrofluidics 2016;10.

CHAPTER 7

Optical biosensor-based detection of *Mycobacterium tuberculosis*

Garima Rathee[1], Sahil Kohli[2], Jyotsna Rathee[3], Antonio Puertas-Segura[1], and Pratima R. Solanki[4]

[1]Department of Chemical Engineering, Universitat Politècnica de Catalunya, Barcelona, Spain,
[2]Research and Development Cell, Department of Chemistry, Manav Rachna International Institute of Research and Studies, Faridabad, Haryana, India,
[3]Computer Science & Engineering Department, Deenbandhu Chhotu Ram University of Science and Technology, Sonipat, Haryana, India,
[4]Nanobio Laboratory, Special Centre for Nanoscience, Jawaharlal Nehru University, New Delhi, Delhi, India

7.1 Introduction

Notwithstanding significant advancements in medical knowledge and health care provision, TB continues to be one of the most deadly infectious illnesses, especially in low- and middle-income areas. Prompt identification and immediate intervention are crucial for curtailing disease's transmission and mitigating its health and economic impacts. The protracted period of TB therapy and the rising prevalence of drug-resistant TB strains provide considerable obstacles to disease management and prevention. In this backdrop, innovations in nanotechnology and photonics have begun transforming the realm of TB diagnoses. These developing technologies provide sensitive and specific approaches for early illness detection, facilitating more precise selection of suitable therapy solutions. The increasing demand for cost-effective, accessible, and portable diagnostic instruments is especially urgent in resource-limited environments. Photonic biosensors, distinguished by their compactness, sensitivity, and possibility for remote application, are widely acknowledged as viable solutions to address this gap. In contrast to traditional diagnostic approaches, which are accurate but frequently costly, time-consuming, and dependent on skilled laboratory staff, photonics-based technologies provide quick, economical alternatives ideal for extensive implementation. Integrating novel biosensing systems into clinical and field-based TB diagnostics might expand access to prompt diagnosis and support worldwide efforts to combat this persistent public health concern (Chauke et al., 2024).

Point-of-care (POC) diagnostic devices have become pivotal instruments for the swift and accurate identification of TB, particularly in environments with constrained laboratory facilities. These technologies have substantial benefits compared to conventional diagnostic procedures, including expedited return times, frequently yielding findings within minutes, portability, ease of use, and minimum training prerequisites

Advances in Biosensor Technology for Mycobacterium Tuberculosis Detection
DOI: https://doi.org/10.1016/B978-0-443-44589-7.00012-1

(Steingart et al., 2014). Additionally, many POC systems can identify drug-resistant strains of *Mycobacterium tuberculosis* (MTB), facilitating the prompt selection of suitable treatment regimens (Lawn and Nicol, 2011; Khan et al., 2007). Commonly utilized POC technologies include nucleic acid amplification tests (NAATs), lateral flow assays (LFAs), and microfluidic devices (Boehme et al., 2010). Nucleic acid amplification tests (NAATs) such as GeneXpert and Truenat, exhibit enhanced sensitivity and specificity. GeneXpert, a cartridge-based technology, can detect MTB DNA and rifampicin resistance within two hours by targeting mutations in the rpoB gene (Denkinger et al., 2014). Truenat, a portable NAAT device powered by batteries, provides findings in less than 90 minutes and is recommended by the World Health Organization (WHO, 2020, 2021) for TB detection, especially in decentralized environments.

LFAs, such as the Determine TB-LAM Ag test and TB IgG/IgM fast tests, identify antigens or antibodies in patient specimens (for example, sputum or urine). While often less sensitive than NAATs, they are economical, user-friendly, and provide findings in 10 to 15 minutes. The Determine TB-LAM test is authorized for the diagnosis of TB in HIV-positive patients exhibiting severe immunosuppression. Microfluidic technologies, such as lab-on-a-chip and mChip, improve diagnostic accessibility by consolidating sample processing and detection into tiny, portable formats. These devices can detect TB and identify drug-resistant strains with high sensitivity and specificity, rendering them particularly appropriate for low-resource settings (Engel et al., 2015; De Castro et al., 2022).

7.2 Principles of optical biosensing

Optical biosensors have developed as strong analytical instruments for identifying infectious diseases, including TB, because of their high sensitivity, rapid response, and capability for label-free detection. These biosensors detect modifications or changes in optical characteristics, such as absorbance, fluorescence intensity, or refractive index, prompted by the interaction of target biomolecules with designated biorecognition components. This section delineates the fundamental operational principles of essential optical biosensing systems and highlights their applicability in TB detection.

7.2.1 Fluorescence-based biosensors

Fluorescence biosensors depend on light emission from fluorophores when excited, allowing real-time, and very sensitive detection of target analytes. Fluorescence-based biosensors for TB detection often include fluorophore-labeled antibodies, aptamers, or nucleic acid probes that specifically bind to TB-specific biomarkers, including MTB DNA, lipoarabinomannan (LAM), or specific antigens. Förster Resonance Energy Transfer (FRET) is a very sophisticated method in this domain, characterized by energy transfer between a donor and an acceptor fluorophore when they are in close proximity, often because of target binding. Variations in FRET effectiveness indicate molecular interactions, enabling precise detection at the nanoscale. Moreover, quantum dot (QD)-based fluorescence biosensors provide improved photostability, adjustable emission spectra, and multiplexing functionalities. Nanoscale semiconductors may be coupled with biomolecules to produce sensitive TB biosensors that can detect numerous targets concurrently, even in complicated biological materials such as sputum or serum (Sharma et al., 2018; Wu et al., 2012; H. R, Schiffman and Balakrishna, 2018).

7.2.2 Surface plasmon resonance biosensors

Surface plasmon resonance (SPR) is a sensitive, label-free optical detection method that utilizes the interaction between light and free electrons on a metal surface, usually gold, to monitor real-time molecule interactions (Fig. 7.1). The fundamental idea of SPR is based on the activation of surface plasmons, which are coherent electron oscillations at the interface between a metal (often gold or silver) and a dielectric (such as an aqueous sample medium), triggered by transmitted polarized light at a precise angle. In a conventional SPR arrangement, light is transmitted through a high-refractive-index

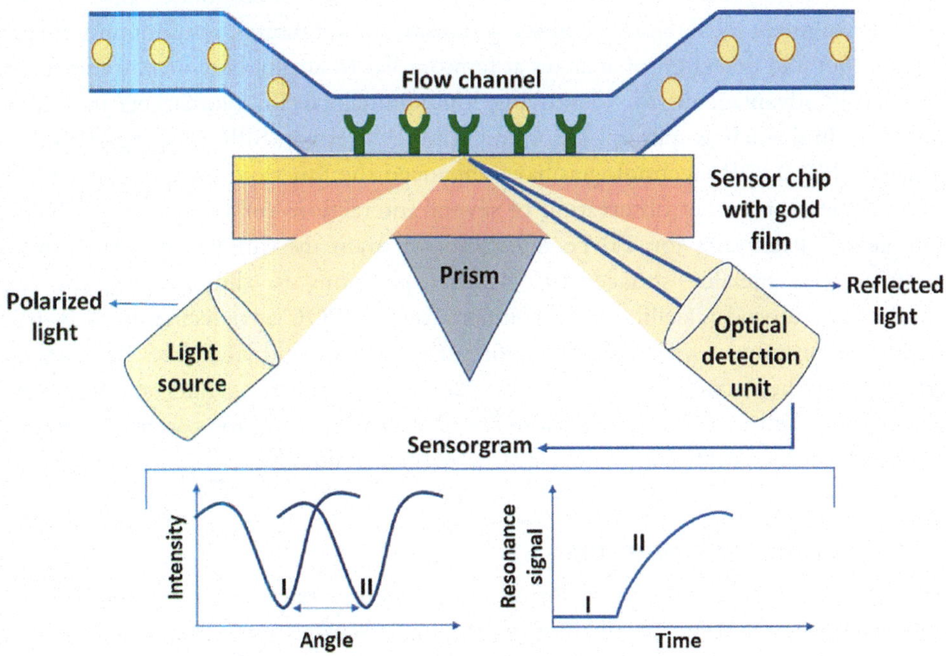

Figure 7.1 Schematic illustration of the surface plasmon resonance (SPR)-based biosensor. *Reprinted with permission (Chaturvedi et al., 2025). Copyright (2025) Elsevier.*

prism (Kretschmann configuration) onto the reverse side of a thin metallic sheet. At a specific angle, termed the resonance angle, the energy of the incident photons aligns with that of the surface plasmons, leading to a decrease in the intensity of the reflected light. This angle is acutely responsive to variations in the local refractive index next to the sensor surface, generally within approximately 200 nm of the interface. Biomolecular interactions on the sensor surface, such as the binding of MTB-specific antigens (e.g., CFP-10, ESAT-6) to immobilized antibodies or aptamers, result in a proportionate rise in the local refractive index corresponding to the mass of the bound analyte. This induces a quantifiable alteration in the SPR angle or resonance wavelength, which may be monitored in real time. The extent of this change is closely correlated with the concentration and kinetics of the analyte, enabling quantitative assessment of binding affinity (KD), association (k_a), and dissociation (k_d) rates.

A primary advantage of SPR is its label-free characteristic, which eliminates the complications and possible interferences linked to fluorescence or enzyme labels. This property and the ability for real-time, high-throughput screening render SPR an optimal instrument for assessing TB biomarkers in fundamental research and clinical diagnosis. Recent advancements in nanomaterial-enhanced SPR have markedly expanded the sensitivity and selectivity of this method. Gold nanoparticles (AuNPs) or graphene oxide (GO) sheets can be functionalized and incorporated into the SPR surface to augment local refractive index variations upon binding, resulting in improved signal output and detection of analytes at femtomolar to attomolar concentrations. This is especially advantageous for identifying low-abundance TB biomarkers in intricate biological fluids such as sputum, blood, or urine. Moreover, SPR imaging (SPRi) and localized SPR (LSPR) technologies have enhanced the functional adaptability of SPR. SPRi enables concurrent observation of several interactions on a single chip, enabling multiplexed diagnostics for TB co-infections or strain discrimination. LSPR utilizes nanostructured metallic surfaces that restrict plasmons to diminutive areas (e.g., nanodots or nanorods), facilitating miniaturization and POC advancements. SPR-based biosensors constitute a viable platform for early, precise, and label-free TB diagnosis, particularly when combined with new surface chemistries, microfluidic sample management, and machine learning-enhanced data processing for pattern recognition and signal interpretation (Zou et al., 2017; Maphanga et al., 2023).

7.2.3 Colorimetric biosensors

Colorimetric biosensors identify color changes caused by particular metabolic events or nanoparticle aggregation, facilitating visual or spectrophotometric analysis. These systems are highly appealing for POC TB diagnosis because of their simplicity, cost-effectiveness, and user-friendliness, particularly in resource-constrained environments. A known method employs the optical characteristics of AuNPs, which have a particular red hue owing to SPR. Aggregation, usually initiated by contact with specific TB biomarkers such as DNA sequences, LAM, or antigens such as CFP-10 and ESAT-6, reduces interparticle distance, causing a spectrum shift and a noticeable color transition from red to blue. Functionalizing AuNPs with particular oligonucleotide probes or antibodies allows for the selective identification of MTB components, enabling quick and sensitive detection without requiring intricate apparatus. Alongside aggregation-based systems, peroxidase-mimicking nanomaterials or nanozymes have exhibited remarkable catalytic efficacy, including iron oxide (Fe_3O_4 NPs), cerium oxide (CeO_2 NPs), and other metal-based catalysts. These nanozymes catalyze the oxidation of chromogenic substrates such as 3,3′,5,5′-tetramethylbenzidine (TMB) in the presence of hydrogen peroxide (H_2O_2), resulting in a colorimetric shift (usually from colorless to blue) in the presence of TB biomarkers. The conjugation of these nanozymes with TB-specific biorecognition components, such as aptamers or antibodies, facilitates target-induced signal production with enhanced sensitivity. Hybrid systems integrating AuNPs with nanozymes have significantly improved assay performance by merging plasmonic and catalytic effects. Sandwich assays employing antibody-functionalized AuNPs and magnetic nanozyme beads enhance colorimetric signal amplification and target separation, enhancing detection limits and reducing background noise. Furthermore, these colorimetric platforms have been effectively incorporated into paper-based analytical devices (PADs) and lateral flow assays (LFAs). These formats have several benefits: they are disposable, need no external power or sophisticated handling, and are compatible with smartphone-based analytical tools for result quantification and remote reporting. Integrating PADs and LFAs in TB diagnostics corresponds with the worldwide demand for rapid, accessible, and scalable testing methods. In conclusion, colorimetric biosensors, utilizing AuNPs, nanozymes, and enzyme-substrate chromogenic reactions, provide an effective and robust method for detecting TB biomarkers. Their receptiveness to POC diagnostics, along with continuous advancements in material chemistry and device integration, establishes them as pivotal contributors to the future of accessible TB detection (Huang et al., 2021; Kooti et al., 2022, 2023; Qian et al., 2022).

7.2.4 Raman and surface-enhanced raman scattering-based biosensors

Raman spectroscopy is a potent vibrational spectroscopy method that identifies the inelastic scattering of incident photons during their interaction with molecular vibrations, rotations, or other low-frequency modes inside a sample. Each molecule generates a unique spectral fingerprint, facilitating accurate chemical identification. The intrinsically weak character of traditional Raman signals, resulting from the low likelihood of inelastic scattering, constrains its sensitivity in practical biosensing applications. To address this constraint, SERS has developed as a revolutionary enhancement technology. SERS significantly enhances Raman signals by many orders of magnitude (enhancement factors ranging from 10^6 to 10^8 or greater) when analytes are situated near nanostructured metallic surfaces, predominantly AuNPs or AgNPs. This amplification results from two primary mechanisms: electromagnetic enhancement, attributed to localized SPR (LSPR), and chemical enhancement, stemming from charge transfer between the analyte and the metal surface. SERS-based platforms in TB biosensing provide exceptional sensitivity and specificity for detecting ultralow quantities of TB-related analytes, including particular nucleic acid sequences, lipids, and protein antigens (e.g., LAM, ESAT-6, CFP-10). Functionalizing metallic SERS substrates with molecular recognition components such as aptamers, antibodies, or peptide probes enables target-specific binding. This facilitates both label-free and tagged detection methodologies based on the assay design.

A primary feature of SERS biosensors is their capacity for quick, multiplexed analysis with little sample preparation. Recent advancements have unveiled SERS nanotags, AuNPs or AgNPs linked with Raman reporter molecules and targeting ligands, that provide highly consistent and distinct spectral signatures for concurrently detecting several TB biomarkers. These attributes are especially beneficial for early diagnosis and strain-specific distinction, essential for effective treatment planning and antibiotic resistance monitoring. Moreover, the interoperability of SERS with portable Raman spectrometers facilitates POC implementation, particularly in resource-limited environments. Incorporating SERS onto microfluidic platforms, lateral flow strips, and smartphone-assisted detection devices has shown significant promise for real-time, field-deployable TB testing. SERS-based biosensors integrate exceptional sensitivity, molecular specificity, and fast detection capabilities, positioning them as a promising advancement in TB diagnosis. Ongoing progress in nanomaterial engineering, substrate consistency, and multiplex assay innovation will augment their dependability and applicability for clinical and epidemiological purposes (Qiao et al., 2024; Hendricks-Leukes et al., 2022).

7.3 Advancements in optical biosensors for *Mycobacterium tuberculosis*

In recent years, substantial advancements have been witnessed in the development of optical biosensors designed to improve the detection of MTB (Table 7.1). These developments have concentrated on enhancing the fundamental performance metrics of sensitivity, specificity, speed, and portability, which are essential for facilitating early and precise TB diagnosis, particularly in resource-limited environments. Significant technical advancements have been propelled by including innovative nanomaterials for signal amplification, compact microfluidic systems for point-of-care applications, and using artificial intelligence (AI) to analyze intricate optical data.

Table 7.1 Summary of optical biosensor platforms for detecting MTB and their analytical characteristics.

Optical biosensor type	Sample type and biomarkers	Material type/electrode	Linear range	Detection limit	Response time	References
SPR	Culture, sputum/ Ag85 B	SPR platform with S2k chips	-	10 ng/ml	15 min	Trzaskowski et al. (2018)
SPR	CFP-10	Gold surface	100 ng/ml - 1 ng/ml μg/ml	100 ng/ ml	200 s	Hawn et al. (2014)
SPR	Culture/ DNA MTB	Gold coated portable-multichannel sensor system	0.05–2 μM	0.05 μM	20 min	Duman and Piskin (2010)
SPR	Culture/ DNA MTB	DNA/Au and PNA/Au (peptide nucleic acid)	5–50 ng/ml	10 ng/ml	10 min	Prabhakar et al. (2008)
FRET	Culture/ IS6110 sequence	Upconversion nanoparticles	10^2–10^6 copies/ μl	10^2 copies/ μl	15 min	He and Zhang (2002)
FRET	Sputum/ ESAT-6	Cadmium-telluride quantum dots/ AuNPs	10 fg – 200 fg	10 fg	4 min	Shojaei et al. (2014)
Electrochemiluminescence	Culture/ rpoB gene	Ruthenium (II) complex/GO	0.1–100 nM	0.04 nM	30 min	Li et al. (2014)
Electrochemiluminescence	Serum/ IS6110 sequence	N-(aminobutyl)-N-(ethylisoluminol) (ABEI) functionalised gold nanoparticles (ABEIAuNPs)	1 fM–1 pM	0.33 fM	5 min	Yu et al. (2012)
Electrochemiluminescence	Culture/ IS6110 sequence	luminol–AuNPs	10 fM–1 pM	6.7 fM	60 min	Jiang, Chai and Cui (2011)

SPR, surface plasmon resonance; *MTB*, *Mycobacterium tuberculosis*; *FRET*, Förster Resonance Energy Transfer

7.3.1 Surface plasmon resonance-based biosensors

SPR-based biosensors may be classified into two primary categories: immunosensors and genosensors. SPR immunosensors identify immunocomplexes generated by the interaction between immobilized antibodies and their corresponding antigens on the sensor surface. Conversely, SPR genosensors detect changes in the local refractive index resulting from the hybridization of single-stranded probe DNA (ssDNA) with its corresponding target DNA sequence. Mukundan et al. (2012) illustrated the efficacy of a waveguide-based optical biosensor utilizing sandwich immunoassay formats for the detection of critical MTB biomarkers, including LAM, early secretory antigenic target-6 (ESAT-6), and the antigen 85 complex (Ag85) (Mukundan et al., 2012). The biomarkers were effectively recognized in both spiked samples and urine, demonstrating the biosensor's efficacy in identifying active TB infection. LAM was measured in urine samples from TB-positive persons (without HIV co-infection) with great sensitivity and specificity, reaching a detection limit as low as 1 picomolar (pM) Fig. 7.2.

Significant work by Hong et al. (2011) created an SPR-based optical immunosensor for detecting Culture Filtrate Protein-10 (CFP-10), a secreted recombinant TB antigen (Hong et al., 2011). A monoclonal antibody against CFP-10 was directly mounted onto a bare gold surface via cysteine-mediated blocking chemistry. The resultant biosensor demonstrated a detection limit of 100 ng/mL, determined by the association between SPR angle shifts and escalating concentrations of CFP-10 (from 0.1 to 1.0 μg/mL). Kimuda et al. (2018) employed SPR technology to evaluate the binding avidity of anti-Ag85A antibodies, an immunological marker pertinent to both latent and active TB infection. The study utilized a Biacore 3000 platform equipped with a CM5 chip functionalized with recombinant Ag85A, revealing that people without TB infection

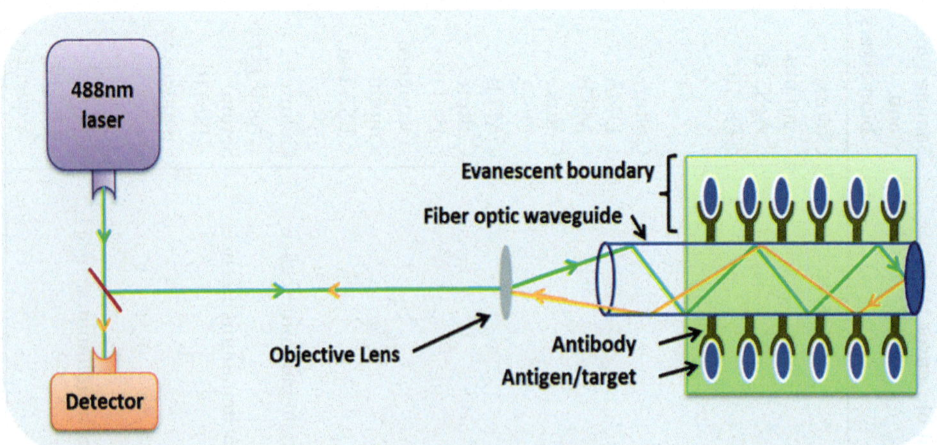

Figure 7.2 Schematic illustration of the surface plasmon resonance (SPR)-based biosensor. *Reprinted with permission (Srivastava et al., 2016). Copyright (2016) Royal Society of Chemistry.*

had somewhat greater antibody avidity than those with active or latent TB (Cose et al., 2018). Xiang et al. (2015) created an SPR-based DNA biosensor utilizing target-primed rolling circle amplification for the genetic detection of the MTB complex and the Mycobacterium avium complex (MAC) (Xiang et al., 2015). A species-specific padlock probe targeting the ITS (Internal transcribed spacer) region of the 16S–23S rRNA gene was affixed to a gold nanoparticle-coated sensor surface. The biosensor exhibited exceptional sensitivity, identifying MTB DNA at concentrations as low as 0.0005 ng/mL and 20 attomoles within a swift amplification period of 40 minutes. Subsequent research by Rinaldi et al. (2019) utilized SPR analysis to qualitatively and quantitatively examine interactions between the MTB antigen Ag85B and antibodies from various sources, facilitating accurate identification of epitope areas on the antigen (Rinaldi et al., 2019). A recent work by Kaur et al. (2022) emphasized material innovation by investigating the application of two-dimensional (2D) nanomaterials, namely carbon nanotubes (CNTs), molybdenum disulfide (MoS_2), and antimonene, as probe materials in SPR-based biosensors for the detection of MTB (Kaur, Kumar and Kaushik, 2022). Antimonene exhibited the greatest sensitivity of the three materials, identifying MTB within a refractive index unit (RIU) range of 1.343 to 1.351, positioning it as a viable contender for prospective biosensor applications.

7.3.2 Fiber-optic biosensors

Light going through an optical fiber may be categorized into two components: the guided light within the fiber's core and an evanescent field that diminishes exponentially outside the core. Owing to total internal reflection (TIR), light is predominantly contained within the core, minimizing energy loss, as only a minor fraction of the evanescent field penetrates beyond the fiber walls. This region, generally extending 100–200 nm beyond the fiber surface, engages insignificantly with the external environment unless deliberately designed. Reducing the cladding layer's thickness enhances the interaction between the evanescent wave and the external medium, rendering this arrangement optimal for biosensing applications. In these systems, energy from the evanescent wave can stimulate fluorophores affixed to the external surface of the fiber, facilitating optical detection with instruments such as spectrophotometers. These sensors may be modified using molecular probes to detect specific bacterial targets. MTB secretes niacin, which interacts with cyanogen bromide and aniline to produce a characteristic yellow molecule. This colorimetric response can be employed alongside fiber-optic sensors to detect MTB. Another method is sandwich immunoassays conducted on waveguide-based fiber-optic systems to detect TB biomarkers, including LAM, early secretory antigenic target-6 (ESAT-6), and the antigen 85 complex (Ag85), providing a sensitive technique of diagnosing current TB infections. Significant research by Pariwono et al. (2007) illustrated the application of fiber-optic biosensors for on-site TB screening. Their technology markedly decreased the diagnostic duration, identifying MTB in as few as five days, compared to the 6–8 weeks generally needed for culture-based techniques (Pariwono et al., 2007).

A novel and promising category of optical biosensors for TB diagnosis is optical ring resonance (ORR) detectors. The fundamental concepts of ORR are intricate and exceed the limits of this examination, yet these systems essentially depend on total internal reflection and constructive interference for their operation. Recent improvements have facilitated the incorporation of ORR sensors into lab-on-a-chip systems featuring many parallel microchannels. Generally, one channel is modified with TB-specific antibodies (test channel), and the remaining channels are coated with nonspecific antibodies that act as reference controls. Antigen binding in the test channel induces changes in total internal reflection and interference patterns, which are then detected and evaluated using a CCD camera. The alterations in the signal are subsequently compared to those in the reference channels to validate particular interactions. The efficacy of ORR biosensors for TB has considerable potential, especially for early detection; nonetheless, advancement relies on the accessibility of highly specific antibodies targeting MTB antigens. Moreover, the surface functionalization of ORR platforms poses a technological barrier akin to that faced by nanowire-based sensors, primarily due to the prevalent usage of silicon as the core material in these systems (Zhu et al., 2007; White, Gohring and Fan, 2007; Baaske and Vollmer, 2012).

7.3.3 Breathalyzer biosensors

Breathalyzer sensors are an innovative method for the noninvasive detection of pulmonary TB. These devices are engineered to gather and evaluate inhaled biological substances after patients cough into a specific, mask-like collection apparatus (McNerney et al., 2010; Begg, Hill and Nickolls, 1964). This device generally comprises a glass tube about 10 cm in length and 3.5 cm in diameter, into which patients exhale aerosols after receiving a nebulized saline treatment. The gathered sample is aimed at the surface of a prism situated at the base of the tube. This prism is adorned with fluorescent peptide epitopes and antibodies unique to MTB. The luminous peptides are synthetic copies of T-cell epitopes from the MTB antigen Ag85B, whereas the antibodies have a greater binding affinity for MTB cells than the peptides. This design enables a competitive binding assay: in the presence of MTB, the bacteria displace the fluorescent peptides adhered to the surface, modifying the fluorescence signal. A diode laser in the gadget identifies these variations by analyzing the displacement process. The biosensor has remarkable sensitivity and can identify MTB concentrations as low as 50–75 colony-forming units (CFU) per milliliter. Furthermore, the system is portable, efficient, and economical, rendering it appropriate for implementation in resource-constrained and outdoor health care environments.

The disposable collecting tube ensures sanitary, and practical single-use for each patient. A commercial prototype, the RBS-TB Breathalyzer, is now being developed by Rapid Biosensor Systems Ltd. in the UK. Preliminary clinical validation trials in India and Ethiopia have produced promising outcomes; however, few false positives were noted. With additional development and validation, this breathalyzer-based biosensor possesses significant potential to function as a sensitive and accessible diagnostic alternative to traditional TB detection techniques (McNerney et al., 2010; Mustafa et al., 2000).

7.3.4 Microfluidic-based photonics

Microfluidics pertains to the exact control of fluids inside microscale channels, often measuring less than 1000 μm in diameter (Nath et al., 2020). These systems provide considerable benefits compared to traditional diagnostic methods by facilitating the execution of many tests on a single platform while necessitating minimal quantities of samples and reagents. Recently, microfluidic photonic systems have garnered interest as efficient point-of-care technologies for bacterial identification. These platforms are engineered to regulate fluid dynamics and light propagation concurrently, providing superior sensitivity and specificity in pathogen detection relative to conventional POC systems. Microfluidic systems have been effectively incorporated into commercial enzyme-linked immunosorbent assay (ELISA) equipment for TB diagnosis, utilizing the interferon-gamma release assay (IGRA) (Lalvani and Pareek, 2010; Pai et al., 2014). Pai et al. and Lalvani et al. indicate that the assay employs interferon-gamma (IFN-γ), a proinflammatory cytokine secreted by immune cells of infected patients in response to TB-specific antigens, as the principal biomarker (Lalvani and Pareek, 2010; Pai et al., 2014). Following the incubation of blood samples with MTB antigens, IFN-γ is produced and may be measured using the ELISA platform. This method, while economical and efficient, frequently requires laborious sample preparation and processing procedures. Researchers have downsized the ELISA test into a lab-on-a-chip system incorporating microfluidic channels. The sensor chip in this platform is functionalized using anti-IFN-γ Fab'$(Cys)_3$ antibodies, which are anchored to gold electrodes. As blood samples containing IFN-γ traverse the microfluidic channels at a velocity of 25 μL/min, electrochemical interactions between the cytokine and the electrode surface elicit variations in current, which are quantified to ascertain the IFN-γ concentration. The technology exhibited a detection limit between 10 and 2000 pg/mL.

7.3.5 Artificial intelligence and machine learning in optical biosensing

Incorporating AI and ML into optical biosensing technologies has created new opportunities for the swift and precise detection of infectious illnesses, such as TB.

Intelligent systems are widely utilized to augment biosensor performance by advanced processing of intricate optical data, enhancing diagnostic accuracy, speed, and dependability. Optical biosensors, including those utilizing fluorescence, SPR, and imaging modalities, can generate extensive and intricate datasets that pose challenges for interpretation through conventional analytical methods. AI and ML techniques, such as support vector machines, random forest classifiers, and deep learning neural networks, are proficient in discerning significant patterns and signals from noisy or high-dimensional data. This is particularly significant in TB diagnostics, where early detection depends on nuanced and occasionally unclear optical signals produced by biomarkers such as LAM, ESAT-6, and Ag85. Recent studies have shown AI's effectiveness in enhancing the classification accuracy of TB-positive and TB-negative samples by analyzing fluorescence intensities or spectral changes. In SPR-based biosensors, ML algorithms may be developed to differentiate between particular binding events and non-specific interactions, therefore minimizing false positives and enhancing specificity. Likewise, AI-based image processing algorithms have been utilized in microfluidic optical devices to automate the identification and quantification of fluorescence-tagged MTB biomarkers from digital pictures. Convolutional neural networks (CNNs) have been utilized to advance smartphone-integrated optical biosensors. These technologies provide on-device analysis and immediate diagnostic input, rendering them exceptionally appropriate for POC testing in resource-constrained environments. Moreover, AI facilitates real-time monitoring, adaptive calibration, and ongoing learning from fresh data, augmenting the sensor's diagnostic efficacy over time (Singh et al., 2022; Ullah et al., 2022).

Notwithstanding the encouraging promise, specific problems persist. This encompasses the necessity for extensive, meticulously annotated datasets for model training, safeguarding data privacy, and addressing variability in sensor performance attributable to environmental or sample-related issues. Integrating AI with optical biosensing signifies a revolutionary method expected to be pivotal in the forthcoming generation of TB diagnostic instruments. In conclusion, AI and ML substantially enhance the functionality of optical biosensors, facilitating swift, sensitive, and intelligent detection of MTB. As the technology advances, it is anticipated to enhance diagnostic procedures and facilitate prompt clinical decision-making in the worldwide battle against TB.

7.4 Challenges and limitations

Notwithstanding the considerable progress in optical biosensing for TB diagnosis, several obstacles impede the extensive use of these technologies. A primary concern is sensitivity and specificity. Precisely differentiating MTB from nontuberculous mycobacteria (NTM) and other respiratory illnesses continues to be a significant challenge. Cross-reactivity and nonspecific interactions may produce false positives or negatives, undermining diagnostic accuracy. Researchers are creating high-affinity recognition elements to overcome these constraints, including monoclonal antibodies, aptamers, and MTB-specific DNA probes. Furthermore, advancements in surface chemistry, nanomaterial-based signal amplification, and sophisticated readout systems are being investigated to enhance biosensor selectivity and reduce background interference.

A significant difficulty is the expense and scalability of these devices. The construction of optical biosensors frequently depends on costly materials such as gold or other precious metals, sophisticated detecting elements, and bespoke microfluidics. These variables result in elevated production costs and create accessibility obstacles, especially in low-resource environments where TB is most common. Moreover, the necessity for specialized apparatus and skilled staff contributes to the obstacles. Efforts are currently directed toward discovering cost-effective alternatives, employing scalable production methods such as roll-to-roll printing, and creating biosensors from readily accessible and biodegradable materials while maintaining performance standards.

Translating optical biosensors from the laboratory to practical clinical and field applications poses considerable challenges. Although several systems have significant potential in controlled experimental environments, they frequently encounter performance challenges when implemented in varied environmental contexts. Effective execution at the point of care necessitates comprehensive clinical validation, adherence to regulatory requirements, and consistent performance in field settings. Moreover, for use in resource-limited regions, biosensors must be intuitive, portable, and able to operate independently of intricate gear. Integration with smartphone interfaces, battery-powered displays, and streamlined sample preparation methods is crucial for practical implementation. Addressing these gaps necessitates technical innovation and interdisciplinary collaboration, including public health, engineering, and regulatory science.

7.5 Prospective outlook

The domain of optical biosensing for TB diagnosis is set for significant advancement. Future advancements will likely focus on next-generation biosensor designs that use various sensing modalities to enhance diagnostic reliability. Hybrid systems that integrate optical and electrochemical detection, nanophotonic materials, plasmonic nanostructures, and microring resonators provide potential for ultrasensitive and quick detection of TB biomarkers. Advancements in label-free biosensing and real-time visual feedback systems are anticipated to improve the efficiency and ease of diagnosis. A potential avenue is the advancement of portable and wearable optical biosensors. These devices are intended to deliver real-time, on-site TB screening, especially in distant or impoverished areas where access to traditional diagnostics is restricted. Wearable devices utilizing flexible materials and soft electronics provide user comfort and enable continuous monitoring, creating new opportunities for proactive TB surveillance and control. The amalgamation of optical biosensors with cutting-edge technologies such as AI, cloud computing, and the Internet of Things (IoT) is anticipated to transform diagnostic processes. Intelligent biosensors with automated data processing, wireless connection, and cloud-based reporting systems can deliver real-time diagnostic feedback and enable centralized disease surveillance. These improvements will enhance the scalability and responsiveness of TB diagnostics while considerably aiding worldwide initiatives to eradicate the disease through early diagnosis and effective tracking.

7.6 Conclusions

Notwithstanding significant progress in the prevention, detection, and treatment of TB, the disease continues to pose a formidable global health burden. The ongoing prevalence of the disease is primarily due to diagnostic delays, the rise of drug-resistant strains, and socioeconomic obstacles that impede access to competent health care. A significant barrier exists due to the scarcity of accessible, swift, and user-friendly diagnostic instruments, especially in resource-constrained environments where the burden of TB is most pronounced. Optical biosensing technologies present a viable approach to address the diagnostic gap in this context. These technologies demonstrate significant potential for the rapid, sensitive, and specific detection of MTB, often within minutes to hours, in contrast to traditional methods such as NAATs and LFAs, which typically necessitate well-equipped laboratories and extended processing durations. The versatility and scalability of optical biosensors enable the simultaneous detection of several MTB indicators, rendering them an invaluable instrument for thorough TB screening. Despite several optical biosensors being in developmental or precommercial phases, their amalgamation with microfluidics, AI, and portable platforms establishes them as advanced diagnostic instruments. These technologies can improve diagnostic efficiency and broaden access to low-resource testing, facilitating earlier detection, prompt treatment, and enhanced TB control. Advancing and investing in developing, validating, and extensively using optical biosensing technologies will be crucial. A multidisciplinary strategy, incorporating cooperation among engineers, doctors, public health specialists, and politicians, will be essential for converting these breakthrough diagnostic tools into practical solutions in the global battle against TB.

References

Baaske M, Vollmer F. Optical resonator biosensors: molecular diagnostic and nanoparticle detection on an integrated platform. Chemphyschem 2012;13(2):427–36. Available from https://doi.org/10.1002/cphc.201100757.

Begg TB, Hill ID, Nickolls LC. Breathalyzer and Kitagawa-Wright methods of measuring breath alcohol. British Medical Journal 1964;1(5374):9–15. Available from https://doi.org/10.1136/bmj.1.5374.9.

Boehme CC, Nabeta P, Hillemann D, Nicol MP, Shenai S, Krapp F, et al. Rapid molecular detection of tuberculosis and rifampin resistance. New England Journal of Medicine 2010;363(11):1005–15. Available from https://doi.org/10.1056/NEJMoa0907847.

De Castro N, Mechaï F, Bachelet D, Canestri A, Joly V, Vandenhende M, et al. Treatment with a three-drug regimen for pulmonary tuberculosis based on rapid molecular detection of isoniazid resistance: a noninferiority randomized trial (FAST-TB). Open Forum Infectious Diseases 2022;9(8). Available from https://doi.org/10.1093/ofid/ofac353.

Chandan HR, Schiffman JD, Balakrishna RG. Quantum dots as fluorescent probes: synthesis, surface chemistry, energy transfer mechanisms, and applications. Sensors and Actuators, B: Chemical 2018;258:1191–214. Available from https://doi.org/10.1016/j.snb.2017.11.189.

Chauke SH, Nzuza S, Ombinda-Lemboumba S, Abrahamse H, Dube FS, Mthunzi-Kufa P. Advances in the detection and diagnosis of tuberculosis using optical-based devices. Photodiagnosis and Photodynamic Therapy 2024;45. Available from https://doi.org/10.1016/j.pdpdt.2023.103906.

Chaturvedi I, Vyas S, Mishra R. A review of the applications of fluorescent Schiff-base sensors for the metal ions detection and AIE properties. Dyes and Pigments 2025;239:112748. Available from 10.1016/j.dyepig.2025.112748.

Cose S, Kimuda SG, Biraro IA, Bagaya BS, Raynes JG. Characterising antibody avidity in individuals of varied *Mycobacterium tuberculosis* infection status using surface plasmon resonance. PLoS ONE 2018;13(10). Available from https://doi.org/10.1371/journal.pone.0205102.

Denkinger CM, Schumacher SG, Boehme CC, Dendukuri N, Pai M, Steingart KR. Xpert MTB/RIF assay for the diagnosis of extrapulmonary tuberculosis: a systematic review and meta-analysis. European Respiratory Journal 2014;44(2):435–46. Available from https://doi.org/10.1183/09031936.00007814.

Duman M, Piskin E. Detection of Mycobacterium tuberculosis complex and Mycobacterium gordonae on the same portable surface plasmon resonance sensor. Biosensors and Bioelectronics 2010;26(2):908–12. Available from https://doi.org/10.1016/j.bios.2010.06.071.

Engel N, Ganesh G, Patil M, Yellappa V, Pai NP, Vadnais C, et al. Barriers to point-of-care testing in India: results from qualitative research across different settings, users and major diseases. PLoS ONE 2015;10(8). Available from https://doi.org/10.1371/journal.pone.0135112.

Hawn TR, Day TA, Scriba TJ, Hatherill M, Hanekom WA, Evans TG, et al. Tuberculosis vaccines and prevention of infection. Microbiology and Molecular Biology Reviews 2014;78(4):650–71. Available from https://doi.org/10.1128/mmbr.00021-14.

Hendricks-Leukes NR, Jonas MR, Mlamla ZC, Smith M, Blackburn JM. Dual-approach electrochemical surface-enhanced raman scattering detection of *Mycobacterium tuberculosis* in patient-derived biological specimens: proof of concept for a generalizable method to detect and identify bacterial pathogens. ACS Sensors 2022;7(5):1403–18. Available from https://doi.org/10.1021/acssensors.2c00121.

He F, Zhang L. Rapid diagnosis of *M. tuberculosis* using a piezoelectric immunosensor. Analytical Sciences 2002;18(4):397–401. Available from https://doi.org/10.2116/analsci.18.397.

Hong SC, Chen H, Lee J, Park HK, Kim YS, Shin HC, et al. Ultrasensitive immunosensing of tuberculosis CFP-10 based on SPR spectroscopy. Sensors and Actuators, B: Chemical 2011;156(1):271–5. Available from https://doi.org/10.1016/j.snb.2011.04.032.

Huang J, Xiao Z, Yang X, Chen X, Wang X, Chen Y, et al. Two target genes based multiple cross displacement amplification combined with a lateral flow biosensor for the detection of *Mycobacterium tuberculosis* complex. BMC Microbiology 2021;21(1). Available from https://doi.org/10.1186/s12866-021-02328-6.

Jiang J, Chai Y, Cui H. The electrogenerated chemiluminescence detection of IS6110 of Mycobacterium tuberculosis based on a luminol functionalized gold nanoprobe. RSC Advances 2011;1(2):247–54. Available from https://doi.org/10.1039/c1ra00206f.

Kaur B, Kumar S, Kaushik BK. Antimonene, CNT and MoS2Based SPR-fiber-optic probe for tuberculosis detection. IEEE Sensors Journal 2022;22(15):14903–10. Available from https://doi.org/10.1109/JSEN.2022.3186995.

Khan MS, Dar O, Sismanidis C, Shah K, Godfrey-Faussett P. Improvement of tuberculosis case detection and reduction of discrepancies between men and women by simple sputum-submission instructions: a pragmatic randomised controlled trial. Lancet 2007;369(9577):1955–60. Available from https://doi.org/10.1016/S0140-6736(07)60916-7.

Kimuda SG, Biraro IA, Bagaya BS, Raynes JG, Cose S. Characterising antibody avidity in individuals of varied Mycobacterium tuberculosis infection status using surface plasmon resonance. PLOS ONE 2018;13:e0205102. Available from 10.1371/journal.pone.0205102.

Kooti S, Abiri R, Kadivarian S, Khazayel S, Mohajeri P, Atashi S, et al. Modified gold nanoparticle colorimetric probe-based biosensor coupled with allele-specific PCR for rapid detection of G944C mutation associated with isoniazid resistance. Sensing and Bio-Sensing Research 2022;38:100543. Available from https://doi.org/10.1016/j.sbsr.2022.100543.

Kooti S, Kadivarian S, Abiri R, Mohajeri P, Atashi S, Ahmadpor H, et al. Modified gold nanoparticle colorimetric probe-based biosensor for direct and rapid detection of *Mycobacterium tuberculosis* in sputum specimens. World Journal of Microbiology and Biotechnology 2023;39(5):118. Available from https://doi.org/10.1007/s11274-023-03564-w.

Lalvani A, Pareek M. Interferon gamma release assays: principles and practice. Enfermedades Infecciosas y Microbiologia Clinica 2010;28(4):245–52. Available from https://doi.org/10.1016/j.eimc.2009.05.012.

Lawn SD, Nicol MP. Xpert® MTB/RIF assay: development, evaluation and implementation of a new rapid molecular diagnostic for tuberculosis and rifampicin resistance. Future Microbiology 2011;6(9):1067–82. Available from https://doi.org/10.2217/fmb.11.84.

Li F, Yu Y, Li Q, Zhou M, Cui H. A homogeneous signal-on strategy for the detection of rpoB genes of *Mycobacterium tuberculosis* based on electrochemiluminescent graphene oxide and ferrocene quenching. Analytical Chemistry 2014;86(3):1608–13. Available from https://doi.org/10.1021/ac403281g.

Maphanga C, Manoto S, Ombinda-Lemboumba S, Ismail Y, Mthunzi-Kufa P. Localized surface plasmon resonance biosensing of *Mycobacterium tuberculosis* biomarker for TB diagnosis. Sensing and Bio-Sensing Research 2023;39. Available from https://doi.org/10.1016/j.sbsr.2022.100545.

McNerney R, Wondafrash BA, Amena K, Tesfaye A, McCash EM, Murray NJ. Field test of a novel detection device for *Mycobacterium tuberculosis* antigen in cough. BMC Infectious Diseases 2010;10:161. Available from https://doi.org/10.1186/1471-2334-10-161.

Mukundan H, Kumar S, Price DN, Ray SM, Lee YJ, Min S, et al. Rapid detection of *Mycobacterium tuberculosis* biomarkers in a sandwich immunoassay format using a waveguide-based optical biosensor. Tuberculosis 2012;92(5):407–16. Available from https://doi.org/10.1016/j.tube.2012.05.009.

Mustafa AS, Shaban FA, Abal AT, Al-Attiyah R, Wiker HG, Lundin KEA, et al. Identification and HLA restriction of naturally derived Th1-cell epitopes from the secreted *Mycobacterium tuberculosis* antigen 85B recognized by antigen-specific human CD4+ T-cell lines. Infection and Immunity 2000;68(7):3933–40. Available from https://doi.org/10.1128/IAI.68.7.3933-3940.2000.

Nath P, Kabir A, Doust SK, Kreais ZJ, Ray A. Detection of bacterial and viral pathogens using photonic point-of-care devices. Diagnostics 2020;10(10):841. Available from https://doi.org/10.3390/diagnostics10100841.

Pai M, Denkinger CM, Kik SV, Rangaka MX, Zwerling A, Oxlade O, et al. Gamma interferon release assays for detection of *Mycobacterium tuberculosis* infection. Clinical Microbiology Reviews 2014;27(1):3–20. Available from https://doi.org/10.1128/CMR.00034-13.

Pariwono AM, Lo T, Lim CS, Wang SX, Chan YW. Rapid tuberculosis detection technique for on-site patient screening. Journal of Biomedical & Pharmaceutical Engineering 2007;1:27.

Prabhakar N, Arora K, Arya SK, Solanki PR, Iwamoto M, Singh H, et al. Nucleic acid sensor for M. tuberculosis detection based on surface plasmon resonance. Analyst 2008;133(11):1587–92. Available from https://doi.org/10.1039/b808225a.

Qian S, Cui Y, Cai Z, Li L. Applications of smartphone-based colorimetric biosensors. Biosensors and Bioelectronics: X 2022;11:100173. Available from https://doi.org/10.1016/j.biosx.2022.100173.

Qiao Y, Wang X, Fan Z, Song Y, Zhang J, Han Q. Raman-enhanced sensor based on CRISPR-SERS technology for the rapid and hypersensitive detection of Mycobacterium tuberculosis. Analytical and Bioanalytical Chemistry 2024;416(28):6551–62. Available from https://doi.org/10.1007/s00216-024-05551-y.

Rinaldi F, Lupu L, Rusche H, Kukačka Z, Tengattini S, Bernardini R, et al. Epitope and affinity determination of recombinant *Mycobacterium tuberculosis* Ag85B antigen towards anti-Ag85 antibodies using proteolytic affinity-mass spectrometry and biosensor analysis. Analytical and Bioanalytical Chemistry 2019;411(2):439–48. Available from https://doi.org/10.1007/s00216-018-1466-z.

Sharma A, Khan R, Catanante G, Sherazi TA, Bhand S, Hayat A, et al. Designed strategies for fluorescence-based biosensors for the detection of mycotoxins. Toxins 2018;10(5):197. Available from https://doi.org/10.3390/toxins10050197.

Shojaei TR, Mohd Salleh MA, Tabatabaei M, Ekrami A, Motallebi R, Rahmani-Cherati T, et al. Development of sandwich-form biosensor to detect *Mycobacterium tuberculosis* complex in clinical sputum specimens. Journal of Infectious Diseases 2014;18(6):600–8. Available from https://doi.org/10.1016/j.bjid.2014.05.015.

Singh M, Pujar GV, Kumar SA, Bhagyalalitha M, Akshatha HS, Abuhaija B, et al. Evolution of machine learning in tuberculosis diagnosis: a review of deep learning-based medical applications. Electronics (Switzerland) 2022;11(17):2634. Available from https://doi.org/10.3390/electronics11172634.

Steingart KR, Schiller I, Horne DJ, Pai M, Boehme CC, Dendukuri N. Xpert® MTB/RIF assay for pulmonary tuberculosis and rifampicin resistance in adults. Cochrane Database of Systematic Reviews 2014;2014(1). Available from https://doi.org/10.1002/14651858.CD009593.pub3.

Srivastava SK, van Rijn CJM, Jongsma MA. Biosensor-based detection of tuberculosis. RSC Advances 2016;6:17759–71. Available from 10.1039/c5ra15269k.

Trzaskowski M, Napiórkowska A, Augustynowicz-Kopeć E, Ciach T. Detection of tuberculosis in patients with the use of portable SPR device. Sensors and Actuators B: Chemical 2018;260:786–92. Available from https://doi.org/10.1016/j.snb.2017.12.183.

Ullah U, Tahir Z, Qazi O, Mirza S, Cheema MI. Raman spectroscopy and machine learning-based optical probe for tuberculosis diagnosis via sputum. Churchill Livingstone, Pakistan Tuberculosis 2022;136. Available from https://doi.org/10.1016/j.tube.2022.102251.

White IM, Gohring J, Fan X. SERS-based detection in an optofluidic ring resonator platform. Optics Express 2007;15(25):17433–42. Available from https://doi.org/10.1364/OE.15.017433.

WHO. ReliefWeb. WHO; 2020.

World Health Organization. Global Tuberculosis Report 2021. World Health Organization; 2021.

Wu S, Duan N, Ma X, Xia Y, Wang H, Wang Z, et al. Multiplexed fluorescence resonance energy transfer aptasensor between upconversion nanoparticles and graphene oxide for the simultaneous determination of mycotoxins. Analytical Chemistry 2012;84(14):6263–70. Available from https://doi.org/10.1021/ac301534w.

Xiang Y, Zhu X, Huang Q, Zheng J, Fu W. Real-time monitoring of mycobacterium genomic DNA with target-primed rolling circle amplification by a Au nanoparticle-embedded SPR biosensor. Biosensors and Bioelectronics 2015;66:512–9. Available from https://doi.org/10.1016/j.bios.2014.11.021.

Yu X, Chai Y, Jiang J, Cui H. Sensitive ECL sensor for sequence-specific DNA from *Mycobacterium tuberculosis* based on N-(aminobutyl)-N-ethylisoluminol functionalized gold nanoparticles labeling. Journal of Photochemistry and Photobiology A: Chemistry 2012;241:45–51. Available from https://doi.org/10.1016/j.jphotochem.2012.04.023.

Zhu H, White IM, Suter JD, Dale PS, Fan X. Analysis of biomolecule detection with optofluidic ring resonator sensors. Optics Express 2007;15(15):9139–46. Available from https://doi.org/10.1364/OE.15.009139.

Zou F, Wang X, Qi F, Koh K, Lee J, Zhou H, et al. Magneto-plamonic nanoparticles enhanced surface plasmon resonance TB sensor based on recombinant gold binding antibody. Sensors and Actuators, B: Chemical 2017;250:356–63. Available from https://doi.org/10.1016/j.snb.2017.04.162.

CHAPTER 8

Biophysical approaches for *Mycobacterium tuberculosis* detection

Kajal Tiwari
CSIR-AMPRI, Bhopal, Madhya Pradesh, India

8.1 Introduction

Conventional diagnostic methods frequently miss latent tuberculosis infections, in which the bacillus is dormant and does not exhibit symptoms. As a result, new diagnostic tools that are quick and accurate, suitable for decentralized and resource-constrained situations, are urgently needed (Bartolomeu-Gonçalves et al., 2024). In response to these difficulties in diagnosing, biophysical methods have emerged as useful alternatives or supplementary techniques. These methods identify *Mycobacterium tuberculosis* (MTB) or its biomarkers (such as DNA, RNA, proteins, lipids, and metabolites) in clinical specimens by applying basic physics concepts such as fluid dynamics, electromagnetism, and optics. Biophysical techniques frequently provide advantages like label-free analysis, real-time monitoring, minimal sample preparation, and superior analytical sensitivity and specificity (Thakur et al., 2007).

8.2 Biophysical approaches in *Mycobacterium tuberculosis* detection

Biophysical diagnostics includes a range of advanced technologies. For the detection of tuberculosis, Raman spectroscopy, infrared spectroscopy, mass spectrometry (MS), and lab-on-a-chip (LOC) technologies are especially helpful (Parihar and Mehta, 2024). These methods help with disease staging, drug resistance profiling, and differentiating between live and dead bacilli in addition to enabling the detection of active infections. Additionally, they are being used in conjunction with AI and ML to automate pattern recognition and classification, thereby decreasing human error and increasing diagnostic precision (Parihar and Mehta, 2024; Puravankara et al., 2024).

A number of biophysical approaches are being scaled down and integrated into portable platforms for point-of-care (POC) diagnostics (Wessels and Raad, 2016). The ability to quickly perform on-site testing without the need for centralized laboratories holds great promise for tuberculosis control efforts, particularly in endemic areas with limited healthcare infrastructure (Parsons et al., 2011). This chapter provides a more in-depth discussion of the most important biophysical techniques for identifying

Advances in Biosensor Technology for Mycobacterium Tuberculosis Detection
DOI: https://doi.org/10.1016/B978-0-443-44589-7.00013-3

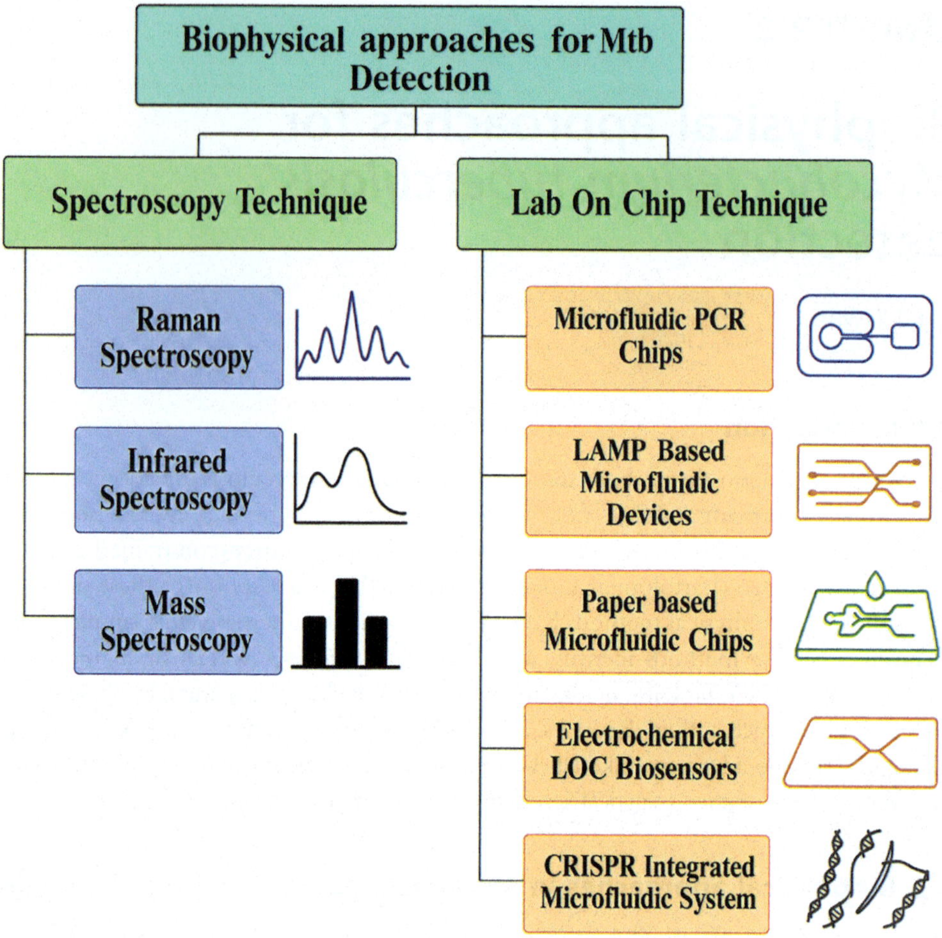

Figure 8.1 Classification of *Mycobacterium tuberculosis*detection techniques.

TB (tuberculosis). It explains the fundamental concepts, workings, diagnostic applications, advantages, and disadvantages of each technique (Bartolomeu-Gonçalves et al., 2024). To emphasize the revolutionary impact of biophysical approaches on TB diagnostics and their potential for use in future global TB control strategies, the chapter highlights recent developments and case studies (Merchant et al., 2022) (Fig. 8.1).

8.3 Key biophysical techniques for tuberculosis detection

8.3.1 Raman spectroscopy

Raman spectroscopy is a vibrational spectroscopic method which probes the inelastic interaction of monochromatic light, generally from a laser source (Keresztury, 2001).

When a sample's molecular bonds absorb light, a minor fraction of photons are scattered at different energy levels, which leads to a Raman spectrum that is like a molecular fingerprint (Keresztury, 2001). Raman spectroscopy involves a series of steps. Initially, a monochromatic laser beam, usually from a near-infrared laser source, is sent to the sample. The laser light interacts with the chemical bonds in the molecules and thus a small number of photons are scattered inelastically. This scattering leads to an energy change (wavelength) of the scattered photons, which is equal to the vibrational frequencies of the molecules (Chu, 2007).

To obtain a Raman spectrum, light that has undergone inelastic scattering is gathered using a lens and then directed through a monochromator and a detector, which is typically a CCD. Each peak in the spectrum corresponds to a vibrational mode that is characteristic of the sample's chemical structures (Liu and Berg, 2012).

In short, the Raman spectrum signatures are responsible for MTB detection. These signatures come from mycolic acids, lipids, peptidoglycans, and nucleic acids. These features allow MTB detection even in difficult biological matrices such as sputum and blood. By using the SERS technique, sensitivity is increased. This is achieved because the sample is on nanostructured metallic substrates, and the Raman signals are amplified by several orders of magnitude (Perumal et al., 2018).

In addition, MTB biomarkers can be selectively captured and ultrasensitively detected by integrating a SERS-based biosensor with functionalized nanostructured metallic substrates (Hendricks-Leukes et al., 2022).

In rural Indian clinics, sputum samples were examined using a portable SERS-based Raman device, which had a sensitivity and specificity of over 90%. The system is an excellent diagnostic tool for environments with limited resources due to its portability and speed (Perumal et al., 2018).

8.3.2 Infrared spectroscopy

Infrared (IR) spectroscopy is a powerful analytical technique that can identify the molecular composition of biological samples through their interaction with infrared radiation. The non-destructive identification of molecular fingerprints, which have a high degree of specificity for distinguishing between various biomolecular structures, is made possible by this method (Theophile, 2012).

The principle of IR spectroscopy is that the molecules are capable of absorbing certain frequencies of infrared light due to which transitions are generated between vibrational energy levels. It turns out that when infrared light falls on a biological sample, the chemical bonds give the energy the form of vibrational excitation such as stretching, bending, or twisting to the energy they have absorbed. Essentially, each bond and functional group is characterized by the absorption of IR radiation at given frequency ranges, which then result in a unique absorption spectrum that is a graphical representation of the sample's molecular composition (Beć et al., 2022).

A typical IR spectroscopy apparatus comprises an infrared light source, a sample holder, a monochromatic or interferometer, and a detector. The sample is passed through the IR radiation, which is then collected and analyzed. Absorption peaks in the resulting spectrum indicate the presence of specific chemical bonds and structures. This spectrum acts as a molecular fingerprint of the sample, allowing identification and comparison of the sample (Coates, 2010).

IR spectroscopy can be carried out in three different ways; transmission (where the IR light goes through the sample), reflection (where the light bounces from the surface of the sample), and attenuated total reflectance (ATR), which not only increases surface sensitivity but also makes sample preparation easier. The last one is particularly good for the analysis of biological samples (Theophile, 2012).

TB detection applications comprise the detection of MTB-specific proteins, lipids, and metabolic indicators in sputum, serum, urine, and exhaled breath condensate (EBC). IR spectroscopy has been employed to trace the changes in biochemical components that signal tuberculosis infection, such as an increased mycolic acid or the MTB-specific lipid profile (Theophile, 2012).

Furthermore, biosensors based on ATR-FTIR have been developed for the detection of tuberculosis. In these biosensors, the evanescent wave produced at the ATR crystal–sample interface interacts with immobilized MTB-specific biomolecules, allowing for rapid, label-free detection in clinical samples through typical infrared absorption changes (Naseer et al., 2021).

8.3.3 Mass spectrometry

Mass spectrometry is an essential tool for accurate chemical determination and measurement that is carried out using the mass-to-charge ratio (m/z) of ionized molecules. It allows for extensive characterization of intricate biological specimens and has a high sensitivity level (Finehout and Lee, 2004).

MS has three main steps: ionization, mass analysis, and detection. The substance is introduced into the mass spectrometer and ionized by one of several methods. Electrospray ionization (ESI) consists of converting a sample into a thin aerosol and then exposing it to a high electric field, which results in the formation of charged droplets that, upon evaporation, leave the gas-phase ions. Matrix-assisted laser desorption/Ionization (MALDI) is a process that consists of a sample that is embedded in a matrix which absorbs the laser thus making the biomolecules desorption and ionization. Atmospheric pressure chemical ionization (APCI) is another method that is quite often used for the analysis of small volatile compounds (Awad et al., 2015).

First, the ions are given some energy and are then sent to the mass analyzer. In Time-of-flight (TOF) analyzers, ions get a push from an electric field and the time it takes for them to arrive at the detector is recorded for their m/z ratio. Quadrupole analyzers make use of the oscillating electric fields to allow the passage of only those ions which have the corresponding m/z. Orbitrap analyzers confine ions in an electrostatic field and track the changes in their oscillation frequencies (Guilhaus et al., 1997).

The detector traces the energy and mass ratio of the particles, generating a mass spectrum. High points in the spectrum indicate the presence of various biomolecules, thus allowing one to conduct qualitative and quantitative investigations (Guilhaus et al., 1997).

In recent years, mass spectrometry has been coupled with biosensor platforms such as MS-based immunosensors, where selective antibody–antigen interactions enable targeted capture of biomolecules prior to ionization, thereby improving specificity and sensitivity (Liu et al., 2022).

In addition, microfluidic MS sensors combine sample preparation, separation, and ionization on a single chip, making it possible to analyze intricate biological samples quickly and efficiently with little sample volume (Ohla and Belder, 2012).

8.4 Applications in tuberculosis diagnosis include

Proteomics: the detection of MTB-specific protein patterns in samples taken from patients (for instance, serum, sputum, and urine) (López-Hernández et al., 2016).

Lipidomics: the pinpointing of MTB lipid signals, which are mycolic acids, phthiocerol dimycocerosates, and trehalose dimycolate (López-Hernández et al., 2016).

Metabolomics: the investigation of both host and bacterial metabolites that can serve as biomarkers for tuberculosis diagnosis (López-Hernández et al., 2016).

LC-MS/MS analysis was employed to detect unique lipid biomarkers in sputum samples that could directly point to active tuberculosis or latent infections. The capacity to create customized treatment protocols and monitor the course of the illness was facilitated by this method (Chen et al., 2021).

MALDI-TOF MS is just one method that is being utilized in the medical labs for the quick detection of the MTB complex from culture isolates. A move toward less complex sample preparation has made it possible to perform a direct analysis of the decontaminated sputum, hence reducing the diagnostic time (Calderaro and Chezzi, 2024) (Fig. 8.2).

8.5 Lab-on-a-chip technology for tuberculosis detection

Lab-on-a-chip (LOC) technology is one of the most innovative ways to detect tuberculosis, that is because it significantly reduces and combines all the laboratory processes across a microfluidic chip (Parihar and Mehta, 2024). Handheld devices of this nature integrate sample preparation, target amplification, and detection into an automated way, hence they are perfect for point-of-care testing in resource-limited

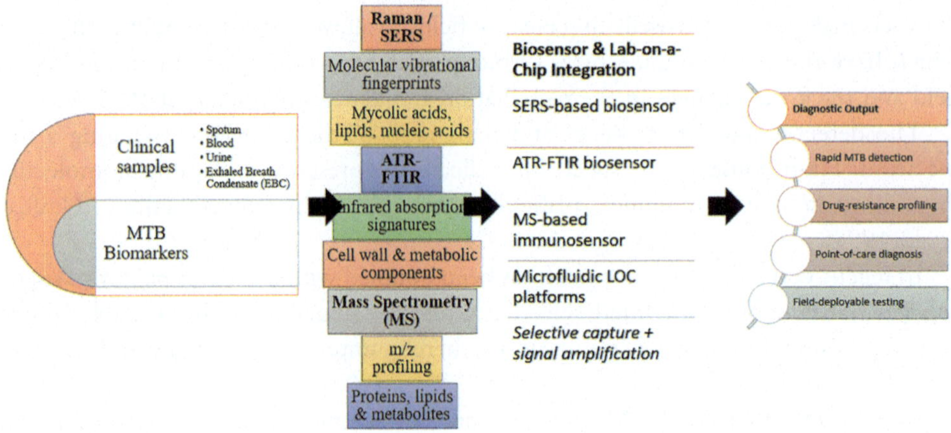

Figure 8.2 Schematic representation of biophysical techniques integrated with biosensor and lab-on-a-chip platforms for Mycobacterium tuberculosis detection.

settings (Zu et al., 2025). LOC gadgets usually contain microscale channels and compartments that are chemically treated with substances such as glass, silicon, or polymers. Here, fluid flow is very accurately regulated by means of capillary action, pressure, or electro kinetic forces (Dawson et al., 2021).

8.5.1 Key types of lab-on-a-chip systems for tuberculosis detection

8.5.1.1 Microfluidic polymerase chain reaction chips

Among their features are small thermal cycling chambers that make for quicker DNA amplification with significantly lower reagent amounts compared to regular polymerase chain reaction (PCR). The use of micro heaters and sensors facilitates precision thermal control and at the same time, the emission of fluorescence in real-time allows the process of amplification to be constantly observed. Current innovations have completed the major areas of the process so that real sample-to-answer systems are now feasible to be used in carrying out the analysis of raw sputum samples (Ahrberg et al., 2016).

Microfluidic PCR chips are especially efficient in detecting MTb because they can quickly amplify and measure bacterial DNA. These chips reduce traditional PCR to a few tiny thermal cycling chambers, preserving high sensitivity while significantly reducing reagent consumption (Wang et al., 2022).

They have the capability to locate MTb-specific genetic markers like the drug resistance gene even in raw sputum samples with very little pre-processing. The micro heaters and temperature sensors, which are integrated, give accurate control of the temperature, and the real-time fluorescence detection enables uninterrupted observation of the amplification process. However, these systems offer the highest level of accuracy but still they rely on complex devices which make them more suitable for laboratories than for use in the field (Lehnert and M. Gijs, 2024).

8.5.1.2 Loop-mediated isothermal amplification-based microfluidic devices

These tools rely on loop-mediated isothermal amplification (LAMP), performing amplification at a constant temperature (60°C–65°C), and thus not requiring sophisticated control of the temperature ramp. They are usually integrated with visual detection (colorimetric) or fluorescence detection (improved sensitivity). Their low cost and high efficiency make them perfect for outdoor applications, and a few models even come with smart phone readout devices (Zhang et al., 2019

LAMP-based microfluidic devices for MTb detection have emerged as a simple alternative to PCR as it runs at a constant temperature (60°C–65°C) without need of complex thermocycling. These devices effectively amplify MTb DNA and can be integrated with colorimetric or fluorescence readout to provide visual or smartphone-based read interpretation (Wang et al., 2021).

LAMP is particularly suited for use in areas with limited resources due to its energy efficiency and light-weight nature. Although it has a little lower sensitivity compared to PCR, it is still a good instrument for rapid diagnosis of MTb, especially in field deployments places, where laboratory infrastructure is not available (Mahato and Kumar, 2024).

8.5.1.3 Paper-based microfluidic chips

These low-cost diagnostic platforms use cellulose or nitrocellulose substrates to transport fluids from capillary actions, which do not require external power. Colorimetric detection systems designed for maximum simplicity often use colors that can be interpreted by the naked eye or basic imaging equipment. Despite having low sensitivity compared to other LOC forms, their disposal and stability in room temperature make them great for large-scale screening operations (Alsaeed and Mansour, 2020).

Paper-based microfluidic chips provide an extremely low-cost alternative for MTb detection, especially in large-scale screening programs. These devices carry fluids via capillary action, which eliminates the need for external power. They often detect MTb antigens in urine samples, such as lipoarabinomannan (LAM), which is effective for detecting tuberculosis in HIV-positive patients. Despite having lower sensitivity than PCR or LAMP, their simplicity, disposability, and durability at ambient temperatures make them excellent for distant and low-resource settings where advanced diagnostics are unavailable (Wu et al., 2024).

8.5.1.4 Electrochemical lab-on-a-chip biosensors

These units sense the tuberculosis biomarkers through changes in electrical signals and provide rapid findings (usually <30 minutes). They are best at detecting non-nucleic acid targets such as LAM antigen and can be established to detect markers of tuberculosis and drug resistance at the same time. Recent developments have increased their efficiency with complex clinical samples, but remain compatible with portable equipment (Mir et al., 2009).

Electrochemical LOC biosensors recognize MTb biomarkers using measurable electrical signals and provide results in less than 30 minutes. These devices are especially good at detecting non-nucleoids targets, such as LAM antigen, and can be organized to detect TB and drug resistance signals at the same time. Recent advances have increased their ability to cope with complex clinical samples while maintaining portability. Their speed and ease make them fanciful for TB tests at the site, especially in situations that require rapid diagnosis (Hendricks-Leukes et al., 2022). (Table 8.1).

Table 8.1 Different biophysical techniques and their role in *Mycobacterium tuberculosis* (MTB) detection.

Biophysical approach	Principle	Role in MTB detection	References
Raman spectroscopy	It quantifies the inelastic scattering of monochromatic light caused by vibrations in molecules.	Uses particular spectral signatures to distinguish TB bacteria; quick identification	Kaewseekhao et al. (2020)
Infrared spectroscopy	It detects molecular vibrations by measuring IR absorption	Identifies MTB's metabolic stages and cell wall components	Pesala et al. (2012), Perumal et al. (2018)
Mass spectrometry	Evaluates the ionized particles' mass-to-charge ratio.	Utilized to quickly identify drug resistance indicators and MTB proteins.	Finehout and Lee (2004), Dang et al. (2015)
Microfluidic PCR chips	Compact heat cycler for small-volume DNA amplification	Enhances MTB-specific genes (such as IS6110); quick diagnosis with a small sample size	Schlanderer et al. (2023)
LAMP-based microfluidic devices	Loop primers and isothermal DNA amplification at a steady temperature	Detects MTB DNA with excellent sensitivity and specificity without the need for complicated equipment.	Yang et al. (2023)

(Continued)

Table 8.1 (Continued)

Biophysical approach	Principle	Role in MTB detection	References
Paper-based Microfluidic Chips	Channels made of porous paper allow fluid to pass through for chemical and biological reactions.	Colorimetric or visual MTB detection is most suited for low-resource environments.	Wu et al. (2024)
Electrochemical LOC Biosensors	Uses microelectrodes to detect electrical signals from biological reactions.	Quantitative identification of MTB antigens or DNA; quick, on-site testing	Golichenari et al. (2019), Joshi et al. (2022)
CRISPR-integrated Microfluidic Systems	It Uses Cas enzymes to identify MTB DNA through cleavage driven by guide RNA.	Enables highly specific MTB gene detection (e.g., using Cas12a/Cas13); detects drug resistance too	Qi et al. (2022)

8.5.1.5 CRISPR-integrated microfluidic systems

These techniques, which combine CRISPR-Cas technology with microfluidics, provide excellent specificity in MTB detection. They can identify different tuberculosis strains and detect drug-resistant mutations with single-nucleobid precision. The integration of sampling preparation procedures allows the processing of original clinical samples, making them potential tools to identify tuberculosis in area conditions resistant to drugs, but adjustment of reactions is still an ongoing research field (Sen et al., 2024).

Every LOC-type provides different benefits for the diagnosis of tuberculosis, ranging from high performance to microfluidic PCR chips and extreme simplicity in the paper-based equipment. Continuous work is focused on increasing sensitivity, reducing costs and improving usability to fulfill the various objectives of global tuberculosis control programs. The combination of smartphone-based reading techniques and cloud connection in next-generation ITC promises to severely revive TB diagnosis in resource-limited environments (Sohail et al., 2022).

Based on CRISPR micro flooding systems provides outstanding, specific MTb discoveries by combining gene examination technology and microfluidics. This technology can identify drug-resistance changes with single-nucleotide precision and separate different TB strains. This is useful for identifying drug-resistant TB to drugs in field conditions, as it combines the stages of sampling preparation that directly examines the original clinical sample. However, it is still difficult to improve the reaction conditions, making them more consistent for general use (Qi et al., 2022).

8.6 Advantages of biophysical methods in tuberculosis diagnostics

Compared to traditional microbiological and molecular techniques, the biophysical diagnostic methods for tuberculosis provide some noteworthy benefits. These benefits include:

In clinical settings, especially those with limited resources, biophysical methods of diagnosing TB offer significant benefits compared to microbiological and molecular methods (Parsons et al., 2011).

Their increased sensitivity and specificity, which enable the identification of MTB at extremely low levels and accurately differentiate it from non-tuberculosis mycobacteria, (NTM) areamongtheir most remarkable advantages. This reduces the likelihood of false-positive results while increasing the likelihood of accurate outcomes (Al-Zamel, 2009).

A short turnaround time, compared to traditional culture-based procedures, which can take days or even weeks, is another major advantage. Data can be obtained in minutes to hours with new innovations. Starting treatment and stopping the spread of the disease require prompt identification. Raman and IR spectroscopy, two biophysical techniques, provide label-free analysis without the use of fluorescent or radioactive dyes, preserving the sample and streamlining the testing procedure (Walla, 2014).

Because sample preparation is typically reduced by biophysical techniques, procedures are made simpler and errors are reduced. This advantage is particularly beneficial in field situations where the laboratory infrastructure is limited. In addition, many methods are self-contained, and some battery-operated, portable devices are ideal for POC diagnostics in remote or resource-limited locations (Walla, 2014).

Their capacity to multiplex provides an additional advantage, allowing for the simultaneous detection of multiple MTB biomarkers. In addition to aiding in diagnosis, this capability also assists in staging the disease and monitoring the treatment. By automating the recognition and classification of intricate patterns in spectral or mass spectrometry data, the application of artificial intelligence AI and ML facilitates automated high-throughput scans and improves diagnosis accuracy (Caulfield and Wengenack, 2016).

Biophysical approaches, most notably those embedded intoLOC systems, are POC ready. LOCs are a crucial link between clinical practice and laboratory-based testing because these integrated systems expand the scope of decentralized healthcare to include sophisticated diagnostics. The outstanding benefits of LOCs strengthen the case for biophysical methods to be used as pioneering solutions in the ongoing global initiatives for enhanced TB diagnosis and control (RSC - Page Load Error, n.d.).

8.7 Challenges and limitations

Although there are significant advantages of using biophysical methods forTB detection, there are still some challenges that need addressing before they can be widely implemented in clinical settings. One of the main challenges is the modern instruments' capital expenditure, which includes mass and Raman spectrometers. These instruments not only require considerable prior investment, but they are difficult to maintain and operate in under-resourced settings. In addition, many of these methods require complex procedures that need trained personnel necessitating availability of skilled workforce which makes them inaccessible (Muthukrishnan, 2021).

A further serious problem is that the biological samples (fluids, e.g., sputum, blood, and serum) are complex and contain a variety of disturbing chemicals that may interfere with or disturb the analytical signal, thereby reducing the accuracy of the detection. Moreover, the absence of standardized methods and reference databases makes replication among laboratories and systems infeasible with consequent inability to establish universal diagnostic cut points (Guzman and Guzman, 2024).

Strict environmental constraints could be relevant, especially for field-deployable instrumentation. Differences in temperature, humidity, and dust levels are potential sources of variation in instrument performance and require refinements of the design to state absolute reliabilities under different circumstances. Reagent stability and implementation are also major technological hurdles in the establishment of LOC technologies, in which most of point-of-care testing applications require long-term storage of reagent and on-chip reagent performance (Garg et al., 2003).

Finally, between regulation and validation they are still barriers that must be overcome before they can be widely accepted. Large, multicenter clinical studies need to be conducted to demonstrate diagnostic accuracy, reproducibility, and usefulness in real world before obtaining permission from health authorities, which is a long-lasting and expensive process. Solving these challenges is critical for translating biophysical breakthroughs into feasible, scalable solutions for global tuberculosis control (Stokes and Wind, 2010).

8.8 Conclusion and future perspectives

In this chapter, we have briefly reviewed the biophysical techniques that have introduced a paradigm shift in the diagnosis of tuberculosis by enabling fast, reliable, and sensitive detection of MTB and its biomarkers. Raman spectroscopy, IR Spectroscopy, mass spectrometry, and LOC systems are among the most promising methods currently under development and clinical evaluation.

Prospects include the development of hybrid devices combining Raman and IR spectroscopy, the integration of advanced ML algorithms for automated interpretation, and further miniaturization to achieve fully handheld diagnostics. With ongoing technological advancements and a growing global health focus, these biophysical methods are poised to significantly enhance TB diagnosis, particularly in underserved and high-burden regions.

The wider implementation of these tools, combined with training programs and regulatory support, will empower healthcare providers to make timely decisions, reduce transmission, and improve treatment outcomes in the fight against tuberculosis.

References

Alsaeed B, Mansour FR. Distance-based paper microfluidics; principle, technical aspects and applications. Microchemical Journal 2020;155:104664. Available from https://doi.org/10.1016/j.microc.2020.104664.

Al-Zamel FA. Detection and diagnosis of*Mycobacterium tuberculosis*. Expert Review of Anti-Infective Therapy 2009;7(9):1099–108. Available from https://doi.org/10.1586/eri.09.92.

Awad H, Khamis MM, El-Aneed A. Mass spectrometry, review of the basics: ionization. Applied Spectroscopy Reviews 2015;50(2):158–75. Available from https://doi.org/10.1080/05704928.2014.954046.

Bartolomeu-Gonçalves G, Souza JMde, Fernandes BT, Spoladori LFA, Correia GF, Castro IMde, et al. Tuberculosis diagnosis: current, ongoing, and future approaches. Diseases 2024;12(9). Available from https://doi.org/10.3390/diseases12090202. Article 9.

Beć KB, Grabska J, Huck CW. Chapter One—Physical principles of infrared spectroscopy. Cozzolino D, editor. Comprehensive Analytical Chemistry, 98. Elsevier; 2022. p. 1–43. Available from https://doi.org/10.1016/bs.coac.2020.08.001.

Calderaro A, Chezzi C. MALDI-TOF MS: a reliable tool in the real life of the clinical microbiology laboratory. Microorganisms 2024;12(2):2. Available from https://doi.org/10.3390/microorganisms12020322.

Caulfield AJ, Wengenack NL. Diagnosis of active tuberculosis disease: from microscopy to molecular techniques. Journal of Clinical Tuberculosis and Other Mycobacterial Diseases 2016;4:33–43. Available from https://doi.org/10.1016/j.jctube.2016.05.005.

Chen J-X, Han Y-S, Zhang S-Q, Li Z-B, Chen J, Yi W-J, et al. Novel therapeutic evaluation biomarkers of lipid metabolism targets in uncomplicated pulmonary tuberculosis patients. Signal Transduction and Targeted Therapy 2021;6(1):22. Available from https://doi.org/10.1038/s41392-020-00427-w.

Chu B. Laser Light Scattering: Basic Principles and Practice. Courier Corporation; 2007.

Coates JP. Infrared spectroscopy for process analytical applications. Process Analytical Technology. John Wiley & Sons, Ltd; 2010. p. 157–94. Available from https://doi.org/10.1002/9780470689592.ch6.

Dang NA, Mourão M, Kuijper S, Walters E, Janssen H-G, Kolk AHJ. Direct detection of *Mycobacterium tuberculosis* in sputum using combined solid phase extraction–gas chromatography–mass spectrometry. Journal of Chromatography B 2015:15–122. Available from https://doi.org/10.1016/j.jchromb.2015.01.045.

Dawson H, Elias J, Etienne P, Calas-Etienne S. The rise of the OM-LoC: opto-microfluidic enabled lab-on-chip. Micromachines 2021;12(12):12. Available from https://doi.org/10.3390/mi12121467.

Finehout EJ, Lee KH. An introduction to mass spectrometry applications in biological research. Biochemistry and Molecular Biology Education 2004;32(2):93–100. Available from https://doi.org/10.1002/bmb.2004.494032020331.

Garg SK, Tiwari RP, Tiwari D, Singh R, Malhotra D, Ramnani VK, et al. Diagnosis of tuberculosis: available technologies, limitations, and possibilities. Journal of Clinical Laboratory Analysis 2003;17(5):155–63. Available from https://doi.org/10.1002/jcla.10086.

Golichenari B, Nosrati R, Farokhi-Fard A, Faal Maleki M, Gheibi Hayat SM, Ghazvini K, et al. Electrochemical-based biosensors for detection of *Mycobacterium tuberculosis* and tuberculosis biomarkers. Critical Reviews in Biotechnology 2019;39(8):1056–77. Available from https://doi.org/10.1080/07388551.2019.1668348.

Guilhaus M, Mlynski V, Selby D. Perfect timing: time-of-flight mass spectrometry† Rapid Communications in Mass Spectrometry 1997;11(9):951–62. Available from https://doi.org/10.1002/(sici)1097-0231(19970615)11:9<951::aid-rcm785>3.0.co;2-h.

Guzman NA, Guzman A. Human sputum proteomics: advancing non-invasive diagnosis of respiratory diseases with enhanced biomarker analysis methods. International Journal of Translational Medicine 2024;4(2). Available from https://doi.org/10.3390/ijtm4020020. Article 2.

Hendricks-Leukes NR, Jonas MR, Mlamla ZC, Smith M, Blackburn JM. Dual-approach electrochemical surface-enhanced raman scattering detection of *Mycobacterium tuberculosis* in patient-derived biological specimens: proof of concept for a generalizable method to detect and identify bacterial pathogens. ACS Sensors 2022;7(5):1403–18. Available from https://doi.org/10.1021/acssensors.2c00121.

Joshi H, Kandari D, Maitra SS, Bhatnagar R. Biosensors for the detection of *Mycobacterium tuberculosis*: a comprehensive overview. Critical Reviews in Microbiology 2022;48(6):784–812. Available from 10.1080/1040841X.2022.2035314.

Kaewseekhao B, Nuntawong N, Eiamchai P, Roytrakul S, Reechaipichitkul W, Faksri K. Diagnosis of active tuberculosis and latent tuberculosis infection based on Raman spectroscopy and surface-enhanced Raman spectroscopy. Tuberculosis 2020;121:101916. Available from https://doi.org/10.1016/j.tube.2020.101916.

Keresztury G. Raman Spectroscopy: Theory. In: Chalmers JM, Griffiths PR, editors. Handbook of Vibrational Spectroscopy. 1st ed.Wiley; 2001. Available from https://doi.org/10.1002/0470027320.s0109.

Lehnert T, M. Gijs MA. Microfluidic systems for infectious disease diagnostics. Lab on a Chip 2024;24(5):1441–93. Available from https://doi.org/10.1039/D4LC00117F.

Liu C, Berg RW. Determining the spectral resolution of a charge-coupled device (CCD) Raman instrument. Applied Spectroscopy 2012;66(9):1034–43. Available from https://doi.org/10.1366/11-06508.

Liu M, Miao D, Qin S, Liu H, Bai Y. Mass tag-based mass spectrometric immunoassay and its bioanalysis applications. TrAC Trends in Analytical Chemistry 2022;157:116745. Available from https://doi.org/10.1016/j.trac.2022.116745.

López-Hernández Y, Patiño-Rodríguez O, García-Orta ST, Pinos-Rodríguez JM. Mass spectrometry applied to the identification of*Mycobacterium tuberculosis*and biomarker discovery. Journal of Applied Microbiology 2016;121:1485–97. Available from https://doi.org/10.1111/jam.13323.

Mahato RP, Kumar S. The future in diagnostic tools for TB outbreaks: a review of the approaches with focus on LAMP and RPA diagnostics tests. Journal of Microbiological Methods 2024;227:107064. Available from https://doi.org/10.1016/j.mimet.2024.107064.

Merchant SA, Shaikh MJS, Nadkarni P. Tuberculosis conundrum - current and future scenarios: a proposed comprehensive approach combining laboratory, imaging, and computing advances. World Journal of Radiology 2022;14(6):114–36. Available from https://doi.org/10.4329/wjr.v14.i6.114.

Mir M, Homs A, Samitier J. Integrated electrochemical DNA biosensors for lab-on-a-chip devices. Electrophoresis 2009;30(19):3386–97. Available from https://doi.org/10.1002/elps.200900319.

Muthukrishnan L. Multidrug resistant tuberculosis – diagnostic challenges and its conquering by nanotechnology approach – an overview. Chemico-Biological Interactions 2021;337:109397. Available from https://doi.org/10.1016/j.cbi.2021.109397.

Naseer K, Ali S, Qazi J. ATR-FTIR spectroscopy as the future of diagnostics: a systematic review of the approach using bio-fluids. Applied Spectroscopy Reviews 2021;56(2):85–97. Available from https://doi.org/10.1080/05704928.2020.1738453.

Ohla S, Belder D. Chip-based separation devices coupled to mass spectrometry. Current Opinion in Chemical Biology, Synthetic Biology / Analytical Techniques 2012;16(3):453–9. Available from https://doi.org/10.1016/j.cbpa.2012.05.180.

Parihar A, Mehta PP. Lab-on-a-chip Devices for Advanced Biomedicines: Laboratory Scale Engineering to Clinical Ecosystem. Royal Society of Chemistry; 2024. Available from https://doi.org/10.1039/9781837673476.

Parsons LM, Somoskövi Á, Gutierrez C, Lee E, Paramasivan CN, Abimiku A, et al. Laboratory diagnosis of tuberculosis in resource-poor countries: challenges and opportunities. Clinical Microbiology Reviews 2011;24(2):314–50. Available from https://doi.org/10.1128/cmr.00059-10.

Perumal J, Dinish U, Bendt A, Kazakeviciute A, Fu CY, Ong ILH, et al. Identification of mycolic acid forms using surface-enhanced Raman scattering as a fast detection method for tuberculosis. International Journal of Nanomedicine 2018;13:6029–38. Available from https://doi.org/10.2147/ijn.s171400.

Pesala B, Gavarna H, Kumar A, Kumaravelu C, Scaria V, Sivasubbu S. Non-invasive detection of *Mycobacterium tuberculosis* using IR and NIR spectroscopy. 2012 37th International Conference on Infrared, Millimeter, and Terahertz Waves, Wollongong, NSW, Australia. IEEE; 2012. p. 1–2. Available from https://doi.org/10.1109/IRMMW-THz.2012.6380079.

Puravankara V, Manjeri A, Ho Kim Y, Kitahama Y, Goda K, Dwivedi PK, et al. Surface-enhanced Raman spectroscopy for point-of-care bioanalysis: from lab to field. Chemical Engineering Journal 2024;498:155163. Available from https://doi.org/10.1016/j.cej.2024.155163.

Qi Y, Li K, Li Y, Guo D, Xu J, Li Y, et al. CRISPR-based diagnostics: a potential tool to address the diagnostic challenges of tuberculosis. Pathogens 2022;11(10). Available from https://doi.org/10.3390/pathogens11101211. Article 10.

RSC - Page load error. (n.d.). Retrieved July 9, 2025, from https://pubs.rsc.org/en/error/pageloaderror.

Schlanderer J, Hoffmann H, Lüddecke J, Golubov A, Grasse W, Kindler EV, et al. Two-stage tuberculosis diagnostics: combining centrifugal microfluidics to detect TB infection and Inh and Rif resistance at the point of care with subsequent antibiotic resistance profiling by targeted NGS. Lab on a Chip 2023;24:74–84. Available from https://doi.org/10.1039/D3LC00783A.

Sen A, Masetty M, Weerakoon S, Morris C, Yadav JS, Apewokin S, et al. Paper-based loop-mediated isothermal amplification and CRISPR integrated platform for on-site nucleic acid testing of pathogens. Biosensors and Bioelectronics 2024;257:116292. Available from https://doi.org/10.1016/j.bios.2024.116292.

Sohail M, Xie S, Zhang X, Li B. Methodologies in visualizing the activation of CRISPR/Cas: the last mile in developing CRISPR-based diagnostics and biosensing – a review. Analytica Chimica Acta 2022;1205:339541. Available from https://doi.org/10.1016/j.aca.2022.339541.

Stokes WS, Wind M. Validation of innovative technologies and strategies for regulatory safety assessment methods: challenges and opportunities. ALTEX Alternativen zu Tierexperimenten - Alternatives to Animal Experimentation 2010;27(3):3. Available from https://doi.org/10.14573/altex.2010.3.198.

Thakur KG, Joshi AM, Gopal B. Structural and biophysical studies on two promoter recognition domains of the extra-cytoplasmic function σ factor σC from *Mycobacterium tuberculosis*. Journal of Biological Chemistry 2007;282(7):4711–8.

Theophile T. Infrared Spectroscopy: Materials Science, Engineering and Technology. BoD – Books on Demand; 2012.

Walla PJ. Modern Biophysical Chemistry: Detection and Analysis of Biomolecules. John Wiley & Sons; 2014.

Wang C-H, Chang J-R, Hung S-C, Dou H-Y, Lee G-B. Rapid molecular diagnosis of live *Mycobacterium tuberculosis* on an integrated microfluidic system. Sensors and Actuators B: Chemical 2022;365:131968. Available from https://doi.org/10.1016/j.snb.2022.131968.

Wang X, Wang G, Wang Y, Quan S, Qi H, Sun L, et al. Development and preliminary application of multiplex loop-mediated isothermal amplification coupled with lateral flow biosensor for detection of *Mycobacterium tuberculosis* complex. Frontiers in Cellular and Infection Microbiology 2021;11. Available from https://doi.org/10.3389/fcimb.2021.666492.

Wessels L, Raad H. Recent advances in point of care diagnostic tools: a review. American Journal of Engineering and Applied Sciences 2016;9(4):1088–95. Available from https://doi.org/10.3844/ajeassp.2016.1088.1095.

Wu T, Shen C, Zhao Z, Lyu M, Bai H, Hu X, et al. Integrating paper-based microfluidics and lateral flow strip into nucleic acid amplification device toward rapid, low-cost, and visual diagnosis of multiple Mycobacteria. Small Methods 2024;8(12):2400095. Available from https://doi.org/10.1002/smtd.202400095.

Yang X, Chen X, Huang J, Chen Y, Zheng W, Chen W, et al. Ultrafast, one-step, label-based biosensor diagnosis platform for the detection of *Mycobacterium tuberculosis* in clinical applications. ACS Infectious Diseases 2023;9(4):762–72. Available from https://doi.org/10.1021/acsinfecdis.2c00475.

Zhang H, Xu Y, Fohlerova Z, Chang H, Iliescu C, Neuzil P. LAMP-on-a-chip: revising microfluidic platforms for loop-mediated DNA amplification. TrAC Trends in Analytical Chemistry 2019;113:44–53. Available from https://doi.org/10.1016/j.trac.2019.01.015.

Zu Y, Chang H, Cui Z. Molecular point-of-care testing technologies: current status and challenges. Nexus 2025;2(2). Available from https://doi.org/10.1016/j.ynexs.2025.100059.

Further reading

Chakaya J, Khan M, Ntoumi F, Aklillu E, Fatima R, Mwaba P, et al. Global Tuberculosis Report 2020 – Reflections on the Global TB burden, treatment and prevention efforts. International Journal of Infectious Diseases 2021;113:S7–12. Available from https://doi.org/10.1016/j.ijid.2021.02.107.

Harries AD, Kumar AMV, Satyanarayana S, Takarinda KC, Timire C, Dlodlo RA. Treatment for latent tuberculosis infection in low- and middle-income countries: progress and challenges with implementation and scale-up. Expert Review of Respiratory Medicine 2020;14(2):195–208. Available from https://doi.org/10.1080/17476348.2020.1694907.

International Journal of Pharmaceutical Research and Development. (n.d.). Available from https://doi.org/10.33545/26646862

Mishra R, Pal R, Mandal RK. Emerging technologies in tuberculosis diagnosis: a comprehensive review. International Journal of Pharmaceutical Research and Development 2025;7(1):265–77. Available from 10.33545/26646862.2025.v7.i1d.119.

Mohammadnabi N, Shamseddin J, Emadi M, Bodaghi AB, Varseh M, Shariati A, et al. *Mycobacterium tuberculosis*: the mechanism of pathogenicity, immune responses, and diagnostic challenges. Journal of Clinical Laboratory Analysis 2024;38(23). Available from https://doi.org/10.1002/jcla.25122.

Tiwari TP, Hattikudur NS, Bharmal RN, Kartikeyan S, Deshmukh NM, Bisen PS. Modern approaches to a rapid diagnosis of tuberculosis: promises and challenges ahead. Tuberculosis (Edinburgh, Scotland) 2007;87(3):193–201. Available from https://doi.org/10.1016/j.tube.2006.07.005.

CHAPTER 9

Nanoparticle-based biosensing of *Mycobacterium tuberculosis*

Anju Nigam[1] and Mansi Chaturvedi[2]
[1]Department of Physics, Govt PG College, Kunidhar, Manila, Almora, Uttarakhand, India,
[2]Academic of Scientific and Innovative Research (AcSIR), CSIR-Advanced Materials and Processes Research Institute, Bhopal, Madhya Pradesh, India

9.1 Introduction

At present times, the terms nanotechnology and nanomaterials sheds light on various research fields, revolutionizing biosensing platforms by offering numerous advantages contributing to their wide applicability (Joudeh and Linke, 2022). Nanoparticles are a number of atoms or molecules fused together with 1–100 nm range of radius. Presently, these nanoparticles have been acquisitioned huge attention because of their unique properties at the nanoscale unlike their bulk counterparts (Sudha et al., 2018). The fascinating change in physical and chemical properties at nanoscale is because of several phenomena such as high aspect ratio, quantum confinement effects, and surface plasmon resonance which revolutionizing the TB diagnostics nowadays. The quantum confinement effect is highly important in the biosensing of *Mycobacterium tuberculosis* (MTB) and it is observed when the size of a material is comparable to the wavelength of electrons which leads to the quantization of energy levels and thus, highly altering the electronic and optical properties depending on the nanoscale range of material. Similarly, another phenomenon, surface plasmon resonance is observed because of the collective vacillation of conduction electrons at the surface of nanoparticles, when exposed to light at specific wavelengths which is of high importance for colorimetric detection of MTB (Maphanga et al., 2023). These effects are highly dependent on the size, shape, and composition of the nanoparticles and thus, different methodologies of nanoparticles play an important role in providing required shapes and sizes to nanoparticles (Anu Mary Ealia and Saravanakumar, 2017). The two primary approaches of synthesizing nanoparticles are "bottom-top" method which involves the flat pack of atom nuclei, resulting in the gradual advancement of nanoscale particles and "top-bottom" approach which involves disintegration of bigger particles into smaller ones by the different methods such as ball milling etc. Thus, the continuous advancements in the synthesis of nanoparticles bring them as most desirable in research because of the requirement of their specific properties in various engineering and biomedical research fields (Kumari and Sarkar, 2021). Particularly, in TB diagnostics, the properties of nanoparticles such as

Advances in Biosensor Technology for Mycobacterium Tuberculosis Detection
DOI: https://doi.org/10.1016/B978-0-443-44589-7.00007-8

large surface to volume ratio, biocompatibility, optical characteristics, magnetic properties, functionalization capability, and electrical conductivity have been extensively explored to enhance sensitivity and reducing detection limits of biosensors. Thus, nanoparticles have emerged as ideal candidates for developing sensitive and rapid diagnostic tools in providing quality health care for many infectious diseases such as tuberculosis (TB) (Khan and Hossain, 2022). As already known, TB has significantly impacted human health for centuries being a fatal and infectious disease associated with significant morbidity, causing deaths of approximately two million people per year and raising serious health concerns (WHO Global Tuberculosis control Report, 2009; Vashishtha, 2009). This disease is caused by some specific species of Mycobacterium such as *M. Tuberculosis Complex* (MTBC) and *M. Lepre*. Among them, MTB is the holoparasite type human pathogen of the MTBC group of *Mycobacteriaceae* family having approximate 20 hours of generation time and grows in oxygen-rich environments. It is a leisurely-growing, aerobic bacterium with complex lipid-rich cell wall structure rich in mycolic acids, which imparts resistance to many conventional antibiotics and contributes to its pathogenicity. Despite peak advancements in the MTB detection techniques, it is still among one of the most death causing pathogens worldwide because of scarcity of reliable and rapid diagnostic tools (Patra et al., 2024; Abrahams and Besra, 2018).

Available conventional methods for the identification of bacteria are generally based on the culture of the microorganisms on agar plates and these techniques were very laborious, slow, and unspecific in some cases (Pornprom et al., 2024; Tigga et al., 2025). Whereas, nanotechnology offers efficient, compact, and simple diagnostic devices suitable for on-the-spot testing in resource limited settings. Thus, in the quench of investigating better tools for the biosensing of MTB, various types of metal nanoparticles synthesized through different techniques have been investigated for providing efficacy in biosensing platforms. The functionalization of nanoparticles with biomolecules such as antibodies, aptamers, or DNA oligonucleotides enhances their specificity, enabling the selective capture of MTB antigens or genetic material (Arora et al., 2006). The ability of nanoparticles bestowing observable changes, such as colorimetric, fluorescence, or electrochemical responses upon target binding enables the facile and accurate detection process, even without the need for complex instrumentation (Biswas et al., 2017). Thus, in today's era nanoparticles are on the way to become most promising materials in the biosensing of not only MTB but for other infectious diseases also.

9.2 Morphology of nanoparticles

The morphology of nanoparticles plays an important role for minimizing or maximizing specific interactions or properties. The high aspect ratio of nanoparticles affords more reaction sites, improving sensitivity of detection, thus providing accurate MTB detection even with low bacterial counts. Detection technologies based on nanomaterials recurrently yield fast outcomes. Variations in nanomorphology are because of thermodynamic or kinetic forces, including size, coatings and surfactants such as parameters, supports, and substrates for applications of nanoparticles in biosensing of many diseases including MTB. The structure of nanoparticles is complex with two or three layers described as, **(1) Surface layer**: This is the outer layer and can be modified with diverse metal ions, tiny molecules, polymers, or surfactants., **(2) The shell layer**: This is the second layer or can be added purposely. It is chemically different, with entirely different structure from the core of the nanoparticles., **(3) The core**: This is the central portion of the nanoparticles or can be considered as nanoparticle itself and thus, responsible for the characteristic properties of the nanoparticles (Christian et al., 2008).

9.3 Properties of nanoparticles

Nanoparticle properties play a crucial role in the biosensing of MTB by enhancing sensitivity and detection accuracy. The nanoscale properties of nanoparticles improve signal amplification, enabling early and low-level detection of the pathogen and the efficient immobilization of biomolecules that specifically recognize MTB targets. Size and shape tunability help in optimizing interactions with MTB antigens or DNA. Surface functionalization increases selectivity and reduces non-specific binding and thus providing rapid response times. Some of the important properties utilized in biosensing of MTB are given in Table 9.1.

Table 9.1 Properties of nanoparticles utilized in biosensing of MTB.

Property	Role in MTB biosensing	Typical nanomaterials	Detection principle
Optical	Direct signal generation	Au nanoparticles, Ag nanoparticles, QDs	LSPR, fluorescence, SERS (Maphanga et al., 2023; Hasan et al., 2024; Cortie and McDonagh, 2011)
Magnetic	Target capture & magnetic readout	Fe_3O_4, $CoFe_2O_4$	Magnetic separation, GMR (Joshi et al., 2022; Gordillo-Marroquín et al., 2018)
Thermal	Signal amplification	Gold nanoparticles, graphene	Photothermal, thermo-resistive (Zhou et al., 2011; Nikolaev et al., 2024)
Mechanical	Nanomechanical signal, stability	CNTs, nanocantilevers	Resonance frequency shift (Ramesh et al., 2022; Hsu et al., 2014)

MTB, Mycobacterium tuberculosis.

9.4 Nano-biosensors

In the present times, the term nano-sensor represent a system measuring physical properties, constituting at least one nanostructure that is used for detecting gases, biological agents, electric fields and chemicals etc. in its construction. The sensitivity of these sensors is pivotally enhanced by the use of nanomaterials such as nanorods, nanotubes, and nanoparticles. Among different kind of available nanosensors, nanobiosensors are highly employed in the medical field, especially in diagnosis of TB (Hsu et al., 2014; Nikolaev et al., 2024). These analytical nanodevices utilize biological components such as enzymes, antibodies, or microorganisms, integrated with physicochemical transducers or bioreceptors (such as electrodes or optical systems), to detect specific analytes, converting their presence or concentration into measurable signals (Yang et al., 2022). The process of signal creation during the contact between the bioreceptor and analyte is termed as biorecognition. Unlike conventional chemical sensors, nanobiosensors work on two principles: (a) biological structures (cells, enzymes, or nucleic acids) will be the sensing elements (b) biological processes or physical changes could be measured (Ma et al., 2021). In nanobiosensors, analyte is attached to a part of the system for the precise detection of a biological element (such as a DNA strand etc.).

9.5 Nanoparticles based biosensing of *Mycobacterium tuberculosis*

Nowadays, nanoparticles have fascinated scientists to meet modern challenges and finding numerous applications in all type of biosensors based on their properties. Nanoparticles have the caliber to be employed in biosensors as biorecognition elements, transducers or both, depending upon the required property of particular type of nanoparticles. The most important parameter in any biosensor is the area interacting with the analyte for enhanced sensitivity and thus, nanoparticles potentially increase biorecognition area, proving themselves the most promising candidates for biosensors in biosensing of MTB (Rabti et al., 2020). However, the basic roles of the nanoparticles in biosensing activity are classified as (1) biomolecules immobilization, (2) catalysis of electrochemical reactions, (3) enhancement of electron transfer, (4) labeling biomolecules, and (5) acting as a reactant. In the biosensing of MTB, the size-dependent properties of nanoparticles play a crucial role for enhancing sensitivity and specificity, utilizing their unique properties. Generally, nanoparticles of size range between 10 to 500 nm have been employed in the field of biotechnology. The large surface to volume ratio makest nanoparticles as extraordinary intrants for biomedical applications as diverse biological processes can occur at nanometer scales, allowing different communications on cells surface with biomolecules in such a way that the various biochemical properties of these cells can be decoded (Zhou et al., 2011; Salimiyan Rizi et al., 2021).

In the biosensing process of MTB, nanoparticles have often been combined with specific targeting molecules, such as antibodies, aptamers, or ligands, which bind selectively to MTB antigens or components. By exposing nanoparticles to a sample, these interact with MTB cells or biomarkers, thus bringing a measurable signal change. Furthermore, various analytes are also employed in nanoparticles-based biosensing for MTB to enhance detection sensitivity and specificity. There are many available analytes such as nucleic acids, proteins, cytokines, and whole bacterial cells contributing significantly in the diagnosis of MTB (Grassian, 2008). The use of nanoparticles enhances the efficiency of detection methods such as hybridization assays and polymerase chain reaction (PCR) techniques. For instance, functionalized gold nanoparticles have been used for colorimetric detection, allowing for the easy and rapid identification of MTB DNA in clinical samples. For differentiating between the MTB-infected individuals and those immunized with BCG vaccine, specific proteins such as primarily secreted antigenic target 6 (ESAT-6) and culture filtrate protein 10 (CFP10) are utilized as biomarkers for the proper treatment of MTB (Zhu et al., 2024).

Interestingly, the whole MTB cells can be directly detected using nanoparticles-based methods. This approach involves the use of magnetic nanoparticles for capturing MTB from complex biological samples such as sputum. The nanoparticles enable downstream applications in bounded state, including PCR or fluorescence-based detection techniques, making it possible to identify the pathogen without the need for extensive sample processing (Seele et al., 2023). For example, miscellaneous metal nanoparticles such as gold, silver, and iron as well as metal oxide nanoparticles like ZnO and MnO_2 have prominent and versatile capabilities (Gordillo-Marroquín et al., 2018; Yang and Shen, 2025). Furthermore, magnetic nanoparticles facilitate magnetic separation and detection of MTB from complex samples using an external magnetic field, followed by assays such as enzyme-linked immunosorbent assays (ELISA) or PCR for quantification. Moreover, polymeric nanoparticles can be engineered for incorporating various sensors or probes, such as fluorescent dyes or electrochemical elements, enhancing the signal output and enabling diverse detection techniques (Burlec et al., 2023).

Quantum dots (QDs) are another protuberant kind of nanoparticles used in bioanalysis which are optical semiconductor nanocrystals which are employed for developing optical biosensors for detecting pharmaceutical analytes, chemicals, and biomolecules. Moreover, the exclusive properties of nanostructured carbons, such as carbon nanotubes (CNTs) are highly exploited in electrical transducers in biosensing devices (Kim et al., 2017; Liang et al., 2021). Their distinct structure, excellent stability, outstanding properties, high thermal conduction, low surface deposition, and low excess voltage (Wang et al., 2024), advances the biosensor's performance for biomedical applications. These nanoparticles also encircle negative effect on cells by causing oxidative stress, cell necrosis, and damage to DNA, mitochondria, and cell membranes (Wang et al., 2013).

At nanoscale, the size, dimensions, and shape of nanoparticles play an important role in their applications such as the absorbance and scattering of light by gold and silver nanoparticles show size and aggregation dependence (Rao et al., 2002; Khan and Hossain, 2022). Thus, the luminescence properties of nanoparticles can be tuned by changing the material's dimensions within the size range of 1–100 nm. For instance, smaller nanoparticles (characteristically with range of 1–10 nm) enhance the binding efficiency of targeting molecules such as antibodies or aptamers to MTB antigens. The amplified surface area provides more active spots for antigen binding, thereby improving sensitivity and allowing detection of lesser concentrations of MTB (Hang et al., 2024), whereas larger nanoparticles (ranging from 20–100 nm) tend to have greater optical or magnetic properties because of their increased volume which is advantageous for signal amplification. For example, gold nanoparticles larger than 20 nm exhibit more pronounced localized surface plasmon resonance effects, leading to stronger colorimetric signals in optical biosensing (Loskutova et al., 2025). However, larger nanoparticles might experience reduced binding efficiency in complex biological matrices because of steric hindrance or aggregation. Some studies also revealed that intermediate-sized nanoparticles (around 10–20 nm) often provide a balance between high sensitivity and optimal signal enhancement, maintaining adequate signal strength and avoiding aggregation issues (Sapsford et al., 2006). The advantage of using nanoparticles is that the size can be chosen willingly with modification in synthesis process for designing required effective biosensing platforms for MTB, as it directly impacts the sensitivity, specificity, penetration depth, and practicality of the diagnostic method (Kaur and Shorie, 2019). The diverse physical properties of the nanoparticles can be obtained by employing different synthesis methods which can also be altered with various chemical functional groups for conjugation of these materials with antibodies, ligands, and drugs of interest. This variation is bringing a range of potential applications in biotechnology, magnetic imparting, the pre-concentration of target analytes, targeted drug delivery, and in diagnostic imaging (Dolai et al., 2021). According to a study report revealed that aptamers, biomolecules or analytes to be recognized should be in the LSPR field penetration depth region of the nanoparticle for the effective sensing (Ramesh et al., 2022).

Moreover, other detection methods based on nanomaterials employing lasers, fluorescence labeling, nuclear magnetic resonance (NMR), and even flow cytometry have been developed for quicker and precise detection of MTB (Semwal et al., 2023; Ckumdee et al., 2016). The general classification of nanomaterials is described in Fig. 9.1 and nanoparticles-based TB detection systems are discussed below.

9.5.1 Metallic nanoparticles in biosensing of *Mycobacterium tuberculosis*

These nanoparticles are composed of metals and attracted huge interest in nanotechnology for over years because of their fascinating properties such as greater surface energies, quantum confinement effect, plasmon excitation, localized surface plasmon

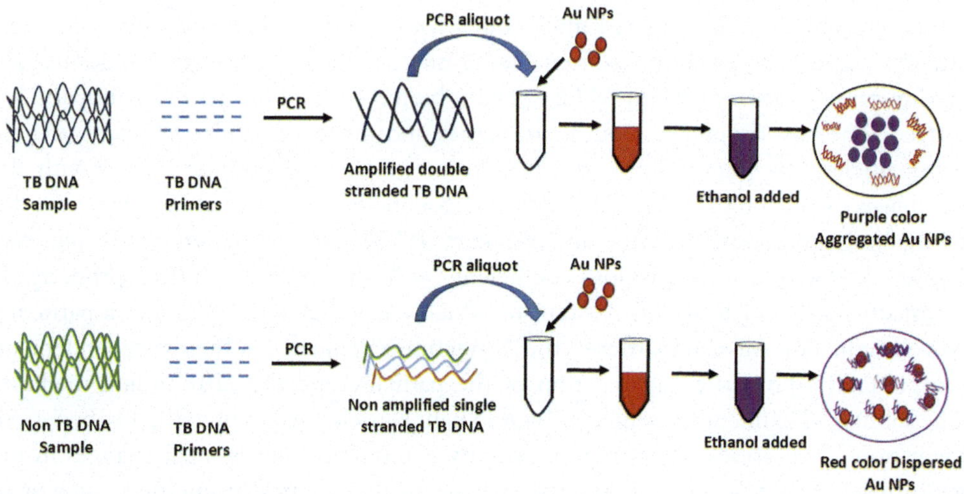

Figure 9.1 Schematic representation of a colorimetric assay for detecting *Mycobacterium tuberculosis* DNA using PCR amplification and gold nanoparticles.

resonance (LSPR), and more kink counts. These nanoparticles enhance signal detection through LSPR, which is a physical phenomenon takes place when light interacts with nanoscale metal particles. When light hits these nanoparticles, the free electrons on the surface begin to vibrate together at a specific frequency, and the collective movement of electrons is called plasmon resonance. Because of the nanoscale particles, much smaller than the wavelength of light, the vibrations are "localized" to the nanoparticles itself and do not spread across a larger surface (Hutter and Fendler, 2004; Rafsan et al., 2025). LSPR-based sensors can monitor multiple binding actions like protein-ligand communications, deoxyribonucleic acid (DNA) hybridizations, and biomarker interactions. Thus, these biosensors enable the biosensing applications by bringing change in the index of refraction of the immediate medium. The absorption and scattering of light by the nanoparticles through LSPR are altered upon binding to MTB, resulting in a detectable color change. These all properties such as enhanced specificity and sensitivity, tiny, and tunable size with emission wavelength, and prompt analyte detection make these nanoparticles prominent candidates for multiple optical bioanalysis (Mathebula et al., 2009).

9.5.1.1 *Gold nanoparticles based* Mycobacterium tuberculosis *detection*

Gold nanoparticles for biosensing MTB have always been a prior interest of researchers from early use till now because of their user-friendly nature. Their facile surface modification leads to effective functionalization with various ligands, antibodies, or

DNA sequences for the targeted interactions with MTB biomarkers. The SPR effect attracted appreciable attention as it enables sensitive detection through changes in light absorption or scattering when MTB-specific biomolecules bind to the nanoparticle's surface. The size-dependent optical properties and the extinction coefficient of gold nanoparticles have significant value in colorimetric sensing methods along with the advantage of biocompatibility, low toxicity, conductivity, and catalytic properties for biomedical applications (Kreibig and Vollmer, 1995). The extinction coefficient, also known as the molar absorption coefficient, is a fundamental optical property of nanoparticles that quantifies the extent to which a material absorbs light at a particular wavelength. For Au nanoparticles, the extinction coefficient is highly dependent on morphological properties, aggregation of the particles, and the surrounding medium. The substantial extinction coefficient of gold nanoparticles makes them highly beneficial for various applications where high sensitivity is required (means small changes in the nanoparticle surface lead to substantial changes in their optical properties), such as in biosensing. Thus, gold nanoparticles are widely employed in precise detection of biomolecular interactions like binding of antigen-antibody, hybridization of DNA, and other molecular recognition events (Karnwal et al., 2024).

The exceptional biocompatibility and electrical conduction of gold nanoparticles enable widespread application of gold nanoparticles in biosensors such as electrochemical, and piezoelectric sensors with improved diagnostic performance as compared to other biosensor designs. Moreover, the immobilization of a large number of biological molecules such as peptides, antibodies or DNA probes specific to MTB antigens is facilitated by the high aspect ratio of gold nanoparticles and possessing ability to significantly intensify the surface plasmon resonance signals, thereby improving the detection limit and reliability of the assay. These nanoparticles also provide a stable platform for biomolecule conjugation, ensuring robust performance under varying environmental conditions encountered in clinical settings (Lin et al., 2016). Functionalized gold nanoparticles have been extensively studied for their minimal cytotoxic effects while effectively enhancing detection signals through signal amplification strategies, such as enzyme-linked immunosorbent assays or electrochemical detection methods (Khan and Hossain, 2022; Patra et al., 2024). By functionalizing gold nanoparticles with specific recognition elements, researchers can tailor biosensors to target unique MTB biomarkers, minimizing cross-reactivity with other microorganisms (Zhao et al., 2006). This specificity is crucial for accurate MTB diagnosis, particularly in regions where TB coexists with other infectious diseases that may present similar clinical symptoms.

Initial studies revealed the application of gold nanoparticles-based bio-barcode analytes for the ultrasensitive detection of proteins and thus, laying the groundwork for their subsequent adaptation in infectious disease diagnostics (Pei et al., 2022; Nam et al., 2003). More advancements with time have been made in nanoparticles synthesis and bioconjugation techniques for the development of highly specific biosensors targeting MTB antigens. These biosensors utilize functionalized gold nanoparticles coupled with antibodies or DNA probes specific to MTB biomarkers, enabling rapid and accurate detection in clinical samples (Maphanga et al., 2023). Many researches were conducted to improve the stability, reproducibility, scalability, and the incorporation of multiplexed detection strategies to simultaneously identify multiple MTB biomarkers in a single assay (Du et al., 2016; Anu Mary Ealia and Saravanakumar, 2017; Kumari and Sarkar, 2021).

In a recent study, a molecular analysis assay proposed that gold nanoparticles derived with thiol modified oligonucleotide improved detection flexibility, efficiency, and cost reduction. This method utilized gold nanoparticles probes for IS6110 and Rv3618 for specific hybridization with target DNAs of MTBC and MTB isolates, respectively. The evaluation of the effectiveness of synthesized gold nanoparticles probes was done by direct and simultaneous detecting MTBC and MTB from various clinical sputum specimens. A sensitivity of approximately 96.6% and specificity of approximately 98.9% towards detection of MTBC was observed, and a sensitivity of 94.7% and specificity of 99.6% for detection of MTB was also observed and thus, proving this method very prompt, facile, and low-cost with accurate detection mechanism (Baptista et al., 2006).

As per recent advancements, a study revealed the development of colorimetric biosensor through simple and cost-effective method using gold nanoparticles, with DNA enzymes functionalization that specifically recognizes MTB DNA sequences, enabling visual recognition of MTB DNA in collected samples. The study demonstrated the potential of AuNP-based biosensors as point-of-care diagnostics for TB, addressing the critical need for rapid and accurate detection methods (Soo et al., 2009).

Another study reported a new electrochemical aptasensor constructed on the basis of dual-signal yield for very sensitive detection of MTB (ESAT-6) antigen. In this study, a new nanocomposite of gold nanoparticles-loaded with molybdenum disulfide was synthesized for generating and amplifying the signal with the detection limit of the aptasensor approximately 2.88 fg mL^{-1} in the Differential pulse voltammetry measurement and 13.50 fg mL^{-1} in effect of incubation time measurement, respectively. The proposed sensor showed better accuracy, with a sensitivity of 97.5% and a specificity of 96.7% (Huang et al., 2023).

A study revealed the development of a rapid detection method by means of spherical shaped modified gold nanoparticles in a colorimetric probe-based biosensor and investigated 52 secretion specimens. The developed biosensor showed 100% sensitivity and specificity for the detection of secretion DNA, with taking minimal time of less than 15 min, and the designed method was claimed to be affordable, reliable, and cost-efficient (Zhang et al., 2024a). Further studies were conducted on SERS biosensors and integration of Au nanoparticles into electrode surfaces and functionalizing them with

MTB-specific probes to detect bacterial biomarkers with high sensitivity had also been studied (Kooti et al., 2023).

Another study revealed potential of gold nanoparticles in multiplexed diagnostic assays, allowing for detecting simultaneous manifold MTB markers. For achieving this, the combination of rolling-circle amplification techniques with AuNP-based biosensors has been utilized, which improved the detection efficiency by amplifying the signal associated with MTB sensing. Thus, these advancements have been providing a platform that is not only sensitive but also adaptable for point-of-care applications (Maphanga et al., 2023).

Another study revealed the development of a LAMP assay for rapid in vitro colorimetric diagnostics of TB with a time of ≤50 min, by the isothermal amplification of target gene 'IS6110' and recognizing it by DNA functionalized gold nanoparticles. The DNA gold nanoparticles support insitu investigation of the amplified DNA fragment and abiding the intrusion in the complex LAMP system, letting for colloidal spreading through the steric effect depending on DNA splicing density. The positive sample was identified through red visuals of the DNA-Au nanoparticles contrasting with the gray visuals in aggregates indicating the negative sample. Additionally, the DNA-gold nanoparticles demonstrate exceptional performance with greater than seven months storage while maintaining the sensitivity. The enhanced specificity of the DNA-gold nanoparticles was observed highly beneficial for MTB diagnosis in urgent situations and can potentially accelerate patient maintenance with primary treatment (Pei et al., 2022).

A facile and prompt detection method for MTB was developed using altered gold nanoparticles with colorimetric visuals probe-based biosensor in specimens. Functionalized gold nanoparticles were manufactured through the citrate reduction method. Au nanoparticles-biosensor and *IS6110* PCR were equated with the gold standard in terms of analytical sensitivity and specificity, positive predictive value (PPV), negative predictive value (NPV), diagnostic odds ratio (DOR), and accuracy in different specimens. The gold standard was defined as a positive result in sputum smear microscopy. The gold nanoparticle-biosensor had sensitivity of approx. 100 percent and specificity for detecting total sputum DNA in less than approximately 15 minutes with ready-to-use gold nanoparticle-biosensor. This method was found to have 100% PPV, 100% NPV, and 2325 DOR with 100% accuracy. The auspicious results of the designed method was a reliable and cost-beneficial way for straight detection of MTB in various clinical specimens (Zhang et al., 2024b).

Another study revealed an alternative engineered paper-based electrochemical biosensor for direct diagnosis of MTB without requiring any pre-enrichment. In this study, a sensitive, low-cost, portable, and disposable biosensor was developed based on gold nanoparticles and GCOOH decorated on the working area. This proposed immunosensor, with requirement of microscopic sample volume of 5 μL, attained better performance as per the limit of detection with the measurement of 0.01 ng/mL and thus can quickly detect blood serum with TB infection without requiring any pre-enrichment, bringing it as a suitable alternative for the primary diagnosis of MTB in resource-scarce countries (Kooti et al., 2023).

A recent study discovered a facile ethanol-induced gold nanoparticles aggregation strategy, a colorimetric method to detect amplified MTB DNA. For the aggregation of Au nanoparticles, ethanol was used in hydrophobic and electrostatic interactions between Au nanoparticles and DNA. 4 μl of 100% ethanol addition to the gold nanoparticles-MTB DNA complex caused aggregation. In the absence of TB DNA, suspensions of gold nanoparticles did not aggregate despite the inclusion of 8 μl of 100% ethanol. This technique was found sensitive to ~340 fM levels for the detection of MTB DNA, and rapid with time taken less than 3 min. As compared to other salt-based aggregation methods, this method has been claimed as more sensitive and reliable for early MTB identification (Pornprom et al., 2024).

Another study reported a facile procedure to form mycobacterial phosphatidylinositol mannoside (PIM) comprising monolamellar liposomes and attaching them on gold exteriors by means of thiol chemistry. The prepared functional fused nanoparticles showed localized surface plasmon resonance (LSPR) swings of approximately up to 2 nanometers upon their interaction with specific antibodies, representing their appropriateness for the optical sensing of antigen–antibody reactions. The pertinence of the system for detecting real-world situation of TB-associated antibodies was verified through sample analysis of human individuals (Tripathi et al., 2023). The process of diagnosis is shown in Fig. 9.2.

Recently, a study revealed a new way of MTB detection using a smartphone camera coupled with a paper-based assay. This method uses wells of magnesium chloride, preventing aggregation of gold nanoparticles functionalized with thiol-modified ssDNA probes. These probes were designed to complement MTB's RNA polymerase β subunit. Change in color of the array wells worked as an indicator, and imaging the color changes in the wells quantified absence or presence of MTB through a simple RGB (red-green-blue color) analysis with a limit of detection of 10 μg/mL reported (Becerra et al., 2024).

9.5.1.2 Role of silver nanoparticles in biosensing of tuberculosis

Silver nanoparticles (Ag nanoparticles) are known to have high reactivity and thus possess high electrochemical oxidation because of their superior electron transfer efficiency, significant scattering to extinction ratio with sharp and clear bands as compared with other metal nanoparticles like gold. Their vulnerable stability and toxicity may be considered as some limitations, and thus, fewer studies are reported on Ag nanoparticles for improvisation in biosensing. To overcome these limitations,

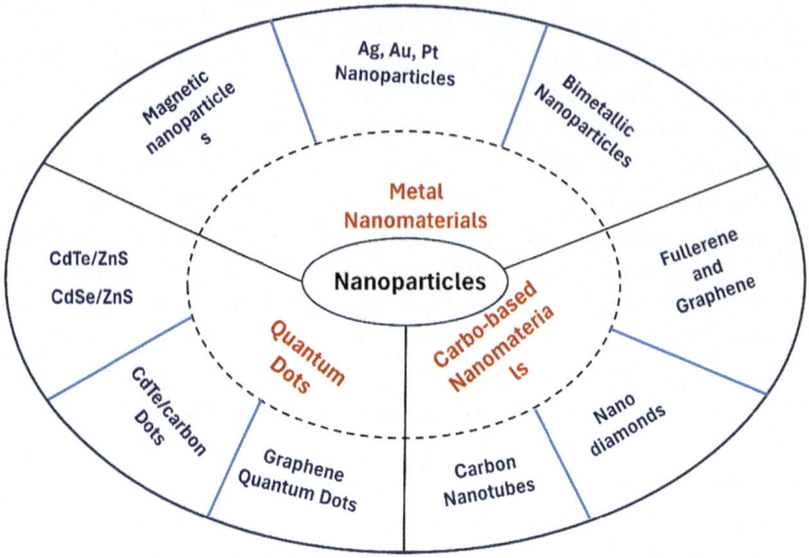

Figure 9.2 General classification of nanomaterials.

various progressive coating techniques are being discovered to enhance the harmfulness, stability, and functionalization issues with Ag nanoparticles (Veigas et al., 2012). As we discussed already, noble metallic nanoparticles such as Ag have been found as potentially capable as detecting color visuals for prompt colorimetric investigation in wide range of molecules, making naked eye visualization of marked species because of LSPR, which enables label-free detection through changes in absorbance or fluorescence upon binding to MTB-specific biomarkers. Moreover, morphological properties of these nanoparticles ease interactions with biomolecules, and probe-conjugation of nanoparticles show high-efficiency in detection of biomolecules, thereby improving sensitivity. Additionally, smooth functionalization with antibodies, aptamers, or other molecules reveals them as most prominent candidates in selective and specific capture of MTB antigens or nucleic acids (Sofi et al., 2022).

Integration of Ag nanoparticles into various biosensing platforms has been studied by many researchers for the improvisation of different biosensor's performance, thus enabling point-of-care testing, a crucial step towards decentralized MTB diagnosis in limited resources (Krutyakov et al., 2008). Furthermore, advancements in nanotechnology have facilitated the development of miniaturized, portable Ag nanoparticles-based biosensors capable of precise detection of MTB antigens in clinical settings (Ibrahim et al., 2021).

From 2016 onwards, the focus shifted towards integrating Ag nanoparticles-based biosensors with emerging technologies such as smartphone-based detection platforms and artificial intelligence for real-time analysis of MTB biomarkers. Additionally, efforts were directed towards overcoming the challenges associated with commercialization and scalability of Ag nanoparticles-based biosensors for MTB diagnosis (Girma et al., 2025).

A study revealed a simple colorimetric sensing approach for the concurrent detection of the antituberculosis drugs such as isoniazid and streptomycin sulfate through maltol-covered silver nanoparticles. This study shows that streptomycin sulfate brought a visible color change from yellowish to pinkish colour with SPR peak at 436 nm, indicating a red shift with peak at 664 nm, whereas isoniazid increased the SPR intensity at 436 nm with minimal colour variation. The method demonstrated good accuracy in biological, and pharmaceutical samples. Additionally, the other results showed dual potential as sensitive chemo-sensors and effective antitubercular agents (Naqvi et al., 2023).

A recent study revealed a rapid, precise, and cost-effective visual molecular biosensing strategy for MTB detection utilizing Ag nanoparticles stabilized via label-free citrate. In this study, unique biomarkers to MTB were used as target using specific primers. In this method, the absence of amplified ssDNA was confirmed by citrate-stabilized Ag nanoparticles aggregates formation on the addition of sodium chloride, thus enabling detection of as few as 4 bacilli. This strategy also discriminated MTB from *M. bovis* with sensitivity approximately 40 bacilli (Tan et al., 2021).

9.5.1.3 *Bimetallic nanoparticles in biosensing of* Mycobacterium tuberculosis

The certain limitations faced by single-component nanoparticles lead to the development of bimetallic nanoparticles, for many advantages over monometallic nanoparticles. These nanoparticles comprised of two different metals with specific atomic ratio, thus exhibiting enhanced unique properties such as electronic, catalytic, and optical properties, arising from the interaction of multiple metals (Patnaik and Dey, 2023).

A recent study revealed a low-cost and facile colorimetric sensing process for the detection of isoniazid, which is an effective and highly selective drug against MTB. An open site solvothermal method wass developed for synthesizing functionalized bimetallic nanozymes (nanozymes are nanomaterials possessing catalytic activity like that of natural enzymes) frameworks consists of one-dimension mixed chains of valence metal-oxygen. Using electrostatic effect, Au nanoparticles were grown in situ on N metal organic framework (MOF) and the results showed noteworthy performance of Au@NMOF in terms of peroxidase-like activity, stability, and substrate attraction. A colorimetric detection platform with integration of smartphone based on hydrogel was created to detect INH, based on manufactured framework which showed a better range of 0.25–120 μM, thus enhancing practical application of detecting INH in human serum and urine (Yu et al., 2023).

A recent study also revealed a method of preparing palladium–platinum composite kind of nanoparticles (Pd@Pt nanoparticles) as the nanozyme, joined with a paper-based multilayer analytical device for MTB biosensing. These nanoparticles were easily prepared with high stability and catalytic activity, which can be an effective alternative for horseradish peroxidase (HRP). For observing catalytic efficiency of Pd@Pt nanoparticles, repetitive IS6110 fragments in MTB strains were employed as target DNA for molecular diagnosis of tuberculosis. The limit of detection for paper-based analytical device (dnPAD) was observed as 0.216 nM for target DNA. The calculated results provide on-site infection prediction with better sensitivity in underprivileged areas (Tung et al., 2024).

9.5.2 Carbon-based nanoparticles

These nanoparticles are comprised of solely carbon atoms and encompass diverse structures such as carbon black nanoparticles, carbon quantum dots, and fullerenes, each possessing unique physicochemical properties because of sp^2-hybridized carbon bonds. Fullerenes which are spherical carbon molecules, exhibit excellent radical scavenging properties and are well investigated for antioxidant and drug delivery applications (Joudeh and Linke, 2022).

A recent study revealed the development of field-effect transistor (FET) based on single-walled carbon nanotube and functionalized to detect the major MTB-secreted antigen 85B (Ag85B). The developed device effectively detected Ag85B in phosphate-buffered saline with 0.05 fg/ml detection limit. Observed results confirmed the effective spikes of Ag85B in artificial sputum. This biosensor showed valuable sensing performance within 10 min, in biological matrices along with Ag85B identification in serum after introduction of bovine serum albumin (BSA) into the obstructive layer, and thus made distinction in TB-positive and negative clinical samples (Le et al., 2024).

A recent study revealed an electrochemical genosensor in which PNE was investigated as coating material for reduced graphene and gold nanoparticles utilizing Biotin-avidin chemistry for the biosensing of MTB. The developed biosensor observations revealed high sensitivity of 2.3×10^{-3} mA μM^{-1}, low detection limit of 0.1×10^{-7} μM, and broad detection range of 0.1×10^{-2} to 0.1×10^{-7} μM of bioelectrode along with low response time of 5 seconds and showed increased enzyme loading and better biosensing performance (Pornprom et al., 2024).

A recent study revealed the potential use of fluorescent nanodiamonds in the early and successful detection of TB infections. These nanodiamonds were conjugated with antibodies and then used as immunosensors for the detection of disease markers such as ESAT-6 and differentiating between TB and NTM infections. Spin-enhanced lateral flow immunoassay technique was used in the prepared sensor. The detection of limit was observed approximately 0.02 ng mL^{-1} with the lateral flow which was found very effective and fast for the biosensing of MTB (Joudeh and Linke, 2022).

Another recent study in this direction developed a low-cost, paper-based disposable electrochemical sensor grounded on gold nanoparticles decorated carboxyl graphene.

The indicator used in this technique was Hsp16.3, a biomarker of TB infection. The developed immunosensor required a sample volume of only 5 μL and showed superior performance, 0.01 ng/mL of limit of detection for the early and rapid detection of MTB infections (Wang et al., 2024).

9.5.3 Magnetic nanoparticles in biosensing of *Mycobacterium tuberculosis*

Magnetic nanoparticles are composed of natural magnetic materials such as iron oxide, cobalt, nickel and magnetite (Fe_3O_4), and maghemite (γ-Fe_2O_3) etc., hold intrinsic magnetic properties with enhanced magnetic coercivity and saturation magnetization, dispersibility with biointeractive stimulus. Contrary to their bulk counterparts, magnetic nanoparticles hold high residual magnetism with significant magnetic sensitivity. Magnetic nanoparticles can be classified as paramagnetic, ferromagnetic, and supermagnetic, depending on their alignment of unpaired electrons with or without external magnetic fields. However, the developed magnetization is insignificant in the absence of magnetic field, but at high values, these tiny particles exhibit immense signals for multiple bio-entities detection and thus, find numerous applications in magnetic resonance imaging, magnetic hyperthermia therapy, and magnetic separation techniques. The morphological properties of magnetic nanoparticles allow facile surface functionalization, which is a critical factor for conjugation of magnetic nanoparticles with target tissues such as antibodies and polysaccharides etc., for expanding their application in improving detection and concentrating biomolecules from complex biological samples. These traits of Magnetic nanoparticles, thus, enable a modulated blood movement without applying external magnetic field and thus, allowing accurate interactions with specific target cells. Moreover, reversible magnetization is exhibited by superparamagnetic iron oxide nanoparticles by applying magnetic field externally, enabling tissue-specific imaging and targeted drug delivery. Therefore, biofunctionalized magnetic nanoparticles have been considered proficient in detecting and concentrating MTB biomarkers such as proteins and nucleic acids in terms of efficiency and allowing specific identification of the pathogen in clinical samples (Li et al., 2021; Gupta et al., 2021).

According to initial study reports, researchers explored various novel synthesis routes to tailor the structural properties of magnetic nanoparticles. The evolution of functionalization techniques initiated using aptamers, peptides, and molecularly imprinted polymers against antibodies for improvisation of specificity and efficiency of MNP-based assays for MTB biomarkers, facilitating accurate and rapid diagnosis (Bisht et al., 2023). The development of microfluidic devices integrated with magnetic nanoparticles for automated sample processing and detection, enabling rapid turnaround times and decentralized testing in resource-limited settings. Concurrently, hybrid nanocomposites combining magnetic nanoparticles with graphene, carbon nanotubes, and other nanomaterials emerged as promising platforms for ultrasensitive and multiplexed TB diagnostics. In a study report, a magnetophoretic immune assay was also designed to detect antigens which are significantly effective in primary stages of MTB. In this study, gold nanoparticles for signaling and magnetic particles for separation were employed, resulting in the detection limit of 0.3 pM. Synthesized nanoparticles showed sturdy and reliable sensing against commercially available MTB growth indicator tube (MGIT) test. Thereafter, efforts were made in overcoming translational barriers and optimizing the clinical utility of MNP-based MTB biosensors (Le et al., 2024; Pornprom et al., 2024; Rotherham et al., 2012).

A study revealed an approach of developing a transportable electrochemical aptasensor for the investigation of MTB secreted protein CFP10-ESAT6 in medical sputum samples. Aptamers (act as bioreceptors) have been synthetically designed as an alternative to antibodies with effective applications in biosensor development. In this study, authors used Fe/Au magnetic nanoparticles coupled with explicit antibodies to detect CFP10-ESAT6 proteins, and the diagnosis showed a linear detection range from 10 to 500 ng/mL through a differential pulse voltammetry technique. The process was authenticated in detecting MTB using smear and culture microscopy from sputum samples, finding 91.7% specificity with 100% sensitivity (Azmi et al., 2021).

Another study proposed a highly precise and sensitive technique based on the principle of GMR (Giant Magnetoresistance) for the point of care diagnosis of the MTB specific ESAT-6 antigen using magnetic nanoparticles as labels. ESAT-6 is an antigenic protein secreted by MTB in the primitive phase of TB infection. Coupling of magnetic nanoparticles with the GMR sensor in the presence and proportional to ESAT-6 protein concentration leads to change in whole GMR sensor resistance. The observations confirmed the effective detection of ESAT-6 with detection range of pg/mL comparing to other available techniques for ESAT-6 detection. Additionally, the signal strength showed increment with rise in ESAT-6 protein concentration (Gupta et al., 2021).

A recent study revealed an original approach of MTB detection based on magnetic separation. The process defined an MTB DNA sequence precisely seized by magnetic nanoparticles and urease-conjugated silica nanoparticles concurrently to form a complex with catalysation of urea into ammonium carbonate by altering it with the prepared complex. By employing silicon nanowire field-effect transistor, the detection of MTB DNA was investigated indirectly by the detection of ammonium carbonate. The detection limit was investigated to be 78.541 fM, and specificity was confirmed by detecting wide range of bacterial species. The efficacy of the biosensor was verified, and the recovery study of DNA was done in the genomic DNA mined from cultured *Mycobacterium tuberculosis*. The biosensor claimed to be an accurate technique for diagnostic purposes (Wei et al., 2023).

Another study revealed a fast method of developing functionalized magnetic nanoparticles combined with silica by immobilizing heterogeneous antibodies on the outer shell of amine-silanized nanoparticles for the early detection of MTB. The observed results confirmed good feasibility as well as the reduced detection time of ~4 hours as compared to the conventional standard diagnostic method, ELISA, taking approximately ~15 hours. Moreover, the prepared magnetic nanoparticles showed higher sensitivity in the samples of infected patients (León-Janampa et al., 2022).

Another study discovered the use of a monoclonal antibody for an electrochemical immunosensor which was immobilized on a chitosan-coated iron oxide biologically derived composite, comprised of magnetic iron oxide and a non-magnetic thin layer for the detection of MTB lipoprotein. A strong connection between the concentration of phosphate-buffered saline solution (LpqH) and the peak current was observed by the electrochemical measurements. LpqH (19-kDa antigen) is a glycollipoprotein attached at outer membrane of MTB that exhibits immunosuppressive functions. The developed sensor exhibited detection limit of approximately 40 μg/mL (2 μM) LpqH for substantial diagnosis of disease (Reaño et al., 2023).

A recent study revealed an innovative magnetic nanoparticle conjugate with antibody and aptamer-grounded technique targeting four different antigens like GlcB, MPT64 and CFP-10 etc. of MTB for the detection of pleural TB, which is much known for its non-specific clinical features. This method was developed by using EDC-NHS chemistry for conjugating heterogeneous antibodies on exterior of magnetic nanoparticles to capture MTB antigens. Among all four, MPT51-based assay executed the best, with 66.6% and 95.4% sensitivity and specificity, respectively (Sharma et al., 2025).

9.5.4 Semiconductor nanoparticles in biosensing of *Mycobacterium tuberculosis*

Materials having electron bandgap and electrical conductivity between insulators and conductors are termed as semiconductors. Usually, the tunable properties of semiconductors, bring them as most desirable for employment in a wide variety of applications such as nanoscaled integrated circuits, biological sensing, laser technology, labeling, and imaging, etc. At the nanoscale range, quantum size effect leads to discrete energy levels in semiconductor nanoparticles, providing wide range of tunable bandgaps, which bring out fascinating optical and electronic characteristics such as size-tunable bandgaps and fluorescence, long luminescence lifetime and heightened photoluminescence (Terna et al., 2021).

Semiconductor nanoparticles, predominantly derived from materials such as cadmium sulfides and titanium dioxides etc. have reformed the biosensing of MTB because of their notable fluorescence and photoelectric characteristics, which are harnessed to develop highly sensitive and selective biosensors for diagnostic purposes. These nanoparticles can be engineered to emit desired fluorescence at required wavelengths on interaction with MTB biomarkers or DNA, enabling low-detection-limit assays with high signal-to-noise ratios. These are considered most appropriate for biosensors because of better interfacial electron transfer, biocompatibility, robust absorption, chemical steadiness and affordable preparation strategies. Corresponding to their metal materials, efficacy of semiconductor nanoparticles-based biosensors is enormously influenced by their morphological properties. Therefore, size-dependent properties evaluate the significant potentials for binding definite antigens and infection endorsing molecules. Several metal oxide nanoparticles and variants of iron oxide are used for rapid adaptation of signal transduction and in sensing analytes, utilizing their electrochemical, optical, magnetic, and electrical mechanisms. For instance, quantum dots (QDs) are also inorganic semiconducting nanoparticles with size range of 1–10 nm emitting size-tunable fluorescence, generally composed of group IV, III–V and II–VI elements (Smith and Nie, 2010; Qureshi et al., 2023; Wang et al., 2023).

Alloys of SC have also emerged as commanding tools in biosensing applications for MTB and thus recently explored for ultrasensitive, multiplexed analysis (Michalet et al., 2005). The composition and size of SC nanoparticles have ability to tune the emission wavelength from blue to near-infrared (Medintz et al., 2005). The narrow emission of quantum Dots and high quantum yield with photostability make them highly effective in multiplexed optical coding and biomedical imaging. Early 2000s focused on the size-tunable fluorescence and high photostability of quantum Dots was advantageous for sensitive detection of biomolecules. Moreover, surface functionalization strategies in quantum dot synthesis paved the way for enhanced sensitivity and specificity in MTB biosensing. Continuous growth in surface chemistry of the quantum Dots leads to their facile bioconjugation with antibodies and aptamers for capturing and detecting MTB biomarkers (Michalet et al., 2005).

Initial studies proposed many novel approaches for the bio-compatibility and stability of QD-based biosensors for MTB detection. The incorporation of Quantum Dots with progressive imaging techniques such as fluorescence resonance energy transfer and Förster resonance energy transfer allowed the real-time monitoring of MTB biomarker interactions and signal amplification for better MTB diagnostics during complete phase of disease. Furthermore, the combination of nanotechnology and artificial intelligence has been explored for data analysis and pattern recognition in quantum dots fluorescence spectra, thus initiating automated and accurate diagnosis of MTB infections (Soni et al., 2024).

Another study revealed the novel approach of preparing CdSe/CdS and $CdSe_{1-x}S_x$/CdS nanorods for the detection of MTB. These nanoprobe were proficient for detecting low-ample targets with sensitive levels of sub-picomolar in artificial sputum within 2.5 hours. Thus, demonstrating significant specificity to differentiate the gene template of MTB from that of non-tuberculosis bacteria (Ibrahim and Chan, 2021).

9.5.5 Polymeric nanoparticles in biosensing of *Mycobacterium tuberculosis*

Nanoparticles that are generally composed of colloidal structures or biocompatible polymers, either synthetic such as polystyrene, polycaprolactone or natural like chitosan or alginate are termed as polymeric nanoparticles. Modification in morphological properties of these nanoparticles can improve stability, interaction with biological environments, and targeting specific cells or tissues and thus specifically designed for a range of applications along with biosensing. Moreover, for the detection of MTB, these nanoparticles are recognized by their easy functionalization, significant pathogen-capturing capability, and ability of controlled release of therapeutic agents with enhanced signal detection. Progress in polymer chemistry has enabled the preparation of polymeric nanoparticles with size control, modification in shapes, and alteration in surface functionalities, crucial for enhanced interactions with MTB-specific biomarkers. Surface functionalization of polymeric nanoparticles can be chemically modified with drug moieties and targeting ligands. Generally, polymeric nanoparticles are of two types, (a) those prepared by di-block copolymers using spontaneous self-assembly with hydrophobic and hydrophilic regions. The hydrophobic used to create a polymer core using biopolymers such as polycaprolactone (PCL) etc., whilst the hydrophilic part prepared by polyethylene glycol (PEG) protects the core against opsonization and destruction (Medintz et al., 2005). Various study reports discovered different types of P nanoparticles such as lactic-co-glycolic acid, chitosan, and dendrimers etc., each one of them offering exclusive advantages like constant drug release and heightened biostability. Surface functionalization of polymeric nanoparticles with aptamers and antibodies enables specific recognition and capture of MTB biomarkers. Further innovations in the production processes of nanoparticles have eased the integration of polymeric nanoparticles with microfluidic devices for MTB diagnostics with reduced processing times and sample volumes.

Furthermore, the integration of polymeric nanoparticles with advanced imaging techniques such as magnetic resonance imaging facilitated real-time monitoring and spatial mapping of MTB infections, thus leading to valuable insights into disease progression and treatment response (Kia et al., 2023; Elmowafy et al., 2023).

Further studies demonstrated the integration of polymeric nanoparticles with biosensors equipped with artificial intelligence algorithms attributed to automated data analysis and predictive modelling and improved diagnostic accuracy (Elsabahy et al., 2015).

In another study, an electrochemical biosensor has been proposed for the specific detection of MTB, based on two-dimension transitional metal carbide, Ti_3AlC_2 phase and decorated with polydopamine (PDA). Moreover, the PDA-conjugated phase contributed to favorable atmosphere for immobilizing specific ssDNA to MTB biomarkers. A widespread detection range of 0.1×10^{-4} μM to 0.1×10^{-12} μM with a detection limit of 0.1 attomolar was confirmed by the prepared biosensor. Thus, showing excellent sensitivity of 1.67×10^{-4} mA μM^{-1} along with greater selectivity requiring low response time in MTB diagnosis (Chou et al., 2025).

A recent study demonstrated a modified electrode for an electrochemical genosensor for the detection of probe DNA specific to MTB. This study discovered polynorepinephrine (PNE) as a novel layer material for reduced graphene oxide (RGO) and gold nanoparticles. The process of electrode modification using biotin-avidin chemistry for the covalent immobilization of the probe DNA on the prepared nanocomposite surface specific to MTB. Establishment of these RGO/PNE electrodes and RGO/PNE/Au nanocomposite was verified using different characterization techniques. Electrochemical studies were done using cyclic voltammetry, which resulted in the significant enhancement of charge transfer rate of prepared electrodes compared with the conventional ones. An enhanced sensitivity of 2.3 ×10 3 mA μM1as well as little detection limit of 0.1 ×10 7 μM with broad detection range of 0.1 ×10 2 to 0.1 ×10 7 μM has been achieved, presenting low response time of 5 seconds with enhanced loading of enzyme and better electrochemical responsiveness (Goel et al., 2024).

9.5.6 Composite nanoparticles in biosensing *Mycobacterium tuberculosis*

The term composite is used for the combination of two or more materials, of which at least one must be in nanodimension range. Typical nanocomposites are comprised of a framework on which nanomaterials are implanted as reinforcements. Composite nanoparticles are gaining constant attention because of their multifunctional properties as they can exhibit strengthened properties and capabilities that cannot be found in the individual components alone and thus making them ideal in various research fields such as electronics, energy storage, environmental protection, and medical applications, notably in the field of biomedical applications such as biosensing for MTB detection. Their remarkable mechanical, electrical, and optical properties have been effectively overcoming the limitations of single-component systems by enhancing intrinsic performance. Integration of Composite nanoparticles with gold nanoparticles, polydopamine and reduced graphene oxide improved their performances in the biosensing of MTB owing to combined properties of their individual components, thereby improving electron transfer kinetics and providing larger surface area, which strengthens the signal outputs during biosensing analysis (Bisht et al., 2023; Perumal et al., 2018).

A recent study highlighted the development of a ternary nanocomposite platform comprising reduced graphene oxide and gold nanoparticles, providing remarkable electrochemical performance because of effective charge transfer and enhanced electron communication. This was demonstrated to facilitate rapid and sensitive detection of MTB proteins, with ability to detect the MPT64 protein with high selectivity. These examples demonstrate how the strategic design of composite nanoparticles can lead to significant advancements in the detection capabilities for tuberculosis, paving the way for more effective diagnostic tools to combat this global health challenge (Mobed et al., 2024).

9.6 Green nanotechnology based nanoparticles in biosensing of *Mycobacterium tuberculosis*

Green nanotechnology refers to creating sustainable solutions with minimal environmental impact in research applications, promoting development of environment friendly design, synthesis of nanoproducts which are low-toxic and low-cost throughout their lifecycle, from production to disposal. The synthesis strategy of nanoparticles plays an important role in categorizing them into organic and inorganic nanoparticles. The conventional methods of synthesizing nanoparticles generally use hazardous chemicals, high energy leads to high cost of fabrication and thus facing certain limitations. A best alternative to physical and chemical methods is the green method for the synthesis of nanoparticles which is gaining attention by utilizing enzymes, fungi, algae and plant extracts etc. as reducing and capping agents for the reduction of metal salts (Chaturvedi et al., 2023). Greenly synthesized nanoparticles have been revolutionizing the biosensing platforms, predominantly in the detection of MTB. Among various natural agents, plant extracts are most reliable, less harmful, and natural bioreducing and capping agents in nanoparticles preparation because of presence of various biomolecules such as enzymes, phenolic acids, flavonoids, amino acids, proteins, vitamins, organic acids, and polysaccharides in plant extracts and thus they restrict the need of chemical reagents required for reduction and capping of nanoparticles from their respective salts (Iravani, 2011).

9.6.1 Limitations of green nanotechnology in biosensing of *Mycobacterium tuberculosis*

Green nanotechnology, which emphasizes the use of environmentally friendly and sustainable methods for synthesizing nanoparticles, has shown promising advantages in the biosensing of MTB. However, despite its advantages, some limitations hinder its various applications in this field which are discussed below:

Scalability and reproducibility: One of the major constraints is the challenge of scaling up the production of biosynthesized nanoparticles while maintaining reproducibility and consistency. Inconsistent results can also be found because of variation in sizes and shapes in green synthesis techniques, affecting their performance in biosensing applications.

Limited functionalization: Sensitivity and specificity of the biosensing of MTB may also vary because surface chemistry of nanoparticles synthesized through green routes is always not same in terms of efficient attachment of ligands or antibodies.
Performance metrics: The performance of green-synthesized nanoparticles can limit their application in context of detecting low concentrations of MTB in biosensing applications, as compared to nanoparticles production through conventional chemical methods (Javed et al., 2023; Ying et al., 2022).

9.6.1.1 Limitations of nanoparticle-based biosensing of Mycobacterium tuberculosis

Although, nanoparticles have shown significant promises in the biosensing of MTB, owing to their unique features. But there are also several challenges accompanying with the use of nanoparticles in MTB biosensing. These challenges need to be addressed to improve the efficiency and applicability of nanoparticle-based sensors and thus stated below:

1. Toxicity concerns

Potential toxicity of nanoparticles to human cells is one of the primary concerns nowadays, especially in biosensing applications. Although, nanoparticles are capable in providing high sensitivity and selectivity in detection of MTB but their interaction with biological systems can lead to cytotoxic effects sometimes. Additionally, the accumulation of nanoparticles in organs and tissues such as liver or kidneys can lead to damage to some extent. These toxic effects can be results from various factors such as size, shape, and surface charge of the nanoparticles, which can be overcome by the cautious selection of materials and surface modifications for improved biocompatibility (Zhou et al., 2011).

2. Non-specific binding and sensitivity

Nanoparticles, when used in biosensors, can result in non-specific binding to other biomolecules present in complex samples, leading to false positives or inaccurate indications. This can be a substantial challenge in the detection of MTB in complex biological matrices such as blood or sputum. However, functionalization of nanoparticles with antibodies/aptamers has improved specificity, but still these methods are facing major challenges in minimizing non-specific interactions.

3. Aggregation of nanoparticles

A general problem with nanoparticles is the aggregation over time because of the presence of van der Waals forces and especially in aqueous mediums. The aggregation of nanoparticles can reduce surface area, resulting in decrease of the sensitivity of the biosensor and thus hindering detection process of MTB in samples. However, researches have been going on to discover methods for the prevention of nanoparticles aggregation such as stabilizing them with surfactants or via polymeric coatings.

4. Scalability and reproducibility

Scaling up the production of nanoparticles for extensive use in diagnostics is still a significant challenge. Synthesis of nanoparticles with maintaining consistent quality and reproducibility can be problematic, especially working with complex biological samples. Additionally, large-scale production processes for ensuring uniformity in size, shape, and functionalization of nanoparticles are still in progress, limiting commercial use of laboratory-based findings (Sangboonruang et al., 2023).

9.7 Limited understanding of nanoparticle-biological interactions

Despite their potential, the interactions between nanoparticles and biological systems are still not fully investigated. The behavior of nanoparticles in particular biological fluids, with their cellular uptake, stability, and their interaction mechanisms with pathogens such as MTB is still an area with continuous need of active research. An improved understanding of these interactions at the molecular level is essential for designing nanoparticle-based biosensors (Ying et al., 2022). Current advancements in the nanoparticle-based biosensing of MTB is summarized in the Table 9.2.

9.8 Future perspectives

The future of nanoparticle-based biosensing for *Mycobacterium tuberculosis* (MTB) detection holds immense potential, with emerging developments in nanotechnology, material science, and molecular diagnostics paving the way for transformative improvements. A key direction is the integration of nanoparticles with multiplexed biosensing systems, enabling the simultaneous identification of a range of MTB biomarkers—such as proteins, lipids, and nucleic acids—thus enhancing diagnostic precision and reducing testing time. Innovations in biocompatible nanomaterials, particularly hybrid nanoparticles conjugated with biomolecules, are expected to enhance both sensitivity and selectivity, making early-stage MTB detection more affordable and efficient, especially in resource-limited environments. Additionally, the development of nanoparticles designed to target specific components of MTB's complex cell wall presents a novel approach to more accurate and direct detection. Moreover, the integration of nanoparticle sensors with wearable technologies or mobile health applications offers the prospect of real-time, on-site diagnostic solutions, fostering decentralized and portable tuberculosis detection capabilities. Collectively, when these advancements combined with artificial intelligence and various modern-era machine learning techniques for enhanced data analysis, promise to significantly enhance the speed, accessibility, and accuracy of MTB diagnostics, playing a pivotal role in the global effort to combat tuberculosis.

Table 9.2 Analysis of performance of various nanoparticles for detection applications.

Sr. no	Nanoparticles	Detection principle	Performance	References
1.	Gold nanoparticles	Colorimetric probe based	96.6% sensitivity and 98.9% specificity	Pei et al. (2022)
2.	Gold nanoparticles	Colorimetric biosensing based	sensitivity of 94.7%, and specificity of 99.6%	Huang et al. (2023)
3.	Gold nanoparticles	Colorimetric probe based	100% sensitivity and specificity	Zhang et al. (2024b)
4.	Au nanoparticles	Colorimetric assay	sensitive to ~340 fM levels	Pornprom et al. (2024)
5.	Ag nanoparticles	Visual molecular biosensing strategy	~40 bacilli	Tan et al. (2021)
6.	(NMOF-CoFe1.8)	Colorimetric sensing method	-	Yu et al. (2023)
7.	Pd@Pt nanoparticles	Paper- based	0.216 nM (LOD)	Tung et al. (2024)
8.	Fluorescent nanodiamonds	Lateral flow immunoassay	0.02 ng mL^{-1}	Joudeh and Linke. (2022)
9.	Au nanoparticles decorated carboxyl graphene.	Paper based technique	limit of detection of 0.01 ng/mL	Wang et al. (2024)
10.	Fe/Au magnetic nanoparticles	Differential pulse voltammetry technique	91.7% specificity with 100% sensitivity	Azmi et al. (2021)
11.	Magnetic nanoparticles	Giant magnetoresistance	-	Gupta et al. (2021)
12.	Magnetic nanoparticles/silica nanoparticles	Magnetic separation	limit of detection~78.541 fM	Wei et al. (2023)
13.	Magnetic nanoparticles	-	detection time ~ 4 hours	León-Janampa et al. (2022)
14.	Chitosan-coated iron oxide nanoparticles	-	detection limit~ 40 μg/mL	Reaño et al. (2023)
15.	Magnetic nanoparticles	Aptamer-based technique	66.6% sensitivity, 95.4% specificity	Sharma et al. (2025)
16.	CdSe/CdS nanorods	-	Detection time ~ 2.5 h	Soni et al. (2024)
17.	Reduced graphene oxide	Biotin-avidin chemistry	detection limit~ 0.1 ×10 7 μM	Goel et al. (2024)
18.	Selenium nanoparticles	Green method	-	Ghaderi et al. (2022)

9.9 Conclusion

Nanoparticles-based biosensing of MTB leads to a significant progression in the field of diagnostic microbiology, offering unmatched sensitivity and specificity for the diagnosis of this pathogenic bacterium. Nanoparticles, possessing unique optical, electronic, and catalytic properties, have transfigured biosensing technologies by allowing fast, accurate, and low-cost detection techniques. The integration of noble metallic nanoparticles and various carbon-based nanoparticles etc., into biosensing platforms leads to the simplified development of extremely sensitive assays able to detect even very low concentrations of MTB biomarkers with extraordinary precision. A great innovation is the functionalization of these nanoparticles, which improved the specificity and stability, crucial for reliable research field applications. In spite of the advancement, issues of scalability, reproducibility, and the need for stringent standardization protocols remain as challenges for the strength of nanoparticle-based biosensors. A new revolution has been brought by researchers by inventing the green synthesis of nanoparticles, which proved themselves as better alternatives in many applications but still owning their own set of challenges such as economic feasibility and reliable performance. Additionally, the development in nanoparticle-based biosensing embraces the promise of renovating MTB diagnostics, leading to more effective health strategies and better patient consequences. Future research must address these limitations while exploring novel nanoparticle materials and synthesis methods to optimize the diagnostic potential of these technologies.

References

Abrahams KA, Besra GS. Mycobacterial cell wall biosynthesis: a multifaceted antibiotic target. Parasitology 2018;145(2):116–33.

Anu Mary Ealia S, Saravanakumar MP. A review on the classification, characterisation, synthesis of nanoparticles and their application. IOP Conference Series: Materials Science and Engineering 2017;263(3):032019. Available from https://doi.org/10.1088/1757-899X/263/3/032019.

Arora K, Chand S, Malhotra BD. Recent developments in bio-molecular electronics techniques for food pathogens. Analytica Chimica Acta 2006;568(1-2):259–74. Available from https://doi.org/10.1016/j.aca.2006.03.078.

Azmi UZM, Yusof NA, Abdullah J, Mohammad F, Ahmad SAA, Suraiya S, et al. Aptasensor for the detection of Mycobacterium tuberculosis in sputum utilising CFP10-ESAT6 protein as a selective biomarker. Nanomaterials 2021;11(9):2446.

Baptista PV, Koziol-Montewka M, Paluch-Oles J, Doria G, Franco R. Gold-nanoparticle-probe–based assay for rapid and direct detection of Mycobacterium tuberculosis DNA in clinical samples. Clinical Chemistry 2006;52(7):1433–4. Available from https://doi.org/10.1373/clinchem.2005.065391.

Becerra AH, Flores MG, Palma-Nicolas JP, Ramírez-García G, López-Marín LM. Mycosome-coated gold nanoparticles for plasmonic detection of tuberculosis-associated antibodies. ACS Applied Nano Materials 2024;7(10):11203–13. Available from https://doi.org/10.1021/acsanm.4c00706.

Bisht N, Patel M, Dwivedi N, Kumar P, Mondal DP, Srivastava AK, et al. Bio-inspired polynorepinephrine based nanocoatings for reduced graphene oxide/gold nanoparticles composite for high-performance biosensing of Mycobacterium tuberculosis. Environmental Research 2023;227:115684. Available from https://doi.org/10.1016/j.envres.2023.115684.

Biswas P, Karn AK, Balasubramanian P, Kale PG. Biosensor for detection of dissolved chromium in potable water: a review. Biosensors and Bioelectronics 2017;94:589–604. Available from https://doi.org/10.1016/j.bios.2017.03.043.

Burlec AF, Corciova A, Boev M, Batir-Marin D, Mircea C, Cioanca O, et al. Current overview of metal nanoparticles' synthesis, characterization, and biomedical applications, with a focus on silver and gold nanoparticles. Pharmaceuticals 2023;16(10):1410.

Chaturvedi M, Patel M, Bisht N, Shruti D, Mukherjee M, Tiwari A, et al. Reduced graphene oxide-polydopamine-gold nanoparticles: a ternary nanocomposite-based electrochemical genosensor for rapid and early Mycobacterium tuberculosis detection. Biosensors 2023;13(3):342.

Chou WC, Canchola A, Zhang F, Lin Z. Machine learning and artificial intelligence in nanomedicine. Wiley Interdiscip Reviews: Nanomed Nanobiotechnology 2025;17(4):e70027.

Christian P, Von der Kammer F, Baalousha M, Hofmann T. Nanoparticles: structure, properties, preparation and behaviour in environmental media. Ecotoxicology 2008;17:326–43.

Ckumdee J, Kaewphinit T, Chansiri K, Santiwatanakul S. Development of Au-nanoprobes combined with loop-mediated isothermal amplification for detection of isoniazid resistance in Mycobacterium tuberculosis. Journal of Chemistry 2016;2016(1):3474396. Available from https://doi.org/10.1155/2016/3474396.

Cortie MB, McDonagh AM. Synthesis and optical properties of hybrid and alloy plasmonic nanoparticles. Chemical Reviews 2011;111(6):3713–35. Available from https://doi.org/10.1021/cr1002529.

Dolai J, Mandal K, Jana NR. Nanoparticle size effects in biomedical applications. ACS Applied Nano Materials 2021;4(7):6471–96. Available from https://doi.org/10.1021/acsanm.1c00987.

Du P, Jin M, Chen G, Zhang C, Jiang Z, Zhang Y, et al. A competitive bio-barcode amplification immunoassay for small molecules based on nanoparticles. Scientific Reports 2016;6(1):38114. Available from https://doi.org/10.1038/srep38114.

Elmowafy M, Shalaby K, Elkomy MH, Alsaidan OA, Gomaa HAM, Abdelgawad MA, et al. Polymeric nanoparticles for delivery of natural bioactive agents: recent advances and challenges. Polymers 2023;15(5):1123.

Elsabahy M, Heo GS, Lim SM, Sun G, Wooley KL. Polymeric nanostructures for imaging and therapy. Chemical Reviews 2015;115(19):10967–1011. Available from https://doi.org/10.1021/acs.chemrev.5b00135.

Ghaderi RS, Adibian F, Sabouri Z, Davoodi J, Kazemi M, Amel Jamehdar S, et al. Green synthesis of selenium nanoparticle by Abelmoschus esculentus extract and assessment of its antibacterial activity. Materials Technology 2022;37(10):1289–97. Available from https://doi.org/10.1080/10667857.2021.1935602.

Girma A, Kassawmar F, Kassa Y, Asrat Y. Nanoparticles as an alternative strategy for the rapid detection of Mycobacterium tuberculosis Complex (MTBC): a systematic literature review of In vitro studies. IET Nanobiotechnology 2025;2025(1):4639233.

Goel H, Patel M, Chaturvedi M, Gupta GK, Khare AK, Mondal DP, et al. Polydopamine functionalized Ti3AlC2 MAX based electrochemical biosensor for early and sensitive detection of Mycobacterium tuberculosis. Microchemical Journal 2024;197:109899. Available from https://doi.org/10.1016/j.microc.2024.109899.

Gordillo-Marroquín C, Gómez-Velasco A, Sánchez-Pérez HJ, Pryg K, Shinners J, Murray N, et al. Magnetic nanoparticle-based biosensing assay quantitatively enhances acid-fast bacilli count in paucibacillary pulmonary tuberculosis. Biosensors 2018;8(4):128.

Grassian VH. When size really matters: size-dependent properties and surface chemistry of metal and metal oxide nanoparticles in gas and liquid phase environments. The Journal of Physical Chemistry C 2008;112(47):18303–13. Available from https://doi.org/10.1021/jp806073t.

Gupta S, Bhatter P, Kakkar V. Point-of-care detection of tuberculosis using magnetoresistive biosensing chip. Tuberculosis 2021;127:102055. Available from 10.1016/j.tube.2021.102055.

Hang Y, Wang A, Wu N. Plasmonic silver and gold nanoparticles: shape-and structure-modulated plasmonic functionality for point-of-caring sensing, bio-imaging and medical therapy. Chemical Society Reviews 2024;53(6):2932–71. Available from https://doi.org/10.1039/D3CS00793F.

Hasan MR, Anzar N, Sharma P, Singh S, Hassan H, Rawat C, et al. Mycobacterium tuberculosis diagnosis from conventional to biosensor-a systematic review. International Journal of Environmental Analytical Chemistry 2024;104(18):6519–34. Available from https://doi.org/10.1080/03067319.2022.2147427.

Hsu SH, Hung SC, Chen YK, Jian ZH. Surface plasmon resonator using high sensitive resonance telecommunication wavelengths for DNA sensors of Mycobacterium tuberculosis with thiol-modified probes. Sensors 2014;15(1):331–40. Available from https://doi.org/10.3390/s150100331.

Huang H, Chen Y, Zuo J, Deng C, Fan J, Bai L, et al. MXene-incorporated C60NPs and Au@ Pt with dual-electric signal outputs for accurate detection of Mycobacterium tuberculosis ESAT-6 antigen. Biosensors and Bioelectronics 2023;242:115734. Available from https://doi.org/10.1016/j.bios.2023.115734.

Hutter E, Fendler JH. Exploitation of localized surface plasmon resonance. Advanced Materials 2004;16(19):1685–706. Available from https://doi.org/10.1002/adma.200400271.

Ibrahim N, Jamaluddin ND, Tan LL, Mohd Yusof NY. A review on the development of gold and silver nanoparticles-based biosensor as a detection strategy of emerging and pathogenic RNA virus. Sensors 2021;21(15):5114. Available from https://doi.org/10.3390/s21155114.

Ibrahim SA, Chan Y. Fluorescent semiconductor nanorods for the solid-phase polymerase chain reaction-based, multiplexed gene detection of Mycobacterium tuberculosis. ACS Applied Materials & Interfaces 2021;13(30):35294–305. Available from https://doi.org/10.1021/acsami.1c05312.

Iravani S. Green synthesis of metal nanoparticles using plants. Green Chemistry 2011;13(10):2638–50. Available from https://doi.org/10.1039/C1GC15386B.

Javed MN, Bangash SAK, Abbas M, Ahmed S, Kaplan A, Iqbal S, et al. Potential and challenges in green synthesis of nanoparticles: a review. Journal of Xi'an Shiyou University (Natural Science Edition) 2023;19:1155–65.

Joshi H, Kandari D, Maitra SS, Bhatnagar R. Biosensors for the detection of Mycobacterium tuberculosis: a comprehensive overview. Critical Reviews in Microbiology 2022;48(6):784–812. Available from https://doi.org/10.1080/1040841X.2022.2035314.

Joudeh N, Linke D. Nanoparticle classification, physicochemical properties, characterization, and applications: a comprehensive review for biologists. Journal of Nanobiotechnology 2022;20:262. Available from https://doi.org/10.1186/s12951-022-01477-8.

Karnwal A, Kumar Sachan RS, Devgon I, Devgon J, Pant G, Panchpuri M, et al. Gold nanoparticles in nanobiotechnology: from synthesis to biosensing applications. ACS Omega 2024;9(28):29966–82. Available from https://doi.org/10.1021/acsomega.3c10352.

Kaur H, Shorie M. Nanomaterial based aptasensors for clinical and environmental diagnostic applications. Nanoscale Advances 2019;1:2123–38. Available from https://doi.org/10.1039/C9NA00153K.

Khan S, Hossain MK. 2 - Classification and properties of nanoparticles. Nanoparticle-Based Polymer Composites. Woodhead publishing; 2022. p. 15–54. Available from https://doi.org/10.1016/B978-0-12-824272-8.00009-9.

Kia P, Ruman U, Pratiwi AR, Hussein MZ. Innovative therapeutic approaches based on nanotechnology for the treatment and management of tuberculosis. International Journal of Nanomedicine 2023;18:1159–91. Available from https://doi.org/10.2147/IJN.S364634.

Kim EJ, Kim EB, Lee SW, Cheon SA, Kim HJ, Lee J, et al. An easy and sensitive sandwich assay for detection of Mycobacterium tuberculosis Ag85B antigen using quantum dots and gold nanorods. Biosensors and Bioelectronics 2017;87:150–6. Available from https://doi.org/10.1016/j.bios.2016.08.034.

Kooti S, Kadivarian S, Abiri R, Mohajeri P, Atashi S, Ahmadpor H, et al. Modified gold nanoparticle colorimetric probe-based biosensor for direct and rapid detection of Mycobacterium tuberculosis in sputum specimens. World Journal of Microbiology & Biotechnology 2023;39(5):118. Available from https://doi.org/10.1007/s11274-023-03564-w.

Kreibig U, Vollmer M. Optical Properties of Metal Clusters. Berlin, Heidelberg: Springer; 1995. Available from https://doi.org/10.1007/978-3-662-09109-8.

Krutyakov YA, Kudrinskiy AA, Olenin AY, Lisichkin GV. Synthesis and properties of silver nanoparticles: advances and prospects. Russian Chemical Reviews 2008;77(3):233. Available from https://doi.org/10.1070/RC2008v077n03ABEH003751.

Kumari S, Sarkar L. A review on nanoparticles: structure, classification, synthesis & applications. Journal of Scientific Research 2021;65(8):42–6. Available from https://doi.org/10.37398/JSR.2021.650809.

Le TN, Descanzo MJN, Hsiao WWW, Soo PC, Peng WP, Chang HC. Fluorescent nanodiamond immunosensors for clinical diagnostics of tuberculosis. Journal of Materials Chemistry B 2024;12(14):3533–42. Available from https://doi.org/10.1039/D3TB03038E.

León-Janampa N, et al. Biorecognition and detection of antigens from Mycobacterium tuberculosis using a sandwich ELISA associated with magnetic nanoparticles. Journal of Pharmaceutical and Biomedical Analysis 2022;215:114749. Available from https://doi.org/10.1016/j.jpba.2022.114749.

Li Z, Zhao L, Diao L, Wang P, Che C, Tian Y, et al. Detection of mycobacterium tuberculosis by magnetic nanoparticle capture-polymerase chain reaction. Science of Advanced Materials 2021;13(11):2116–22. Available from https://doi.org/10.1166/sam.2021.4137.

Liang L, Chen M, Tong Y, Tan W, Chen Z. Detection of Mycobacterium tuberculosis IS6110 gene fragment by fluorescent biosensor based on FRET between two-dimensional metal-organic framework and quantum dots-labeled DNA probe. Analytica Chimica Acta 2021;1186:339090. Available from https://doi.org/10.1016/j.aca.2021.339090.

Lin DZ, Chuang PC, Liao PC, Chen JP, Chen YF. Increasing the spectral shifts in LSPR biosensing using DNA-functionalized gold nanorods in a competitive assay format for the detection of interferon-γ Biosensors and Bioelectronics 2016;81:221–8. Available from https://doi.org/10.1016/j.bios.2016.02.071.

Loskutova A, Seitkali A, Aliyev D, Bukasov R. Quantum dot-based sensors: review from analytical perspective. International Journal of Molecular Sciences 2025;26(14):6674. Available from https://doi.org/10.3390/ijms26146674.

Ma J, Du M, Wang C, Xie X, Wang H, Li T, et al. Rapid and sensitive detection of mycobacterium tuberculosis by an enhanced nanobiosensor. ACS Sensors 2021;6(9):3367–76. Available from https://doi.org/10.1021/acssensors.1c01227.

Maphanga C, Manoto S, Ombinda-Lemboumba S, Ismail Y, Mthunzi-Kufa P. Localized surface plasmon resonance biosensing of Mycobacterium tuberculosis biomarker for TB diagnosis. Sensing and Bio-Sensing Research 2023;39:100545. Available from https://doi.org/10.1016/j.sbsr.2022.100545.

Mathebula NS, Pillay J, Toschi G, Verschoor JA, Ozoemena KI. Recognition of antimycolic acid antibody at self-assembled mycolic acid antigens on a gold electrode: a potential impedimetric immunosensing platform for active tuberculosis. Chemical Communications 2009(23):3345–7. Available from https://doi.org/10.1039/B905192A.

Medintz IL, Uyeda HT, Goldman ER, Mattoussi H. Quantum dot bioconjugates for imaging, labelling and sensing. Nature Materials 2005;4(6):435–46. Available from https://doi.org/10.1038/nmat1390.

Michalet X, Pinaud FF, Bentolila LA, Tsay JM, Doose SJJL, Li JJ, et al. Quantum dots for live cells, in vivo imaging, and diagnostics. Science 2005;307(5709):538–44. Available from https://doi.org/10.1126/science.1104274.

Mobed A, Darvishi M, Kohansal F, Dehfooli FM, Alipourfard I, Tahavvori A, et al. Biosensors; nanomaterial-based methods in diagnosing of Mycobacterium tuberculosis. Journal of Clinical Tuberculosis and Other Mycobacterial Diseases 2024;34:100412. Available from https://doi.org/10.1016/j.jctube.2023.100412.

Nam JM, Thaxton CS, Mirkin CA. Nanoparticle-based bio-bar codes for the ultrasensitive detection of proteins. Science 2003;301(5641):1884–6.

Naqvi SS, Anwar H, Noori MY, Siddiqui A, Ali Z, Shah MR, et al. Silver nanoparticles as a nanoprobe for trace level simultaneous detection of streptomycin sulfate and isoniazid and antiTB activity of their nanoparticles complexes. Journal of Molecular Structure 2023;1275:134634. Available from https://doi.org/10.1016/j.molstruc.2022.134634.

Nikolaev VV, Lepekhina TB, Alliluev AS, Bidram E, Sokolov PM, Nabiev IR, et al. Quantum dot-based nanosensors for In vitro detection of Mycobacterium tuberculosis. Nanomaterials 2024;14(19):1553.

Patnaik N, Dey RJ. Label-free citrate-stabilized silver nanoparticles-based, highly sensitive, cost-effective, and rapid visual method for the differential detection of Mycobacterium tuberculosis and mycobacterium bovis. ACS Infectious Diseases 2023;10(2):426–35. Available from https://doi.org/10.1021/acsinfecdis.3c00390.

Patra S, Pareek D, Gupta PS, Wasnik K, Singh G, Yadav DD, et al. Progress in treatment and diagnostics of infectious disease with polymers. ACS Infectious Diseases 2024;10(2):287–316. Available from https://doi.org/10.1021/acsinfecdis.3c00528.

Pei X, Hong H, Liu S, Li N. Nucleic acids detection for mycobacterium tuberculosis based on gold nanoparticles counting and rolling-circle amplification. Biosensors 2022;12(7):448. Available from https://doi.org/10.3390/bios12070448.

Perumal V, Saheed MSM, Mohamed NM, Saheed MSM, Murthe SS, Gopinath SC, et al. Gold nanorod embedded novel 3D graphene nanocomposite for selective bio-capture in rapid detection of Mycobacterium tuberculosis. Biosensors and Bioelectronics 2018;116:116–22. Available from https://doi.org/10.1016/j.bios.2018.05.042.

Pornprom T, Phusi N, Thongdee P, Pakamwong B, Sangswan J, Kamsri P, et al. Toward the early diagnosis of tuberculosis: a gold particle-decorated graphene-modified paper-based electrochemical biosensor for Hsp16. 3 detection. Talanta 2024;267:125210.

Qureshi A, Shaikh T, Niazi JH. Semiconductor quantum dots in photoelectrochemical sensors from fabrication to biosensing applications. Analyst 2023;148(8):1633–52.

Rabti A, Raouafi A, Raouafi N. DNA markers and nano-biosensing approaches for tuberculosis diagnosis. Nanotechnology based approaches for tuberculosis treatment. Academic Press; 2020. p. 207–30.

Rafsan AA, Das S, Roy KK, Mahdi MI, Sakib S, Maraz MFA, et al. Surface plasmon resonance–based biosensor for accurate and rapid detection of Mycobacterium tuberculosis. Microchimica Acta 2025;192(12):1–21.

Ramesh M, Janani R, Deepa C, Rajeshkumar L. Nanotechnology-enabled biosensors: a review of fundamentals, design principles, materials, and applications. Biosensors 2022;13(1):40.

Rao CNR, Kulkarni GU, Thomas PJ, Edwards PP. Size-dependent chemistry: properties of nanocrystals. Chemistry – A European Journal 2002;8(1):28–35. Available from https://doi.org/10.1002/1521-3765(20020104)8:1<28::AID-CHEM28>3.0.CO;2-B.

Reaño RL, Panghulan GR, Hernandez CAT, Tamayo JP. Bioengineered monoclonal antibody chitosan–iron oxide bio-composite for electrochemical sensing of *mycobacterium tuberculosis* lipoprotein. Engineering Proceedings 2023;58(1):23. Available from https://doi.org/10.3390/ecsa-10-16065.

Rotherham, L.S., Maserumule, C., Dheda, K., Theron, J., & Khati, M. (2012). Selection and application of ssDNA aptamers to detect active TB from sputum samples.

Salimiyan Rizi K, Aryan E, Meshkat Z, Ranjbar G, Sankian M, Ghazvini K, et al. The overview and perspectives of biosensors and Mycobacterium tuberculosis: a systematic review. Journal of Cellular Physiology 2021;236(3):1730–50. Available from https://doi.org/10.1002/jcp.30007.

Sangboonruang S, Semakul N, Suriyaprom S, Kitidee K, Khantipongse J, Intorasoot S, et al. Nano-delivery system of ethanolic extract of propolis targeting Mycobacterium tuberculosis via aptamer-modified-niosomes. Nanomaterials 2023;13:269. Available from https://doi.org/10.3390/nano13020269.

Sapsford KE, Pons T, Medintz IL, Mattoussi H. Biosensing with iuminescent semiconductor quantum dots. Sensors 2006;6(8):925–53.

Seele PP, Dyan B, Skepu A, Maserumule C, Sibuyi NRS. Development of gold-nanoparticle-based lateral flow immunoassays for rapid detection of TB ESAT-6 and CFP-10. Biosensors 2023;13(3):354.

Semwal V, Jensen OR, Bang O, Janting J. Investigation of performance parameters of spherical gold nanoparticles in localized surface plasmon resonance biosensing. Micromachines 2023;14(9):1717.

Sharma P, Gupta RK, Aittan S, Anthwal D, Dass M, Yadav R, et al. Development and assessment of a novel magnetic nanoparticle antibody-conjugate and aptamer-based assay (MNp-Ab-Ap assay) for the rapid diagnosis of pleural tuberculosis. Nanotheranostics 2025;9(1):20.

Smith AM, Nie S. Semiconductor nanocrystals: structure, properties, and band gap engineering. Accounts of Chemical Research 2010;43(2):190–200. Available from https://doi.org/10.1021/ar9001069.

Sofi MA, Sunitha S, Sofi MA, Pasha SK, Choi D. An overview of antimicrobial and anticancer potential of silver nanoparticles. Journal of King Saud University - Science 2022;34(2):101791. Available from https://doi.org/10.1016/j.jksus.2021.101791.

Soni A, Nehra K, Dahiya B, Rais A, Prasad T, Gahlaut A, Mehta PK. Detection of MPT-64 protein in pleural tuberculosis cases by magnetic bead-gold nanoparticle-PCR amplified immunoassay. Future Microbiology 2024;20(2):107–15. Available from https://doi.org/10.1080/17460913.2024.2432179.

Soo PC, Horng YT, Chang KC, Wang JY, Hsueh PR, Chuang CY, et al. A simple gold nanoparticle probes assay for identification of Mycobacterium tuberculosis and Mycobacterium tuberculosis complex from clinical specimens. Molecular and Cellular Probes 2009;23(5):240–6. Available from https://doi.org/10.1016/j.mcp.2009.04.006.

Sudha PN, Sangeetha K, Vijayalakshmi K, Barhoum A. Chapter 12 - Nanomaterials history, classification, unique properties, production and market. Emerging applications of nanoparticles and architecture nanostructures. Elsevier; 2018. p. 341–84. Available from https://doi.org/10.1016/B978-0-323-51254-1.00012-9.

Tan P, Li H, Wang J, Gopinath SC. Silver nanoparticle in biosensor and bioimaging: clinical perspectives. Biotechnology and Applied Biochemistry 2021;68(6):1236–42. Available from https://doi.org/10.1002/bab.2045.

Terna AD, Elemike EE, Mbonu JI, Osafile OE, Ezeani RO. The future of semiconductors nanoparticles: synthesis, properties and applications. Materials Science and Engineering: B 2021;272:115363. Available from https://doi.org/10.1016/j.mseb.2021.115363.

Tigga JG, Sahu VK, Patel MK. Next-generation metal-based tuberculosis biosensing: from conventional methods to nanoscale innovations. Microchemical Journal 2025;219:116134. Available from https://doi.org/10.1016/j.microc.2025.116134.

Tripathi A, Jain R, Dandekar P. Rapid visual detection of Mycobacterium tuberculosis DNA using gold nanoparticles. Analytical Methods 2023;15(20):2497–504. Available from https://doi.org/10.1039/D3AY00195D.

Tung CY, Tsai TT, Chiu PY, Viter R, Ramanavičius A, Yu CJ, et al. Diagnosis of mycobacterium tuberculosis using palladium–platinum bimetallic nanoparticles combined with paper-based analytical devices. Nanoscale 2024;16(12):5988–98.

Vashishtha VM. WHO Global Tuberculosis Control Report 2009: Tuberculosis elimination is a distant dream. Indian pediatrics 2009;46(5).

Veigas B, Jacob JM, Costa MN, Santos DS, Viveiros M, Inácio J, et al. Gold on paper–paper platform for Au-nanoprobe TB detection. Lab on a Chip 2012;12(22):4802–8. Available from https://doi.org/10.1039/C2LC40739F.

Wang J, Shao W, Liu Z, Kesavan G, Zeng Z, Shurin MR, et al. Diagnostics of tuberculosis with single-walled carbon nanotube-based field-effect transistors. ACS Sensors 2024;9(4):1957–66. Available from https://doi.org/10.1021/acssensors.3c02694.

Wang S, Inci F, De Libero G, Singhal A, Demirci U. Point-of-care assays for tuberculosis: role of nanotechnology/microfluidics. Biotechnology Advances 2013;31(4):438–49. Available from https://doi.org/10.1016/j.biotechadv.2013.01.006.

Wang Z, Han D, Wang H, Zheng M, Xu Y, Zhang H. Organic semiconducting nanoparticles for biosensor: a review. Biosensors 2023;13(4):494.

Wei S, Dou Y, Yu Y, Yang J, Yu F, Sha W, et al. A novel biosensor based on a bio-barcode for the detection of Mycobacterium tuberculosis. Analytical Methods 2023;15(30):3683–91. Available from https://doi.org/10.1039/D3AY00772C.

Yang X, Fan S, Ma Y, Chen H, Xu JF, Pi J, et al. Current progress of functional nanobiosensors for potential tuberculosis diagnosis: the novel way for TB control? Frontiers in Bioengineering and Bbiotechnology 2022;10:1036678. Available from https://doi.org/10.3389/fbioe.2022.1036678.

Yang Z, Shen J. A review: metal and metal oxide nanoparticles for environmental applications. Nanoscale 2025;17:15068–85. Available from 10.1039/D5NR01973G.

Ying S, Guan Z, Ofoegbu PC, Clubb P, Rico C, He F, et al. Green synthesis of nanoparticles: current developments and limitations. Environmental Technology & Innovation 2022;26:102336. Available from https://doi.org/10.1016/j.eti.2022.102336.

Yu J, Sun Q, Sun J, Wang X, Niu N, Chen L. Development of gold nanoparticles composite functionalized bimetallic metal-organic framework nanozymes and integrated smartphone to detection isoniazid. Sensors and Actuators B: Chemical 2023;390:134024. Available from https://doi.org/10.1016/j.snb.2023.134024.

Zhang S, Zhang Y, Jiang J, Charconnet M, Peng Y, Zhang L, et al. Shape-specific gold nanoparticles for multiplex biosensing applications. ACS Omega 2024;9(35):37163–9.

Zhang X, Tian Y, Shi Y, Liu J, Zhao C, Chang CC, et al. Naked-eye LAMP Assay of M. tuberculosis in sputum by In situ Au nanoprobe identification: for the In vitro diagnostics of tuberculosis. ACS Infectious Diseases 2024;10(8):2668–78. Available from https://doi.org/10.1021/acsinfecdis.4c00013.

Zhao J, Zhang X, Yonzon CR, Haes AJ, Van Duyne RP. Localized surface plasmon resonance biosensors. Nanomedicine 2006;1(2):219–28. Available from https://doi.org/10.3390/s150715684.

Zhou L, He X, He D, Wang K, Qin D. Biosensing technologies for Mycobacterium tuberculosis detection: status and new developments. Journal of Immunology Research 2011;2011(1):193963. Available from https://doi.org/10.1155/2011/193963.

Zhu J, Wang H, Chen L. Recent advances in nanomaterials for the detection of Mycobacterium tuberculosis. International Journal of Molecular Medicine 2024;55(3):36. Available from https://doi.org/10.3892/ijmm.2024.5477.

CHAPTER 10

Role of 2D materials in the detection of *Mycobacterium tuberculosis*

Archana and Pratima R. Solanki
Nanobio Laboratory, Special Centre for Nanoscience, Jawaharlal Nehru University, New Delhi, Delhi, India

10.1 Introduction

Mycobacterium tuberculosis (MTB) is a slow-growing bacterium that takes up to 6–8 weeks to be detected even after using the sophisticated cell culture methods (Mobed et al., 2024). Other diagnostic methods for the detection of MTB such as tuberculin skin test, sputum microscopy test, and chest X-ray. However, a large gap exists between treatment and diagnosis of tuberculosis (TB) which leads to misdiagnosis. Therefore, the timely detection of the MTB can help in early treatment of TB and will help in reducing the fatality rate. Several biomarkers have been reported that are associated to the detection of MTB and it mainly comprises of immunological molecules such as interferon gamma (IFN-γ), MTB secretory protein MPT64, early secreted antigenic target 6 (ESAT-6) antigen, and nucleic acid targets such as IS6110 sequence of MTB genome, *rpoB* gene, DNA sequences of MTB, etc (Yin et al., 2013; Cavalcanti et al., 2012; Passos et al., 2024; Hwang et al, 2015; Pai et al, 2016; Huang et al., 2023; Gou et al., 2018).

These biomarkers and their respective recognition elements have been used to fabricate biosensing platforms using different nanomaterials such as noble metal nanoparticles, 2D-nanomaterials, nanorods, nanotubes, etc (Thiruppathiraja et al., 2011; Das et al., 2011; Costa et al., 2014). Among these nanomaterials, 2D-nanomaterials has been used extensively for the fabrication of sensors for the detection of biomarkers related to MTB (Chen et al., 2017). The 2D-nanomaterials has various advantages than other dimension nanomaterials such as it has higher surface area to volume ratio, ease of surface functionalization, tunability, biocompatibility, targeted functionality, electro-optical properties, easy to miniaturize, *etc*. The 2D-nanomaterials have excellent durability also owing to the presence of covalent bonds within the plane aiding it to tolerate high level of strain (Manoharan et al., 2024). Various 2D-nanomaterials have been reported in the previous years among them we will discuss the application of mainly three categories of 2D-nanomaterials that is, carbon-based 2D-nanomaterials, transition metal chalcogenides, and MXenes (Soleymaniha et al., 2019; Xiao Li et al., 2017; Wang et al., 2008). Here Fig. 10.1 shows an illustration depicting

Advances in Biosensor Technology for Mycobacterium Tuberculosis Detection
DOI: https://doi.org/10.1016/B978-0-443-44589-7.00003-0

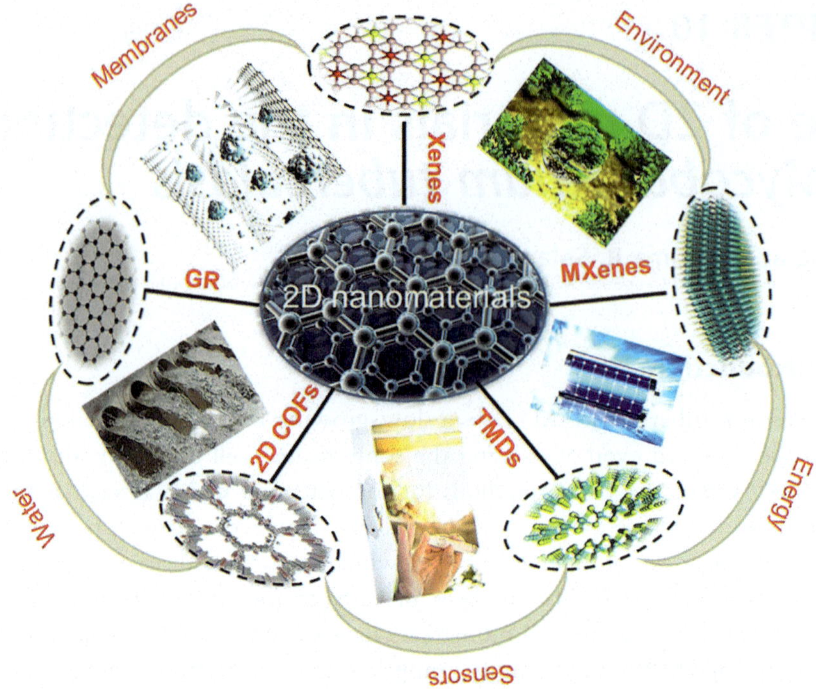

Figure 10.1 Highlighting various applications of 2D-nanomaterials. *Baig N. Two-dimensional nanomaterials: aA critical review of recent progress, properties, applications, and future directions. Composites Part A: Applied Science and Manufacturing 2023;165107362. https://doi.org/10.1016/j.compositesa.2022.107362.*

different applications of 2D-nanomaterials (Baig, 2023). This chapter will further dwell deeper into the applications of various 2D-nanomaterials family for the fabrication of biosensor for the detection of MTB.

10.2 Carbon 2D-nanomaterial based biosensor for the detection of *Mycobacterium tuberculosis*

The carbon-based 2D-nanomaterial family are generally known for its two most excellent members called as graphene oxide (GO) and reduced graphene oxide (rGO) nanosheets (Rattana et al., 2012; Teymourian, Salimi and Khezrian, 2013; Shamsaei et al., 2018). The GO nanosheets is derived from the bulk graphite synthesized using a well-known chemical method called as Hummers method (Chen et al., 2022). This method uses harsh conditions provided by concentric acid such as sulfuric acid and oxidizing agents such as potassium permanganate to exfoliate graphite flakes into thinner layers yielding hydrophilic GO nanosheets. Moreover, when GO nanosheets are further exfoliated using techniques such as ultrasonication, hydrothermal method, chemical method (using reducing agent), green method, hydrophobic rGO nanosheets are produced which have the interplanar distance lower than that of GO nanosheets (Hafeez et al., 2019; Saini, Mishra and Kumar, 2024; Archana et al., 2024). Both the top-down as well as the bottom up approach can be used for the synthesis of graphene-based nanosheets. Also, rGO have higher *Id:Ig* (intensity of defects:intensity of graphene content) ratio than GO as during exfoliation period several defects are introduced in the rGO nanosheets and due to fewer layers than GO, it has lesser graphene content (Archana et al., 2024).

Moreover, these nanomaterials have excellent electrical conductivity properties alongwith the optical properties. The electrical conductivity properties of these nanosheets are due to the presence of oxygen based moieties in their structure on basal planes as well as the edges and also larger surface area to volume ratio. As known for graphene it has the highest specific surface area of 2630 m^2/g (Baig, 2023). Here Fig. 10.2A and B shows the structure of GO and rGO, respectively. Here we can observe in Fig. 10.1 that GO has more hydrophilic functional groups such as carboxylic acid, epoxy groups, hydroxy groups, but for rGO these groups have reduced during exfoliation process which leads to its hydrophobic nature in comparison to GO.

Moreover, the band gap of GO is also tunable by modifying its functional groups which aids in bringing variation of their electrical properties (Syama and Mohanan, 2019). Also, the superior charge carrier properties of graphene are due to the presence of two carbon in their atomic unit cell where their respective band (conductance and valence band) touches each other and results in the formation of a point, also known as

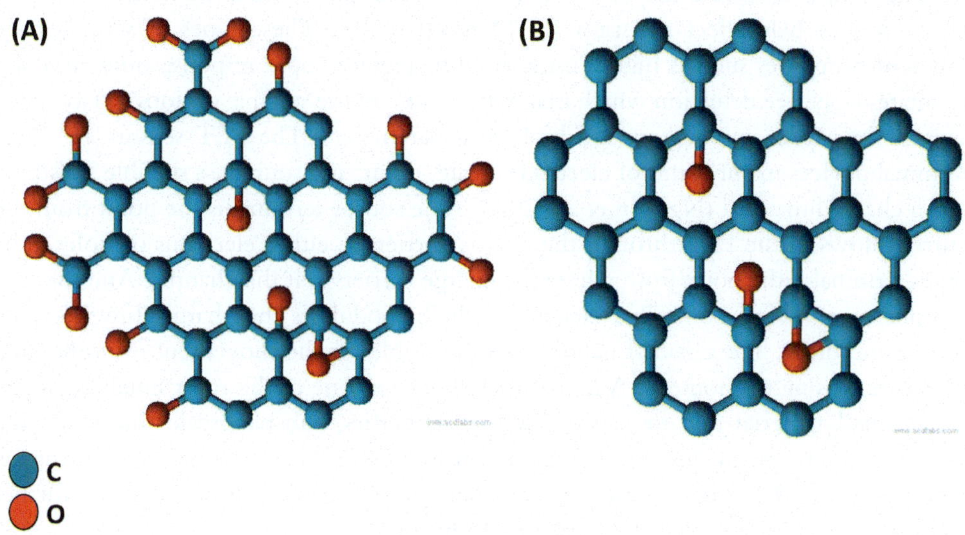

Figure 10.2 Structure of (A) GO and (B) rGO. *GO*, graphene oxide; *rGO*, reduced graphene oxide.

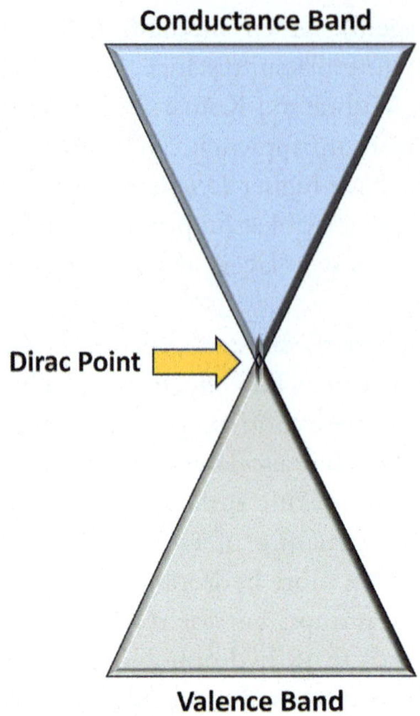

Figure 10.3 Dirac point of the conductance and valence band.

Dirac point (Jiang et al., 2007; Geim and Novoselov, 2007). The Fig. 10.3 shows a schematic representation of Dirac point.

The graphene-based 2D-nanomaterials has also been used immensely for the fabrication of field-effect transistor (FET) based sensor. These types of sensors have inherent properties such as high sensitivity, integrability, rapid response time, and also facilitates label free detection which makes them one of the suitable options to fabricated sensors for several biomarkers and biological parameters. The FET sensors are three terminal devices mainly built of electrodes (gate, drain, and source), a suitable substrate, and a channel material (Nguyen et al., 2023). The source terminal is the point from the current flows in the FET through the charge carriers as either electrons or holes. The drain terminal is the point from where the charge carriers exit the channel. And the gate terminal is a voltage-controlled electrode which modulates the charge flow between source and drain. The channel material is responsible for the movement of probes and monitors sensing phenomena. And graphene has been one of the most suitable choices for channel material due to its conducting properties, sub-nanoscale thickness, and biocompatibility with the biological receptors. Here Fig. 10.4 shows an overall fabrication of FET based biosensor. Overall, these electrical properties are further explored to fabricate electrochemical based sensors.

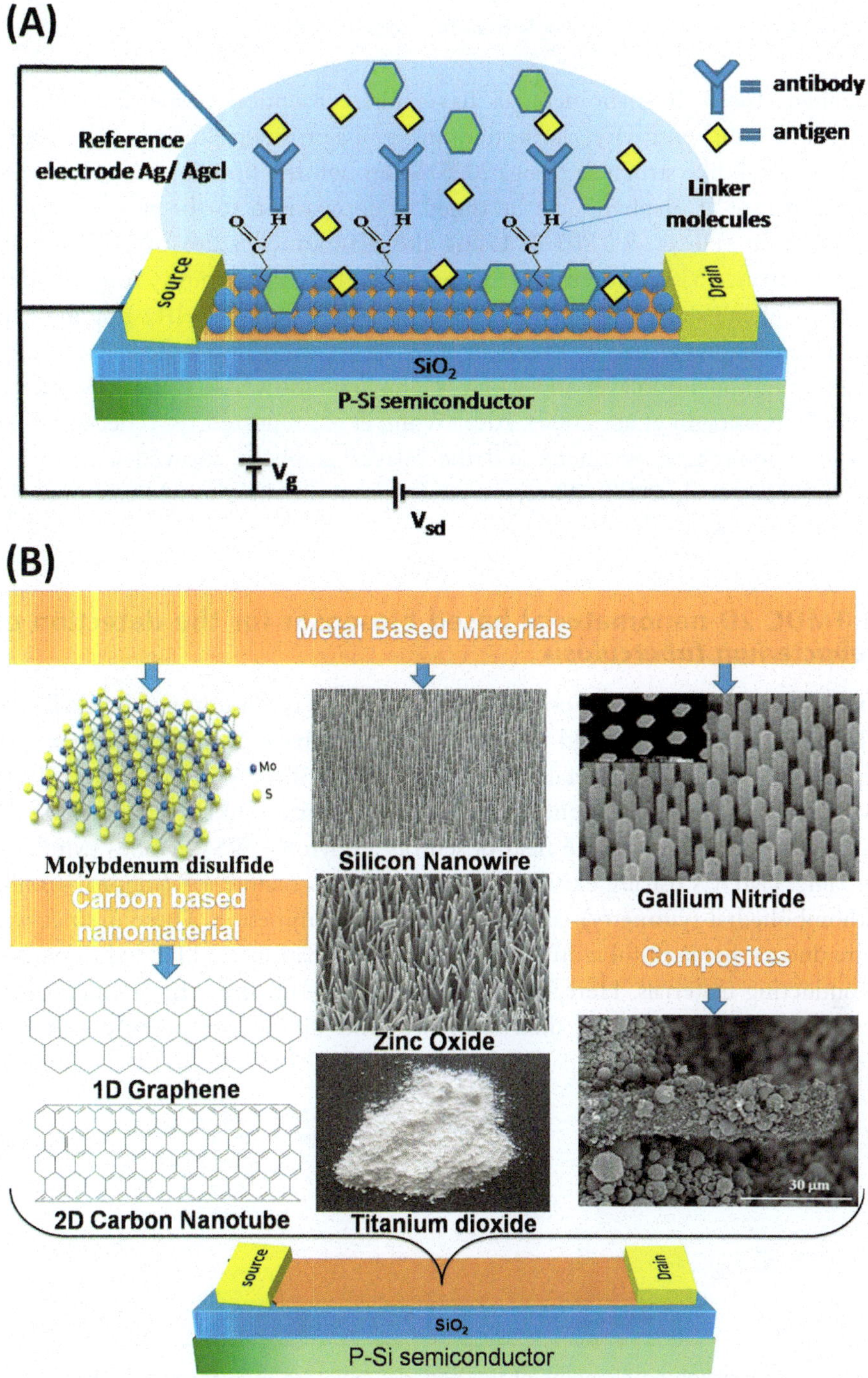

Figure 10.4 (A) Illustration of FET based sensing platform with its major component namely source, drain, semiconducting layer for immobilization of bioreceptors molecules. (B) Types of materials suitable for the gated region of FET. *From Nehra, A, Pal Singh, K. Current trends in nanomaterial embedded field effect transistor-based biosensor. Biosensors and Bioelectronics 2015;74: 731–43. https://doi.org/10.1016/j.bios.2015.07.030.*

Graphene-based 2D-nanomaterials have inherent optical properties such as plasmonic effect, light sensitivity, absorption and emission owing to its single atom thick 2D-honey-comb like structure (Baig, 2023). Through the findings it was reported that plasmonic nature of graphene can be tuned in comparison to the plasmonic range of thin metal film (He et al., 2016). Using the plasmonic nature of graphene surface plasmon resonance based biosensors have been developed previously. Also, the different layers of graphene show distinct reflectance as well as infrared (IR) responses. It was reported that graphene has an excellent 97.7 % optical transparency. As reported by Casiraghi et al., the monolayer and 10-layers graphene showed 0.1% and 2% reflectance, respectively (Casiraghi et al., 2007). Also, Wang et al., reported IR reflectance was not observed in monolayer graphene and the bilayer graphene showed a sharp peak at 350 meV (Wang et al., 2008). Therefore, these properties help to fabricate optical based sensor for detection of the biomarkers.

10.3 TMDC 2D-nanomaterial based biosensor for the detection of *Mycobacterium tuberculosis*

Further, another class of 2D-nanomaterials are called as transition metal dichalcogenides (TMDC) (Chen and McDonald, 2016). The TMDC have MX_2 stoichiometry, where M represents the transition metals from 4 to 7 group of the periodic table *i.e.*, W, Ta, Ti, Mo, and Nb and X represents the chalcogenides group like Te, Se, and S. Several well-known are TMDCs in the field of sensors are MoS_2, WS_2, $MoSe_2$ (Muller et al., 2023). The TMDCs usually exists in three symmetries mainly 1T (trigonal symmetry), 3R (rhombohedral symmetry), and 2H (hexagonal symmetry). These distinct symmetries are due to the plentiful atomic phases and combinations and has led to synthesis of semiconducting materials. Here Fig. 10.5A and (B) (i–iii) show the elements used for the synthesis of TMDCs and their structures and different geometries, respectively (Zhao et al., 2022).

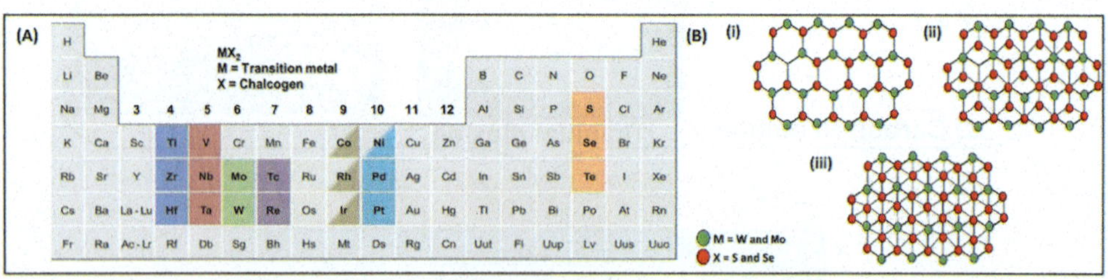

Figure 10.5 (A) Highlighted elements of the periodic table are used for the synthesis of TMDCs generally and (B) Structural units of the TMDC cells (i) 2H, (ii) 1T, and (iii) 3R. *From Zhao Y., Wang S.B., Chen A.Z., Kankala R.K. Nanoarchitectured assembly and surface of two-dimensional (2D) transition metal dichalcogenides (TMDCs) for cancer therapy. Coordination Chemistry Reviews 2022;472:214765. https://doi.org/10.1016/j.ccr.2022.214765.*

The usual synthesis methods followed for the synthesis of TMDCs are hydrothermal technique, chemical vapor deposition, physical vapor transport (Archana, Kumar and Solanki, 2024; Tedstone et al., 2016; Archana, Kumar and Solanki, 2025). The structure of TMDCs is very similar to that of graphene-based 2D-nanomaterials as the strong covalent bonds bring together the transition metals and chalcogens followed by the presence of weak van der Waal forces that hold its sheet-like morphology. And by overcoming the weak forces the 3D-sheet of TMDCs can be changed to 2D-nanosheets via the exfoliation process. There are several defects available in the TMDC structures due to the substitutional impurities, vacancies, and grain boundaries which, therefore, implies significant impacts on their opto-electronic properties (Tang and Jiang, 2015).

The TMDCs based nanomaterials have a versatile chemistry which helps them to tune the band gap of the nanomaterials by controlling the number of layers in their nanosheets as well as the defects. The band gaps of TMDC switches from indirect to direct band when the number of layers in the nanomaterial changes from bulk to few layers. The monolayer of MoS_2 has been reported with the direct band gap of 1.8 eV. These changes in the electrical properties are attributed to the change due to quantum confinement and the electron shell. These change in the band gaps also effects the optical properties of the TMDCs such as emission and light absorption (Hu et al., 2018). It has also been reported that the monolayer of the TMDCs have higher light emission capability which can be harnessed for the optoelectrical devices (Dutta et al., 2023). Also, the functionalization of TMDC due to the higher surface area to volume ratio of its planar structure helps to utilize the unique properties of these materials to a greater extent. However, owing to the inert nature of TMDCs it is preferred to opt for covalent modification methods which does not introduce too many defects on the surface of the nanomaterials. This inert nature of the TMDC is due to the saturated sites of chalcogen and the surrounded transition metals, overall making it ineffective for functionalization(Di Giorgio et al., 2021). In conclusion, the various properties of TMDCs make them one of the excellent choices for the fabrication of biosensor platforms.

And the other class of 2D-nanomaterials as well as the most emerging ones are known as MXenes, the 2D-nanomatertials being used in almost all the fields of chemistry. The term MXenes is used for the 2D transition metal nitrides (TiN and CrN), carbides (Ti_3AlC_2), and carbonitride (Ti_4C_3N) (Quirós-Ovies et al., 2020; Wen et al., 2011; Wu et al., 2012). The bulk MAX phase is used to produce the respective nanolaminates MXenes. In MAX phase "M" represents the transition metals, "A" represents the p-elements of IIIA and IVA group of the periodic table, and "X" represents C, N, or both the elements. The elements found generally in MAX phase is shown in Fig. 10.6A. The structure of MAX phase and MXenes are shown in Fig. 10.6 (i and ii). The method to obtain MXenes focuses on removing the "A" layer which is weakly intercalated. The exfoliation of Ti_3AlC_2 yields Ti_3C_2 in the hydrofluoric acid

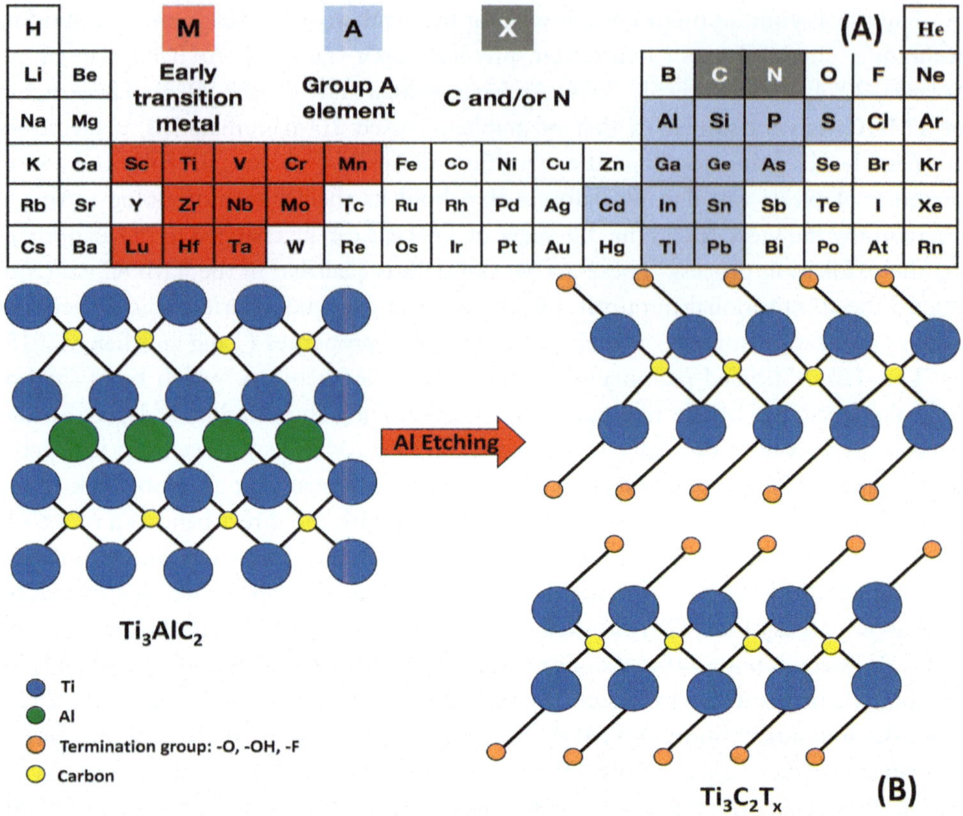

Figure 10.6 (A) Highlighted elements in the periodic table are found in MAX phase and (B) Basic structure of MAX phase and MXene. *From Magnuson M., Mattesini M. Chemical bonding and electronic-structure in MAX phases as viewed by X-ray spectroscopy and density functional theory. Thin Solid Films 2017;621:108–130. https://doi.org/10.1016/j.tsf.2016.11.005.*

where the acid also acts as termination group. The formula for MAX phase and MXenes is $M_{n+1}AX_n$ (n=1–3) and $M_{n+1}X_nT_x$, where T represent the termination groups (-F, -O, and -OH). Therefore, several properties of MXene are dependent on the available terminal group.

The flexibility of the MXene make them a suitable candidate for flexible electronics. The aqueous solution of MXene showcases excellent conductivity up to 6500 S cm^{-1} and it can also increase up to 15,000 S cm^{-1} (Magnuson and Mattesini, 2017). MXenes are generally metallic or semimetallic in nature, and only the few MXene are of semiconducting nature (Zhang et al., 2017). The bare MXene is all metallic in nature because d orbital of M element releases all the fermi energy. If F is the termination group, then the downward shifting of the fermi energy is observed because each F gains one electron from the system. Moreover, if -O is the termination group, then the energy level decreases more leading to formation of semiconducting MXene (Dillon et al., 2016). The conductivity and stability of the MXene is also affected by its precursor such as Shuck et al., used graphite based precursor as well as the TiC-based carbon precursor for synthesis of $Ti_3C_2T_x$. Here the $Ti_3C_2T_x$/graphite based precursor showed the conductivity of ≈4400 S cm^{-1} and $Ti_3C_2T_x$/TiC-based carbon precursor ≈3480 S cm^{-1} (Wang et al., 2020). Also, due to the excellent sheet-like morphology of MXene it makes it a suitable 2D-nanomaterial for immobilization of recognition molecules such as antibodies, DNA, *etc.* Hui Gu et al., reported the 3D-porous $Ti_3C_2T_x$ MXene-graphene nanosheets hybrid for the fabrication of glucose oxidase enzymatic sensor for the detection of glucose on glassy carbon electrodes (GCE) (Shuck et al., 2019). The rich presence of functional groups on MXene provides hydrophilicity and tunable properties without sacrificing its conducting nature (Gu et al., 2019). In another work Shankar et al., used 2D-Ti_2CT_x MXene for the detection of adrenaline for the first time on MXene/graphite composite paste electrode. Here, they observed that with the increasing concentration of adrenaline the differential pulse voltammetry (DPV) current increase subsequently, owing to the electrochemically active nature of adrenaline (Wu et al., 2017).

However, the optical properties of MXene nanosheets have been a lesser explored topic. The different optical properties of MXenes such as reflectivity, transmittance, absorbance has been theoretically analyzed by the researchers. It has been observed that -F and -OH terminated MXenes (Ti_2C and Ti_3C) have lower in-plane absorption capability in comparison to the -O functionalized and bare MXenes counterpart (Shankar et al., 2018). Also, Wu et al., reported the application of the non-linear optical response of $Ti_3C_2T_x$ for the fabrication of femtosecond laser generator (Bai et al., 2016). Overall, MXene 2D-nanosheets can be used for fabrication of various electrochemical as well as optical based sensing platforms.

10.4 Application of 2D-nanomaterials for detection of *Mycobacterium tuberculosis* related nucleic acid biomarkers

There are several known MTB related biomarkers which are nucleic acid based such as IS6110, *dnaJ*, *rpoB* gene, PPE24, *etc.* The choice of nucleic acid targets for MTB detection using sensing platforms is based on their abundance in the microorganism which further determines the limit of detection of the assay.

Among them IS6110 has been most explored DNA biomarker for the detection of MTB owing to its higher copy number leading to enhanced selectivity and sensitivity. IS6110 is an insertion sequences (IS), these IS are mobile genetic elements. They have the capability to transpose and multiply at different sites in the genome. Also, this IS sequence is responsible for rearrangement, deletion, duplication in the bacterial genome which provides them strain identity. Moreover, IS6110 has been established as the gold standard for studies related to TB epidemiology. After considering the significance of the IS6110 sequence several electrochemical sensors have been developed in synergism of 2D-nanomaterials. In a similar context, Hwang et al., group reported the detection of IS6110 DNA sequence of MTB complex using polymerase chain reaction (PCR) amplified sequences which were further incubated with the fluorophores labeled primer and GO to determine its resultant fluorescence. Here, they amplified the IS6110 sequence and further incubated it with 6-carboxyfluorescein (FAM) tagged primer sequence functionalized on GO. If IS6110 sequence is present in the sample then the FAM tagged primer sequence specific for IS6110 will detach from GO and will increase the fluorescence produced by FAM molecule. The overall schematic for this work is shown in Fig. 10.7(Hwang et al., 2015).

And if IS6110 sequence is not available in the sample the FAM labelled primer sequence will remain bound to GO which will result in no change of the fluorescence intensity. Here, GO has been used for its quenching capabilities against many fluorescent molecules as it acts as a universal quencher. Also, GO have preferential binding for single stranded nucleic acids by π-π stacking interactions, and has lower inherent affinity for double stranded nucleic acid (Wu et al., 2019; Hwang et al., 2015).

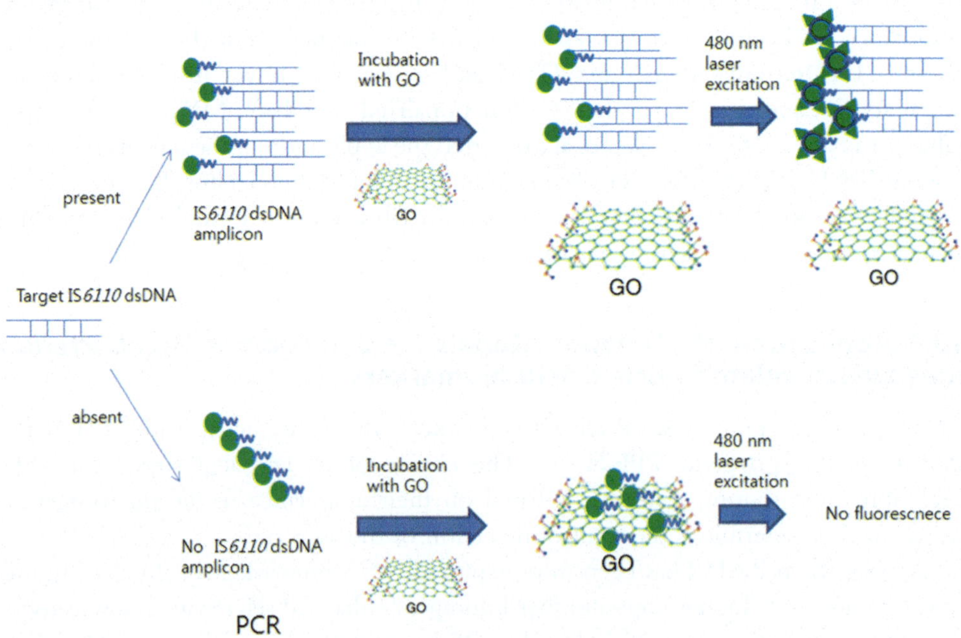

Figure 10.7 Illustration describing the GO based biosensing of IS6110 using FAM tagged primer sequence. *From Hwang S.H., Kim D.E., Sung H., Park B.M., Cho M.J., Yoon O.J. et al. Simple detection of the IS6110 sequence of Mycobacterium tuberculosis complex in sputum, based on PCR with graphene oxide. PLoS ONE 2015;10(8): https://doi.org/10.1371/journal.pone.0136954.*

In another work Rizi et al., reported a genosensor for the detection of IS6110 DNA sequence using MXene nanosheets and polypyrrole composites on the GCE in a PCR free technique (Rizi et al., 2022). The GCE surface was modified with MXene and polypyrrole and further covalently immobilized with IS6110 DNA probe (5'-CAAA GTGTGGCTAACCCTGAA-3') also containing methylene blue (MB) as the redox probe. Here, when the DNA probe present on the modified electrode was incubated with the IS6110 DNA sequence obtained from the sputum samples, they observed a decrease in the peak current of MB recorded with increasing concertation of the target DNA. The change in peak current was observed with differential pulse voltammetry technique from -0.4 to -0.1 V in the presence of the PBS (pH 7.0, 0.05 M). The range of DNA concentration detected using the modified ssDNA/PPY/MXene/GCE was 100 fM to 25 nM with limit of detection (LOD) and limit of quantification of 11.24 fM and 37.09 fM, *respectively*. The MXene nanosheets were used for the significant properties it imparts such as surface functionalization, high metallic conductivity, and excellent ionic conductivity and its further combination with polypyrrole helped in the enhancement of the sensitivity of the fabricated sensor (Park, Goo and Kim, 2014; Chakravorty et al., 2024). The overall schematic for this work is shown in Fig. 10.8.

The IS6110 fragments of MTB was also detected using gold nanoparticles decorated fullerene nanoparticles/nitrogen-doped graphene nanosheet (Au-nano-C_{60}/NGS) sandwich-type electrochemical sensor as reported by (Bai et al., 2019). In this work they immobilized the capture probe (CP) on the GCE using the biotin–avidin system. Furter, using NGS with nano-C_{60} providing helped in obtaining an improved surface area ratio, conductivity, as well as excellent redox activity. Moreover, the decoration of AuNPs on nano-C_{60}/NGS helped on better immobilization of the signal probes (SP) through the Au-S bond and also increased the electron transfer. The sandwich kind of system was followed in the manner of: CP:Target (IS6110):SP. The electrochemical response was captured through DPV in the presence of phosphate buffer (0.1 M/pH 7.0) with tetraoctylammonium bromide (TOAB) treatment resulting in a clear correlation between the increasing DPV current response and increasing concentration of target DNA (IS6110) (on a logarithmic scale) due to enhanced intrinsic electro-activity *via* TOAB functionalized nano-C_{60}.

Liu et al., reported an electrochemical biosensor using rGO-AuNPs as the biosensor matrix on a GCE to detect the sequence IS6110 (Liu et al., 2014). Here, they designed a capture probe (5'-GGTGAGGTCT(SH)-3') immobilized on rGO-Au NPs and the target DNA of IS6110 (5'-AGACCTCACCTATGTGTCGA-3') is sandwiched between the former nanocomposite and the signal probe (50-SH-CCCCCCAAAAA TCGACACATA-3') immobilized Au–PANI nanocomposite. It was observed that DPV reduction peak current increases with increasing concentration of the target DNA from 1.0×10^{15} and 1.0×10^{9} M in the presence of PBS (0.1 M/pH 7.0). However, the IS6110 helps in distinguishing the MTB members from other mycobacteria but does

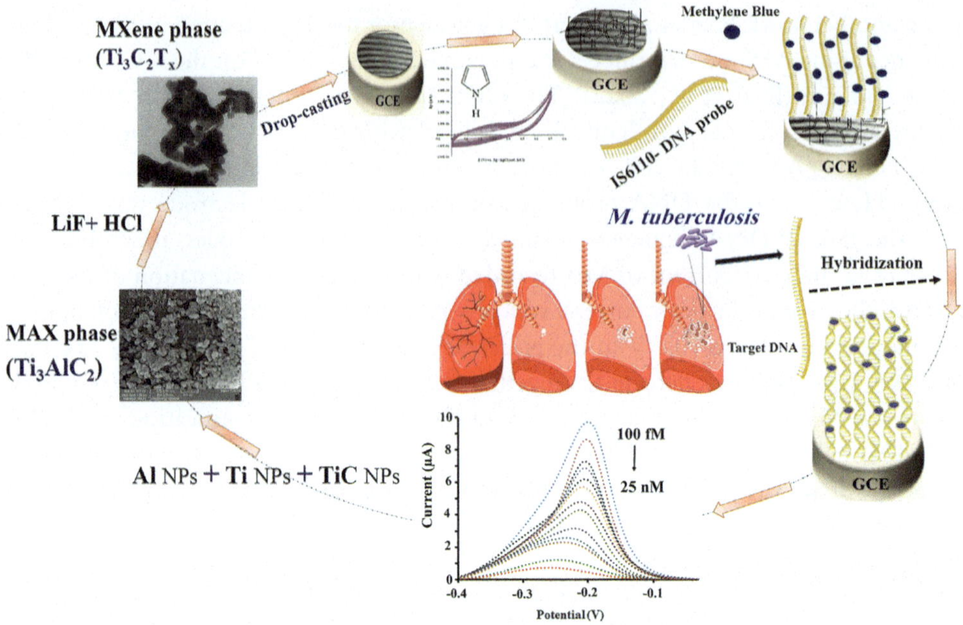

Figure 10.8 Schematic illustration for the detection of IS6110 DNA using PPY/MXene nanocomposite on glassy carbon electrode. *From Rizi K.S. et al. PCR-free electrochemical genosensor for Mycobacterium tuberculosis complex detection based on two-dimensional Ti3C2 Mxene-polypyrrole signal amplification. Microchemical Journal 2022;179:107467. https://doi.org/10.1016/j.microc.2022.107467.*

not have efficiency to distinct the members within the MTB complex. Therefore, other nucleic acid biomarkers such as *rpoB* gene have been used as the detecting marker for MTB.

In the similar context Li et al., reported an electrochemiluminescence (ECL) sensor for the detection of *rpoB* gene of multiple drug resistant (MDR)-TB using ruthenium complex functionalized GO (Ru-GO). *rpoB* gene encodes for DNA dependent-RNA polymerase β subunit, a target for anti-TB drug rifampicin, and the mutation in this gene leads to evolving of MDR-TB species (Li et al., 2021). Therefore, the detection of rpoB gene in patient samples can help in achieving correct drug dosage for their treatment. Here ferrocene (Fc) labelled Fc-ssDNA has been used as controller for ECL intensity. The fabricated sensor was highly sensitive to detect a recognition of single nucleotide polymorphisms in codon 531 of *rpoB* gene. The Fc-ssDNA immobilized on Ru-GO through strong π-π stacking interactions showed decrease in ECL signal as Fc acts as a quencher as shown in Fig. 10.9.

After the formation of dsDNA complex between the complementary mutated ssDNA target (containing the mutated codon 531 TCG→TTG) (TM) and Fc-ssDNA, the complex gets detached form Ru-GO as the dsDNA has lesser affinity towards GO

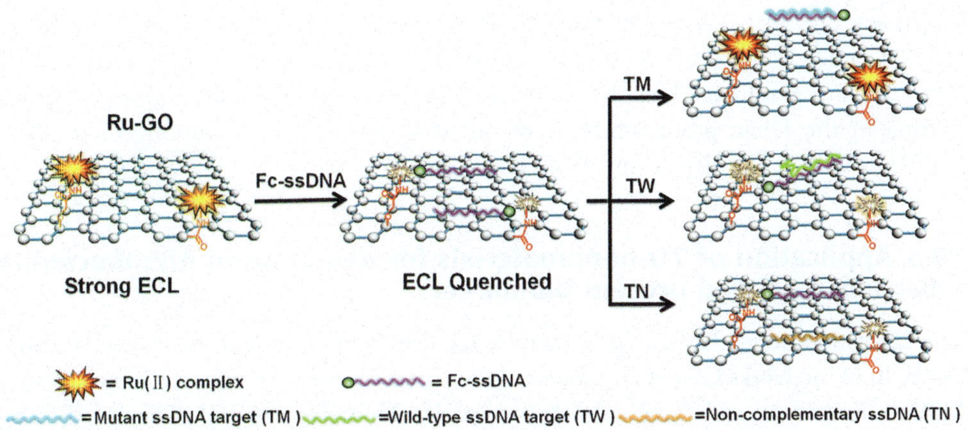

Figure 10.9 Detection of mutated *rpoB* gene using GO supported electrochemiluminescence (ECL) sensor. *From Li F., Yu Y., Li Q., Zhou M., Cui H. A homogeneous signal-on strategy for the detection of rpoB genes of Mycobacterium tuberculosis based on electrochemiluminescent graphene oxide and ferrocene qQuenching. Analytical Chemistry 2014;86(3):1608–1613. https://doi.org/10.1021/ac403281g.*

in comparison to ssDNA and, therefore, the ECL intensity of Ru-GO is restored back. In contrast when wild-type ssDNA target (TW) was introduced then a mismatched heteroduplex with Fc-ssDNA was formed which detached lesser Fc-ssDNA sequence the surface of Ru-GO and resulting no lesser restoration of ECL. Moreover, upon the addition of non-complementary ssDNA sequence (TN) was introduced in Fc-ssDNA, there was no recovery of ECL signal as quenching remained as it is. Therefore, the fabricated sensor showed utmost sensitivity and selectivity towards the TM ssDNA sequence (Li et al., 2014).

In another study carried out by Wang et al., a graphene-based field-effect transistor (G-FET) was developed to detect MTB using Clustered Regularly Interspaced Short Palindromic Repeats (CRISPR/Cas) system for the first time. In this work the DNA sequence IS1081was detected using the CRISPR/Cas system (Wang et al., 2025). They fabricated the system on a gold electrode by depositing graphene by chemical vapour deposition to connect the source and drain. The polydimethoxysilane chamber is placed on the sensing channel and the Ag/AgCl reference electrode is inserted into the solution to provide gate voltage. Further, the device is connected to a printed circuit board. The Cas12a-crRNA RNP (functional ribonucleoprotein) complex was immobilized on G-FET using the linker molecule *i.e.*, 1-pyrenebutonic acid succinimidyl ester which consists of aromatic pyrene group and N-hydroxysuccinamide (NHS). The pyrene group bind to graphene of FET by -Π bonds and the remaining NHS group binds to the amine groups of the biomolecules. Further, when the target DNA of MTB is added, the Cas12a-crRNA RNP identifies the protospacer adjacent motifs (PAM) sequence, due to which the target DNA is repositioned according to the alignment of crRNA which leads to further hybridization of crRNA-target DNA. The formation of hybrid crRNA-target DNA leads to the formation of N doping due to the changes in the Fermi level of graphene. The change in the Fermi level of graphene leads to a negative shifting in the Dirac point which is recorded as ΔV_{Dirac}. They reported a LOD of 2.42×10^{-8} M for the detection of MTB.

10.5 Application of 2D-nanomaterials for detection of *Mycobacterium tuberculosis* related protein biomarkers

The 2D-nanomaterials has also been used for the detection of protein biomarkers related to MTB. In of the works Gou et al., fabricated the sensor for the detection of secretory protein MPT64 antigen, an early stage as well as a specific biomarker for TB. They used GO@ Fe_3O_4@Pt nanocomposite for the immobilization of MPT64 antibodies alongwith protein G. Here, protein G ensures the maintenance of the proper orientation of the antibody binding site *i.e.*, the fragment of crystallization region. Moreover, the presence of GO provides a large surface area to anchor Fe_3O_4@Pt and the presence of Pt and Fe_3O_4 nanoparticles helped in better catalytic activity and reusability of the material. Also, as GO shows higher affinity towards ssDNA, here, rolling circle amplification technique has been used to amplify the DNA probe consisting of tandem repeats which eventually amplifies the output signal. The sensing platform construction beings from the immobilization of Protein G followed by anti-MPT64 antibody on the gold electrode. Further, when MPT64 bind to the antibody it is sandwiched from the other side by APP which was specific towards MPT64. Later, the RCA amplified ssDNA probe is incubated with the previous set-up alongwith GO@Fe_3O_4@Pt nanocomposite and the current response it later recorded using DPV in a working solution of H_2O_2. The concentration range detected for MPT64 is 5.0 fg mL^{-1} to 1.0 ng mL^{-1} in 4 hours and a LOD of 0.34 fg mL^{-1} was obtained (Gou *et al.*, 2018).

Similarly, Huang et al., reported an electrochemical aptasensor for the detection of another secretory protein ESAT-6 antigen using MXene/C_{60}NPs/Au@Pt nanocomposite as shown in Fig. 10.10. The high specific surface area and biofunctionalization of MXene allows functionalization of the aptamer as well as the two electroactive materials (Au and Pt) in order to improve its sensing efficiency. Here, the aptamer specific towards ESAT-6 antigen was immobilized on Au@Pt to synthesize the tracer labels. In order to fabricate the electrode, the GCE was first immobilized by MoS_2_Au to act as a sensor interface. Further the aptamer 1 (APt1) was immobilized on MoS_2_Au using biotin-streptavidin chemistry. Then, ESAT-6 antigen was added on the fabricated electrode further incubated with MXene/C_{60}NPs/Au@Pt/APt2. Here, a sandwich assay has been fabricated where ESAT-6 is present between APt1 and APt2. The DPV peak current showed increase with increase the concentration of the target ESAT-6 antigen yielding the LOD of 2.88 fgm^{-1} in a linear range of 100 fg mL^{-1} to 50 ng mL^{-1} (Huang et al., 2023).

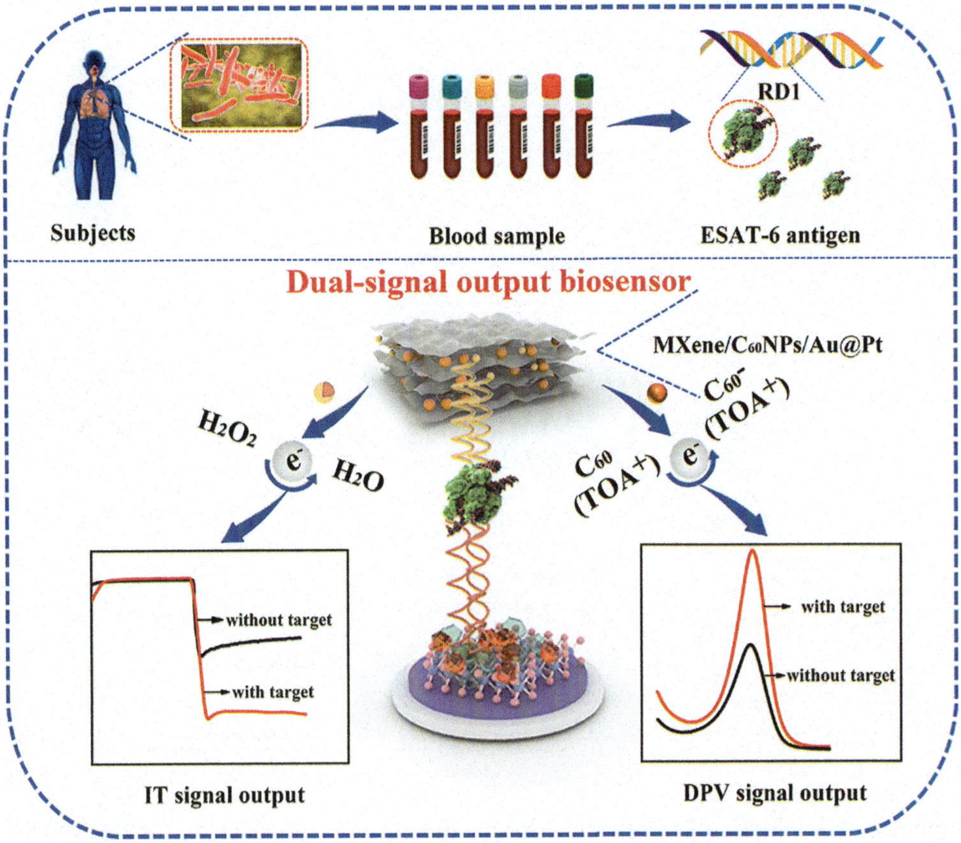

Figure 10.10 Detection of ESAT-6 antigen using an MXene/C_{60}NPs/Au@Pt based aptasensor. *From Huang H., Chen Y., Zuo J., Deng C., Fan J., Bai L. et al. MXene-incorporated C60NPs and Au@Pt with dual-electric signal outputs for accurate detection of Mycobacterium tuberculosis ESAT-6 antigen. Biosensors and Bioelectronics 2023;242115734. https://doi.org/10.1016/j.bios.2023.115734.*

Another electrochemical sensor for protein-based biomarker, IFN-γ, a macrophage activating factor, one of the detection biomarkers known for MTB infection. Here, Jin et al., fabricated an aptasensor on nafion treated GCE using AuNPs-poly(amidoamine) dendrimer (PAMAM)/MoS_2 composite as shown in Fig. 10.11. For this work, MoS_2 nanosheets were used because of its enhanced surface area to volume ration which facilitates in the immobilization of AuNPs-PAMAM. The AuNPs-PAMAM/MoS_2 provides a synergistic effect of MoS_2 and AuNPs-PAMAM leading to more charge transfer rate and enhanced effective electrode area on the GCE surface. Further, using the AuNPs through Au–S coupling HS-terminated hairpin-DNA aptamer for IFN-γ was immobilized on it. The immobilized methylene blue (MB) through hydrogen bond, intermolecular force and non-covalent interactions transmitted weak

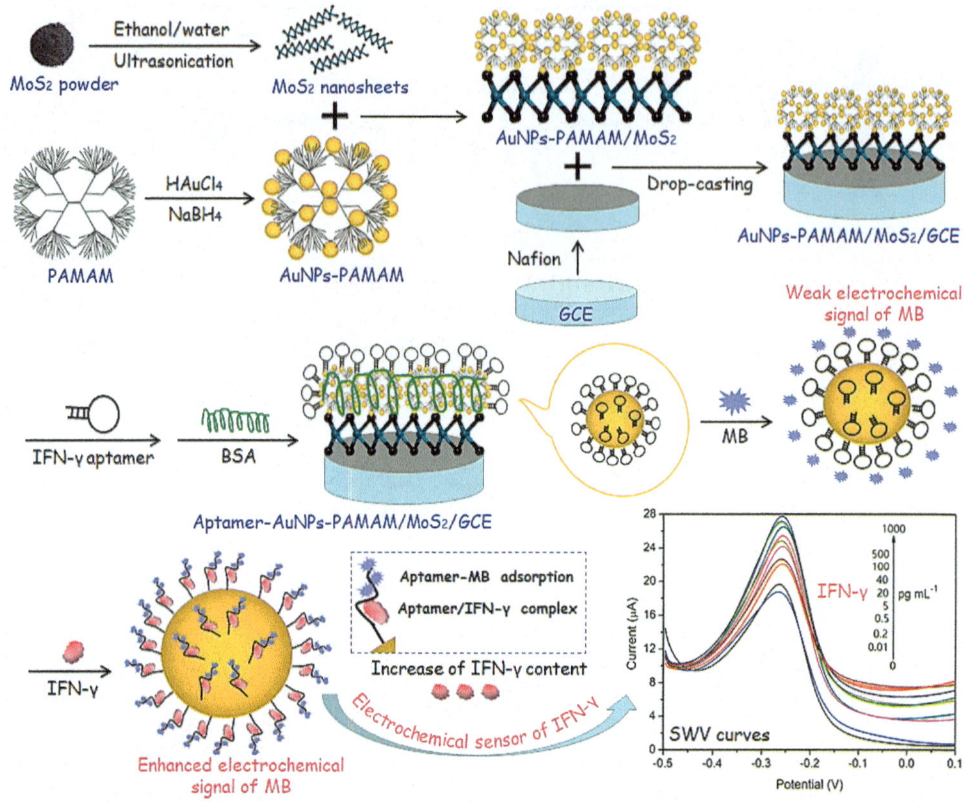

Figure 10.11 Fabrication of an aptasensor for the detection of IFN-γ using AuNPs-PAMAM/MoS_2/GCE. *From Jin H., Gui R., Gao X., Sun Y. An amplified label-free electrochemical aptasensor of γ-interferon based on target-induced DNA strand transform of hairpin-to-linear conformation enabling simultaneous capture of redox probe and target. Biosensors and Bioelectronics 2019;145111732. https://doi.org/10.1016/j.bios.2019.111732.*

electrochemical signals. However, when IFN-γ was incubated with the hairpin aptamer it got transformed into linear aptamer due to which greater number of MB molecules got adsorbed on the aptamer leading to increased current response as recorded by square wave voltammetry. They reported a LOD of 2 fg mL^{-1} in the linear range of 0.01–1000 pg mL^{-1} for the detection of IFN-γ (Jin et al., 2019).

10.6 Other 2D-nanomaterial for the detection of *Mycobacterium tuberculosis*

Apart from carbon and TMDC based 2D-nanomaterials, black phosphorus (BP) is one of the emerging 2D-nanomaterials owing to its excellent optoelectronic properties. BP has a tunable band gap, high charge mobility, and strong optical properties as well.

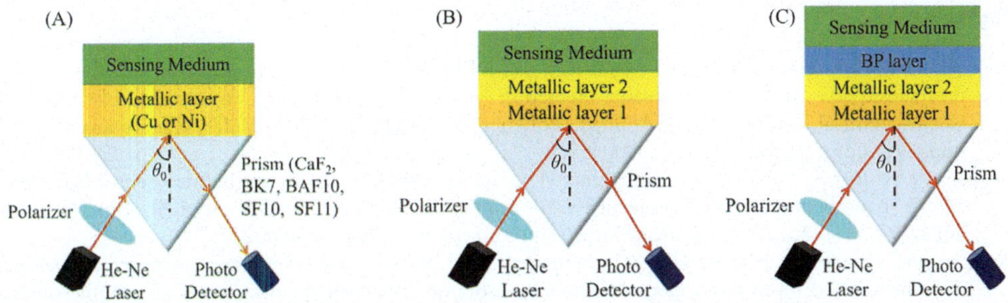

Figure 10.12 Schematic diagrams of the proposed biosensors. (A) Monometallic layer architecture. (B) Bimetallic layer architecture, and (C) BP-capped bimetallic layer architecture. *From Huang X., Peng Y., Wang Y., Zhang L. High-pPerformance sSurface pPlasmon rResonance bBiosensor bBased on cCopper, nNickel, and bBlack pPhosphorus for dDetection of Mycobacterium tuberculosis. Plasmonics 2025;20(12):10937–10953. https://doi.org/10.1007/s11468-025-03158-2.*

Based on these properties Huang et al., fabricated a surface plasmon resonance (SPR) based sensor using BP alongwith a bimetallic layer composed of Cu and Ni as shown in Fig. 10.12 (Huang et al., 2025). The sensor performance was assessed on the transfer matrix method (TMM) alongwith the angular interrogation technique at a specific wavelength of 632.8 nm. The designed sensor was capable for detecting MTB in blood plasma of an infected individual by detecting the change of SPR angle with respect to refractive index in the presence of MTB in the diseased sample in comparison to the healthy blood plasma.

In another work Mondal et al., design and optimized SPR based sensor using N-FK51A prism, titanium dioxide (TiO_2), silicon (Si), barium titanate ($BaTiO_3$), and BP. They also used TMM approach and simulations using finite element method to analyze the result of the sensor (Mondal et al., 2025).

10.7 Conclusion

The 2D-nanomaterials has tremendous capabilities for the fabrication of biosensors in order to detect biomarkers related to MTB. This chapter has overall covered mainly two types of biomarkers (nucleic acids and protein) and for which biosensor has been fabricated using 2D-nanomaterial as one of the prominent parts of the nanocomposite. The major properties explored for any 2D-nanomaterial in these works are based on the enhanced surface area to volume ratio which provides them improved electrical as well as ionic conductivity. The functional groups available in these 2D-nanomaterials also helps in better immobilization of the biological recognition molecules such as capture DNA, signal DNA, aptamers, *etc*. Therefore, the 2D-nanomaterials are the promising option for the fabrication of biosensing platforms for the detection of MTB.

References

Archana N, Hashmi Z, Bhardwaj H, Kumar A, Solanki PR. Reduced graphene oxide nanosheets enhance the sensitivity of p-cresyl sulfate detection using a molecularly imprinted polymer-based electrochemical sensor. ACS Applied Polymer Materials 2024;6(11):6278–89. Available from https://doi.org/10.1021/acsapm.4c00435.

Archana N, Kumar A, Solanki PR. 4-ethylphenyl sulfate detection by an electrochemical sensor based on a MoS2 nanosheet-modified molecularly imprinted biopolymer. ACS Applied Bio Materials 2024;7(6):3841–53. Available from https://doi.org/10.1021/acsabm.4c00227.

Archana A, Kumar P, Solanki R. Chitosan-supported molecularly imprinted polypyrrole@MoSe2 nanosheet-based bioinspired electrochemical sensor for improved quantification of gut microbiota-derived hippuric acid. ACS Applied Bio Materials 2025;8(6):4961–72.. Available from https://doi.org/10.1021/acsabm.5c00329 Available fromhttps://doi.org/10.1021/acsabm.5c00329.

Baig N. Two-dimensional nanomaterials: a critical review of recent progress, properties, applications, and future directions. Composites Part A: Applied Science and Manufacturing 2023;165:107362. Available from https://doi.org/10.1016/j.compositesa.2022.107362.

Bai L, Chen Y, Liu X, Zhou J, Cao J, Hou L, et al. Ultrasensitive electrochemical detection of Mycobacterium tuberculosis IS6110 fragment using gold nanoparticles decorated fullerene nanoparticles/nitrogen-doped graphene nanosheet as signal tags. Analytica Chimica Acta 2019;1080:75–83. Available from https://doi.org/10.1016/j.aca.2019.06.043.

Bai Y, Zhou K, Srikanth N, Pang JHL, He X, Wang R. Dependence of elastic and optical properties on surface terminated groups in two-dimensional MXene monolayers: a first-principles study. RSC Advances 2016;6(42):35731–9. Available from https://doi.org/10.1039/c6ra03090d.

Casiraghi C, Pisana S, Novoselov KS, Geim AK, Ferrari AC. Raman fingerprint of charged impurities in graphene. Applied Physics Letters 2007;91(23). Available from https://doi.org/10.1063/1.2818692.

Cavalcanti YVN, Brelaz MCA, Neves JKDAL, Ferraz JC, Pereira VRA. Role of TNF-alpha, IFN-gamma, and IL-10 in the development of pulmonary tuberculosis. Pulmonary Medicine 2012. Available from https://doi.org/10.1155/2012/745483.

Chakravorty S, Archana, Lakshmi GBVS, Solanki PR, Kumar A. Trimethylamine n-oxide detection by electrochemical sensor based on screen printed electrode modified with molecularly imprinted polypyrrole-molybdenum(III) sulfide nanosheets. Colloids and Surfaces B: Biointerfaces 2024;244:114164. Available from https://doi.org/10.1016/j.colsurfb.2024.114164.

Chen Y, Li Y, Yang Y, Wu F, Cao J, Bai L. A polyaniline-reduced graphene oxide nanocomposite as a redox nanoprobe in a voltammetric DNA biosensor for Mycobacterium tuberculosis. Microchimica Acta 2017;184(6):1801–8. Available from https://doi.org/10.1007/s00604-017-2184-5.

Chen X, McDonald AR. Functionalization of two-dimensional transition-metal dichalcogenides. Advanced Materials 2016;28(27):5738–46. Available from https://doi.org/10.1002/adma.201505345.

Chen X, Qu Z, Liu Z, Ren G. Mechanism of oxidization of graphite to graphene oxide by the hummers method. ACS Omega 2022;7(27):23503–10. Available from https://doi.org/10.1021/acsomega.2c01963.

Costa MP, Andrade CAS, Montenegro RA, Melo FL, Oliveira MDL. Self-assembled monolayers of mercaptobenzoic acid and magnetite nanoparticles as an efficient support for development of tuberculosis genosensor. Journal of Colloid and Interface Science 2014;433:141–8. Available from https://doi.org/10.1016/j.jcis.2014.07.014.

Das M, Dhand C, Sumana G, Srivastava AK, Vijayan N, Nagarajan R, et al. Zirconia grafted carbon nanotubes based biosensor for M. tuberculosis detection. Applied Physics Letters 2011;99(14). Available from https://doi.org/10.1063/1.3645618.

Dillon AD, Ghidiu MJ, Krick AL, Griggs J, May SJ, Gogotsi Y, et al. Highly conductive optical quality solution-processed films of 2D titanium carbide. Advanced Functional Materials 2016;26(23):4162–8. Available from https://doi.org/10.1002/adfm.201600357.

Dutta R, Bala A, Sen A, Spinazze MR, Park H, Choi W, et al. Optical enhancement of indirect band gap 2D transition metal dichalcogenides for multi-functional optoelectronic sensors. Advanced Materials 2023;35(46). Available from https://doi.org/10.1002/adma.202303272.

Geim AK, Novoselov KS. The rise of graphene. Nature Materials 2007;6(3):183–91. Available from https://doi.org/10.1038/nmat1849.

Giorgio CD, Blundo E, Pettinari G, Felici M, Polimeni A, Bobba F. Exceptional elasticity of microscale constrained MoS2 domes. ACS Applied Materials & Interfaces 2021;13(40):48228–38. Available from https://doi.org/10.1021/acsami.1c13293.

Gou D, Xie G, Li Y, Zhang X, Chen H. Voltammetric immunoassay for Mycobacterium tuberculosis secretory protein MPT64 based on a synergistic amplification strategy using rolling circle amplification and a gold electrode modified with graphene oxide, Fe3O4 and Pt nanoparticles. Microchimica Acta 2018;185(9). Available from https://doi.org/10.1007/s00604-018-2972-6.

Gu H, Xing Y, Xiong P, Tang H, Li C, Chen S, et al. Three-dimensional porous Ti 3 C 2 T x MXene–graphene hybrid films for glucose biosensing. ACS Applied Nano Materials 2019;2(10):6537–45. Available from https://doi.org/10.1021/acsanm.9b01465.

Hafeez HY, Lakhera SK, Ashokkumar M, Neppolian B. Ultrasound assisted synthesis of reduced graphene oxide (rGO) supported InVO4-TiO2 nanocomposite for efficient hydrogen production. Ultrasonics Sonochemistry 2019;53:1–10. Available from https://doi.org/10.1016/j.ultsonch.2018.12.009.

He X, Gao P, Shi W. A further comparison of graphene and thin metal layers for plasmonics. Nanoscale 2016;8(19):10388–97. Available from https://doi.org/10.1039/C5NR09061J.

Huang H, Chen Y, Zuo J, Deng C, Fan J, Bai L, et al. MXene-incorporated C60NPs and Au@Pt with dual-electric signal outputs for accurate detection of Mycobacterium tuberculosis ESAT-6 antigen. Biosensors and Bioelectronics 2023;242:115734. Available from https://doi.org/10.1016/j.bios.2023.115734.

Huang X, Peng Y, Wang Y, Zhang L. High-performance surface plasmon resonance biosensor based on copper, nickel, and black phosphorus for detection of Mycobacterium tuberculosis. Plasmonics 2025;20(12):10937–53. Available from https://doi.org/10.1007/s11468-025-03158-2.

Hu Z, Wu Z, Han C, He J, Ni Z, Chen W. Two-dimensional transition metal dichalcogenides: interface and defect engineering. Chemical Society Reviews 2018;47(9):3100–28. Available from https://doi.org/10.1039/c8cs00024g.

Hwang SH, Kim DE, Sung H, Park BM, Cho MJ, Yoon OJ, et al. Simple detection of the IS6110 sequence of Mycobacterium tuberculosis complex in sputum, based on PCR with graphene oxide. PLoS ONE 2015;10(8). Available from https://doi.org/10.1371/journal.pone.0136954.

Jiang Z, Zhang Y, Stormer HL, Kim P. Quantum hall states near the charge-neutral dirac point in graphene. Physical Review Letters 2007;99(10). Available from https://doi.org/10.1103/physrevlett.99.106802.

Jin H, Gui R, Gao X, Sun Y. An amplified label-free electrochemical aptasensor of γ-interferon based on target-induced DNA strand transform of hairpin-to-linear conformation enabling simultaneous capture of redox probe and target. Biosensors and Bioelectronics 2019;145:111732. Available from https://doi.org/10.1016/j.bios.2019.111732.

Liu C, Jiang D, Xiang G, Liu L, Liu F, Pu X. An electrochemical DNA biosensor for the detection of Mycobacterium tuberculosis, based on signal amplification of graphene and a gold nanoparticle–polyaniline nanocomposite. Analyst 2014;139(21):5460–5.. Available from https://doi.org/10.1039/C4AN00976B, http://doi.org/10.1039/C4AN00976B Available from.

Li M-chao, Lu J, Lu Y, Xiao T-yang, Liu H-can, Lin S-qiang, et al. rpoB mutations and effects on rifampin resistance in Mycobacterium tuberculosis. Infection and Drug Resistance 2021;14:4119–28. Available from https://doi.org/10.2147/idr.s333433.

Li F, Yu Y, Li Q, Zhou M, Cui H. A homogeneous signal-on strategy for the detection of rpoB genes of mycobacterium tuberculosis based on electrochemiluminescent graphene oxide and ferrocene quenching. Analytical Chemistry 2014;86(3):1608–13. Available from https://doi.org/10.1021/ac403281g.

Magnuson M, Mattesini M. Chemical bonding and electronic-structure in MAX phases as viewed by X-ray spectroscopy and density functional theory. Thin Solid Films 2017;621:108–30. Available from https://doi.org/10.1016/j.tsf.2016.11.005.

Manoharan AK, Batcha MIK, Mahalingam S, Raj B, Kim J. Recent Advances in two-dimensional nanomaterials for healthcare monitoring. ACS Sensors 2024;9(4):1706–34. Available from https://doi.org/10.1021/acssensors.4c00015.

Mobed A, Darvishi M, Kohansal F, Dehfooli FM, Alipourfard I, Tahavvori A, et al. Biosensors; nanomaterial-based methods in diagnosing of Mycobacterium tuberculosis. Journal of Clinical Tuberculosis and Other Mycobacterial Diseases 2024;34:100412. Available from https://doi.org/10.1016/j.jctube.2023.100412.

Mondal TP, Mahmud RR, Rafi SA, Islam MS, Barua B. Design and optimization of a highly sensitive surface plasmon resonance biosensor for accurate detection of mycobacterium tuberculosis. IEEE Sensors Journal 2025;25(6):9216–23. Available from 10.1109/JSEN.2025.3536976.

Muller SE, Prange MP, Lu Z, Rosenthal WS, Bilbrey JA. An open database of computed bulk ternary transition metal dichalcogenides. Scientific data 2023;10(1). Available from https://doi.org/10.1038/s41597-023-02103-4.

Nguyen TT-H, Nguyen CM, Minh Anh Huynh HHV, Tuan-Khoa Nguyen N-TN. Field effect transistor based wearable biosensors for healthcare monitoring. Journal of nanobiotechnology 2023;21(1):411. Available from https://doi.org/10.1186/s12951-023-02153-1, https://doi.org/10.1186/s12951-023-02153-1.

Pai M, Nicol MP, Boehme CC, Jacobs Jr WR, McShane H, Mizrahi V, et al. Tuberculosis Diagnostics: state of the art and future directions. Microbiology spectrum 2016;4(5). Available from https://doi.org/10.1128/microbiolspec.tbtb2-0019-2016.

Park JS, Goo NI, Kim DE. Mechanism of DNA adsorption and desorption on graphene oxide. Langmuir 2014;30(42):12587–95. Available from https://doi.org/10.1021/la503401d.

Passos BBS, Araújo-Pereira M, Vinhaes CL, Amaral EP, Andrade BB. The role of ESAT-6 in tuberculosis immunopathology. Frontiers in Immunology 2024;15. Available from https://doi.org/10.3389/fimmu.2024.1383098.

Quirós-Ovies R, Sulleiro MV, Vera-Hidalgo M, Prieto J, Gómez IJ, Sebastián V, et al. Controlled covalent functionalization of 2 H-MoS2 with molecular or polymeric adlayers. Chemistry – A European Journal 2020;26(29):6629–34.. Available from https://doi.org/10.1002/chem.202000068, https://doi.org/10.1002/chem.202000068 Available from.

Rattana, et al. Preparation and characterization of graphene oxide nanosheets. Procedia Engineering 2012;32:759–64. Available from https://doi.org/10.1016/j.proeng.2012.02.009.

Rizi KS, et al. PCR-free electrochemical genosensor for Mycobacterium tuberculosis complex detection based on two-dimensional Ti3C2 Mxene-polypyrrole signal amplification. Microchemical Journal 2022;179:107467. Available from https://doi.org/10.1016/j.microc.2022.107467.

Saini R, Mishra RK, Kumar P. Green synthesis of reduced graphene oxide using the tinospora cordifolia plant extract: exploring its potential for methylene blue dye degradation and antibacterial activity. ACS Omega 2024;9(18):20304–21. Available from https://doi.org/10.1021/acsomega.4c00748.

Shamsaei E, de Souza FB, Yao X, Benhelal E, Akbari A, Duan W. Graphene-based nanosheets for stronger and more durable concrete: a review. Construction and Building Materials 2018;183:642–60. Available from https://doi.org/10.1016/j.conbuildmat.2018.06.201.

Shankar SS, Shereema RM, Rakhi RB. Electrochemical determination of adrenaline using MXene/Graphite composite paste electrodes. ACS Applied Materials and Interfaces 2018;10(50):43343–51. Available from https://doi.org/10.1021/acsami.8b11741.

Shuck CE, Han M, Maleski K, Hantanasirisakul K, Kim SJ, Choi J, et al. Effect of Ti3AlC2 MAX phase on structure and properties of resultant Ti3C2Tx MXene. ACS Applied Nano Materials 2019;2(6):3368–76. Available from https://doi.org/10.1021/acsanm.9b00286.

Soleymaniha M, Shahbazi MA, Rafieerad AR, Maleki A, Amiri A. Promoting role of MXene nanosheets in biomedical sciences: therapeutic and biosensing innovations. Advanced Healthcare Materials 2019;8(1). Available from https://doi.org/10.1002/adhm.201801137.

Syama S, Mohanan PV. Comprehensive application of graphene: emphasis on biomedical concerns. Nano-Micro Letters 2019;11(1). Available from https://doi.org/10.1007/s40820-019-0237-5.

Tang Q, Jiang D-en. Stabilization and band-gap tuning of the 1T-MoS2 monolayer by covalent functionalization. Chemistry of Materials 2015;27(10):3743–8. Available from https://doi.org/10.1021/acs.chemmater.5b00986.

Tedstone AA, Lewis DJ, O'Brien P. Synthesis, properties, and applications of transition metal-doped layered transition metal dichalcogenides. A Chemistry of Materials 2016;28(7):1965–74. Available from https://doi.org/10.1021/acs.chemmater.6b00430.

Teymourian H, Salimi A, Khezrian S. Fe3O4 magnetic nanoparticles/reduced graphene oxide nanosheets as a novel electrochemical and bioeletrochemical sensing platform. Biosensors and Bioelectronics 2013;49:1–8. Available from https://doi.org/10.1016/j.bios.2013.04.034.

Thiruppathiraja C, Kamatchiammal S, Adaikkappan P, Santhosh DJ, Alagar M. Specific detection of Mycobacterium sp. genomic DNA using dual labeled gold nanoparticle based electrochemical biosensor. Analytical Biochemistry 2011;417(1):73–9. Available from https://doi.org/10.1016/j.ab.2011.05.034.

Wang W, Du H, Dai C, Ma H, Luo S, Wang X, et al. Amplification-free detection of mycobacterium tuberculosis using CRISPR-Cas12a and graphene field-effect transistors. Nanoscale 2025;17(8):4603–9.. Available from https://doi.org/10.1039/D4NR03852E, http://doi.org/10.1039/D4NR03852E Available from.

Wang Y, Xu Y, Hu M, Ling H, Zhu X. MXenes: focus on optical and electronic properties and corresponding applications. Nanophotonics 2020;9(7):1601–20. Available from https://doi.org/10.1515/nanoph-2019-0556.

Wang G, Yang J, Park J, Gou X, Wang B, Liu H, et al. Facile synthesis and characterization of graphene nanosheets. The Journal of Physical Chemistry C 2008;112(22):8192–5. Available from https://doi.org/10.1021/jp710931h.

Wang F, Zhang Y, Tian C, Girit C, Zettl A, Crommie M, et al. Gate-variable optical transitions in graphene. Science 2008;320(5873):206–9. Available from https://doi.org/10.1126/science.1152793.

Wen Z, Cui S, Pu H, Mao S, Yu K, Feng X, et al. Metal nitride/graphene nanohybrids: general synthesis and multifunctional titanium nitride/graphene electrocatalyst. Advanced Materials 2011;23(45):5445–50. Available from https://doi.org/10.1002/adma.201102772.

Wu Q, Jin X, Chen S, Jiang X, Hu Y, Jiang Q, et al. MXene-based saturable absorber for femtosecond mode-locked fiber lasers. Optics Express 2019;27(7):10159–70. Available from https://doi.org/10.1364/OE.27.010159.

Wu M, Lin X, Wang Y, Wang L, Guo W, Qi D, et al. Economical pt-free catalysts for counter electrodes of dye-sensitized solar cells. Journal of the American Chemical Society 2012;134(7):3419–28. Available from https://doi.org/10.1021/ja209657v.

Wu X, Wang Z, Yu M, Xiu L, Qiu J. Stabilizing the MXenes by carbon nanoplating for developing hierarchical nanohybrids with efficient lithium storage and hydrogen evolution capability. Advanced Materials 2017;29(24). Available from https://doi.org/10.1002/adma.201607017.

Xiao Li JS, Weizhen Zhang SS, Lihui Yuwen LW. Recent advances in synthesis and biomedical applications of two-dimensional transition metal dichalcogenide nanosheets. Small (Weinheim an der Bergstrasse, Germany) 2017;13(5):1602660.. Available from https://doi.org/10.1002/smll.201602660, https://doi.org/10.1002/smll.201602660 Available from.

Yin X, Zheng L, Lin L, Hu Y, Zheng F, Hu Y, et al. Commercial MPT64-based tests for rapid identification of Mycobacterium tuberculosis complex: a meta-analysis. Journal of Infection 2013;67(5):369–77. Available from https://doi.org/10.1016/j.jinf.2013.06.009.

Zhang CJ, Anasori B, Seral-Ascaso A, Park SH, McEvoy N, Shmeliov A, et al. Transparent, flexible, and conductive 2D titanium carbide (MXene) films with high volumetric capacitance. Advanced Materials 2017;29(36). Available from https://doi.org/10.1002/adma.201702678.

Zhao Y, Wang SB, Chen AZ, Kankala RK. Nanoarchitectured assembly and surface of two-dimensional (2D) transition metal dichalcogenides (TMDCs) for cancer therapy. Coordination Chemistry Reviews 2022;472. Available from https://doi.org/10.1016/j.ccr.2022.214765.

CHAPTER 11

Commercially available biosensors for *Mycobacterium tuberculosis* detection

Sapna Kaithwas[1] and Chetna Dhand[2]
[1]Innovative Materials and Processes Division, CSIR-Advanced Materials and Processes Research Institute (AMPRI), Bhopal, Madhya Pradesh, India,
[2]Academic of Scientific and Innovative Research (AcSIR), CSIR-Advanced Materials and Processes Research Institute, Bhopal, Madhya Pradesh, India

11.1 Introduction

Every year approximately 10 million new cases and 1.5 million deaths due to *Mycobacterium tuberculosis* (MTB) per year recorded by WHO. According to recent predictions, only 7.5 million of the 10.6 million people who were expected to have contracted TB in 2022 had a formal diagnosis (Global Tuberculosis Report, 2023). Therefore, the rapid detection and treatment is very essential for the cure of MTB. Clinical diagnostics demand very sensitive detection techniques because of MTB susceptibility and quick dissemination (Albert et al., 2016; Salimiyan Rizi et al., 2021). (Fig. 11.1)

Implementing quick and accurate screening techniques is crucial for successfully controlling and limiting the spread of tuberculosis (World Health Organization, 2013). MTB complex identification in sputum samples using methods such as acid-fast staining and mycobacterial culture has historically been the basis for MTB diagnosis (Bark et al., 2011). Despite being the gold standard, these techniques are frequently labor-intensive, time-consuming, and need specialized laboratory infrastructure, all of which impede prompt diagnosis, particularly in areas with low resources (Boehme et al., 2010; Bisognin et al., 2020). Modern advancement in biosensor technology, especially electrochemical biosensors, have shown great promise for transforming TB diagnosis. Electrochemical biosensors incorporate a variety of detecting techniques, including amperometry, conductometry, potentiometry, and impedimetry. With the help of these technologies, the biochemical interactions can be properly detected as they are converted to electrical signals, which means that electrochemical biosensors are highly sensitive and can be used to make a fast diagnosis (For Health Care Providers, 2024; Cohen et al., 2019; Frieden et al., 2009). Application of biosensors enhances sensitivity and reduces the interval between sample and retention of the result. The devices are sensitive and have fast turnaround times and this is accompanied by the capability to record markers of MTB and drug resistance with a minimum of sample processing.

Advances in Biosensor Technology for Mycobacterium Tuberculosis Detection
DOI: https://doi.org/10.1016/B978-0-443-44589-7.00004-2

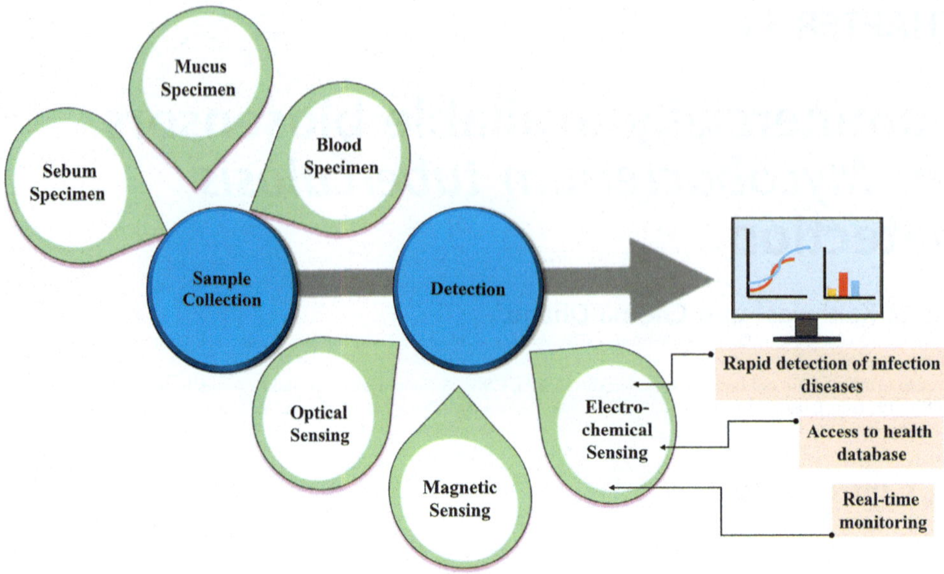

Figure 11.1 Overview of different techniques for the *Mycobacterium tuberculosis* (MTB) detection. This figure explain the traditional and conventional method for MTB detection.

The GeneXpert MTB/rifampicin (RIF) assay (Cepheid Inc.), quantiFERON -TB ELISA (Qiagen), and some lagging genosensor-based amperometric systems with ssDNA probes are examples of these tools, which are commercially available (Frieden et al., 2009). Triage tools such as C-reactive protein (CRP) tests and AI-assisted digital analysis of chest X-rays are also becoming increasingly popular as the tools of set-screening. In spite of such developments, a vast majority of existing technologies do not deliver to the WHO optimum target product profile (TPP) criteria on the deployment of point-of-care devices, especially in aspects of operational feasibility, as well as affordability (GeneXpert® Edge Technical Training, 2022). This chapter offers a comprehensive discussion of the principles, performance, and public health impacts of current and novel biosensor-based technologies in the detection of MTB, along with highlighting the dire necessity of more technological advances in this essential field (Table 11.1).

11.2 GeneXpert *Mycobacterium tuberculosis*/rifampicin MHDC

The GeneXpert test is an integrated, completely computerized nucleic acid amplification assay that identifies MTB and RIF resistance in 90minutes. GeneXpert is a viable test with great accuracy, precision, and a quick processing rate (GeneXpert® Edge Technical Training, 2022; Sasikumar et al., 2020; Horne et al., 2019). The geneXpert

Table 11.1 List of the commercially available biosensors for the detection of MTB.

S. no.	Name of the biosensor	Manufacturing company	References
1.	Truenat^{TM} MTB Rif Dx	Molbio diagnostic, India	Truenat (n.d)
2.	Quantiplus MTB fast detection kit	Huwel lifesciences, India	Made-in-India TB Diagnostic Tests Validated by ICMR (n.d)
3.	GeneXpert^{R} MTB/RIF ultra	Cepheid inc., United States	Xpert® MTB/RIF Ultra Instructions For Use. (2019)
4.	Cobas^{R} 5800- 6800 series	Hoffmann-L roche ltd., Quimica e Industrial, Argentina	About Roche Diagnostics. (2025)
5.	QFT-Plus	Qiagen, United State	QuantiFERON-TB Gold Plus QFT-Plus (n.d)
6.	TB-Feron ELISA	SD biosensor, INC. South Korea, REPUBLIC OF KOREA	Sd Biosensor_ Products (n.d)

MTB, Mycobacterium tuberculosis; RIF, rifampicin.

MTB/RIF (geneXpert) cassette test represents a major advancement in the molecular diagnosis of TB. This technology is particularly essential in the early identification and control of transmissible infections such as TB, where rapid and accurate diagnosis can significantly influence the overall response to treatment and associated human wellness (GeneXpert® Edge Technical Training, 2022; Stevens et al., 2017; Zeka et al., 2011).

In December of 2010, WHO has sanctioned the geneXpert TB/RIF test, a novel diagnostic method, as an alternative to established methods of TB diagnostics. In October 2013, WHO updated its policy to recommend the use of this advanced method for the rapid detection of TB, including in extrapulmonary and pediatric cases (Global Tuberculosis Report, 2023, 2021). The GeneXpert MTB/RIF assay is a sophisticated in vitro diagnostic tool that quickly and precisely detects MTB DNA and Rif resistance straight from clinical samples using semi-nested real-time qPCR technology. Its ability to detect both TB and multidrug resistant TB makes it an essential tool, particularly in environments with limited resources or high burden by facilitating decentralized testing, cutting down on delays, and enhancing results, its incorporation into national TB programs has revolutionized diagnostic methodologies (Golichenari et al., 2018). MTB and Rif resistance are detected by the GeneXpert MTB/RIF assay, a quick molecular test that amplifies MTB-specific DNA and finds mutations in the 81-base pair core region of the rpoB gene. It combines detection, DNA amplification, and specimen processing into one system. To distinguish between wild type and altered sequences associated with RIF resistance, the assay employs probes. Testing at least one sample from people who have a moderate to high suspicion of tuberculosis is advised by the CDC (Hopmeier et al., 2020).

11.2.1 Working principle of geneXpert *Mycobacterium tuberculosis*/rifampicin MHDC

The xpert MTB/RIF test applies a nested real-time PCR method to qualitatively identify members of the Mycobacterium tuberculosis complex and detect RIF resistance. In this assay, specific primers target and amplify an 81-base pair segment of the rpoB gene, referred to as the core region. Distinct fluorescent probes then differentiate between the preserved wild-type sequence and mutations in the core region that are linked to Rif resistance (Saeed et al., 2017; Ioannidis et al., 2011). (Fig. 11.2)

The probe check control (PCC) function ensures the cartridge reagents are properly rehydrated, the PCR chamber is correctly filled, and the probes and associated dyes are performing accurately. The geneXpert platform automates key steps of the analysis, including sample preparation, nucleic acid amplification, and target detection, through integrated rt-PCR and real-time PCR technologies (Jing et al., 2014). According to the test algorithm, MTB detection is confirmed when at least two probes yield cycle threshold (Ct) values

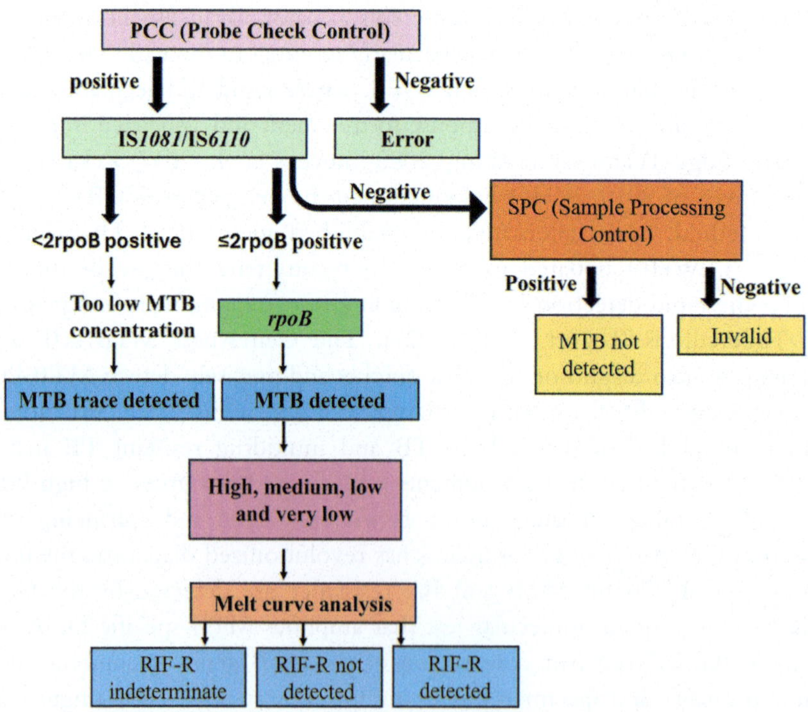

Figure 11.2 Schematic representation of geneXpert MTB/RIF MHDC working mechanism. This figure explain the methodology and working principle. *MTB*, *Mycobacterium tuberculosis*; *RIF*, rifampicin.

within an acceptable range and a delta Ct less than 2.0. RIF resistance is reported when the delta Ct exceeds 4.0 or when the rpoB probe set displays a maximum valid Ct value, with the earliest probe showing a Ct difference greater than 4.0 (Lewinsohn et al., 2017).

11.2.2 Key features of the technology

The geneXpert system automates all major stages of molecular diagnostics, including sample processing, nucleic acid extraction, amplification, and detection, within a single, self-contained disposable cartridge. This integrated platform provides diagnostic results in approximately 45–90minutes and is suitable for both raw and concentrated sputum samples from patients suspected of tuberculosis, including those with HIV coinfection (Zeka et al., 2011; Ong et al., 2025). The test is able to identify Mycobacterium tuberculosis genetic material at the same time identifying mutations in the rpoB gene involved in RIF resistance. The test has sensitivity and specificity of approximately 8998 and 99, respectively depending on the study parameters and the type of specimen used. In processing a sample, a tuberculocidal reagent is used to render viable pathogens inert, and thus making the sample much safer to work with in laboratories and permits the collection of diagnostic samples in the immediate proximity of patients (Ioannidis et al., 2011; General Information, n.d). Initial sample processing is done using a tuberculocidal reagent that reduces the biosafety hazard to the laboratory staff and facilitates its use in a near-patient diagnostic environment (Xpert MTB_RIF and Xpert MTB_RIF Ultra assays_ TB Knowledge Sharing, n.d). The cartridges have inbuilt controls to ascertain that the cartridge is getting the correct sample processing and correct functioning of the probe to ensure accuracy and reproducibility.

The WHO and CDC suggest the use of the geneXpert system as an initial diagnostic instrument of tuberculosis and RIF resistance, especially in the areas where the disease is prevalent (Interpreting Xpert MTB/RIF Assay Results, n.d). It will be particularly useful in the diagnosis of smear-negative, culture-positive, and extrapulmonary TB cases where the traditional method is less effective. GeneXpert MTB/RIF system is also well-liked because of its automation, prompt results, dependability, and two-diagnostic capacity (Chakravorty et al., 2017). Based on its enclosed cartridge, it can be operated without the need to have the advanced laboratory infrastructure (Sample to Insight, 2023).

11.3 Cepheid xpert *Mycobacterium tuberculosis*/rifampicin

Cepheid xpert MTB/RIF or xpert MTB/RIF Ultra (xpert ultra) assay was an automated, on-demand test that was an enhanced version of the original xpert MTB/RIF assay, which had been found to be not quite sensitive enough to detect TB in some circumstances (Lewinsohn et al., 2017). To measure the strength of the same, systematic review and meta-analysis was done in order to find out the accuracy in terms of detecting TB and Rif resistance (Torati et al., 2016). The xpert MTB/RIF assay utilizes nested real-time PCR technology to simultaneously detect DNA from the MTB complex and identify Rif resistance associated with mutations in the rpoB gene, all within approximately 2 hours (Uppe et al., 2018).

11.4 Working principle of cepheid xpert *Mycobacterium tuberculosis*/ rifampicin system

The working principle of this system is identical to that of the xpert MTB/RIF assay. The kit includes all necessary reagents for the detection of MTB and Rif resistance. It also incorporates a sample processing control (SPC), which verifies proper bacterial lysis and detects the presence of any PCR inhibitors during amplification and melt peak analysis. Additionally, the probe check control (PCC) ensures reagent rehydration, correct filling of PCR tubes within the cartridge, optimal probe performance, and dye stability (The New GeneXpert® System, n.d). The xpert MTB/RIF Ultra assay employs primers that amplify targeted regions of the *rpoB* gene, including the critical 81-base pair Rif resistance determining core region, along with segments of the multicopy insertion elements IS1081 and IS6110. The system which identifies the conserved wild-type sequence and the mutations in the region which is replenished by rpoB probes by using four rpoB probe sets differentiated RIF resistance mutations in the core region and

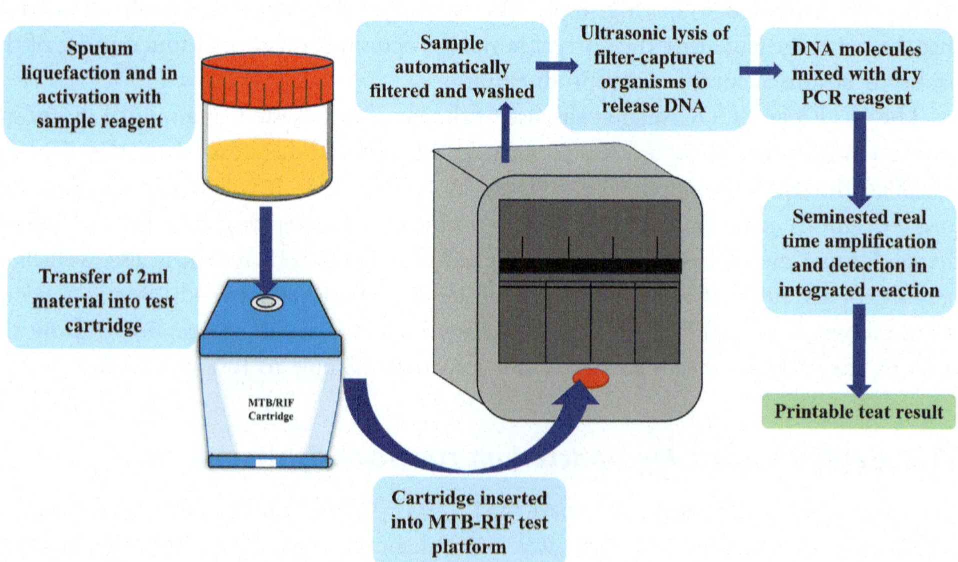

Figure 11.3 Schematic representation of cepheid xpert MTB/RIF system working mechanism. This figure explain the methodology and working principle. *MTB*, *Mycobacterium tuberculosis*; *RIF*, rifampicin.

especially in the bacterial direct repeat transcripts by using RNase II probes (Ferguson and Novak E, 2003). The assay is further made sensitive by the addition of probes that attack IS1081 and IS6110 which are present in multi copies in the vast majorities of the strains of the MTB complexes (Xpert® MTB/RIF Ultra Instructions For Use, 2019; Ng et al., 2018). (Fig. 11.3)

11.4.1 Key features of the cepheid xpert *Mycobacterium tuberculosis*/rifampicin

The cepheid xpert MTB/RIF assay is a representation of one of the significant innovations in molecular diagnostics and the automated technology of RTPCR in the rapid and accurate detection of MTB and Rif resistance (Vitro, I. GeneXpert® Dx Reference Guide, 2017). The system drastically reduces turnaround time by making the entire PCR process automated into a single, closed cartridge, resulting in regular production of results within less than two hours. Such speed is groundbreaking in clinical decision-making and enables one to start proper treatment on time and to better patient outcomes (Haas et al., 2025). The assay has a convenient design and has an easy to use nature. Usser-friendly interface and simple workflow, which needed only a little training of operators and the few manual operations, and is therefore most suitable in decentralized and resource limited laboratories. Self contained cartridge format which also enhances Laboratory biosafety. It lowers the contact of infectious materials and diminishes the impact of cross contamination (Trademark, 2020). Collectively, these benefits make Xpert MTB/RIF one of the most transparent and useful diagnostic platforms in the high- and low-resource environments that can ensure timely tuberculosis diagnosis and detect drug resistance (Xpert® MTB/RIF Instructions for Use For Use with GeneXpert® System with Touchscreen, 2013). As the global TB burden continues, the xpert MTB/RIF remains a cornerstone of modern diagnostic strategies, advancing efforts in detection, management, and disease control (Xpert MTB/RIF Implementation Manual, 2014; Phd et al., 2011).

11.5 ELISA (quantiFERONVR-TB)

The quantiFERON-TB ELISA represents a significant advancement in tuberculosis diagnostic technology. This commercially available assay facilitates the rapid and precise identification of latent TB infection (Yancey, 2021). Developed to assess immune reactivity to the same antigenic components purified protein derivative used in the tuberculin skin test (TST) (Bisognin et al., 2020). QFT-plus has become the most widely adopted interferon-gamma release assay for detecting latent TB infection. QFT-plus results obtained through the ELISA on the skyLab platform with those generated

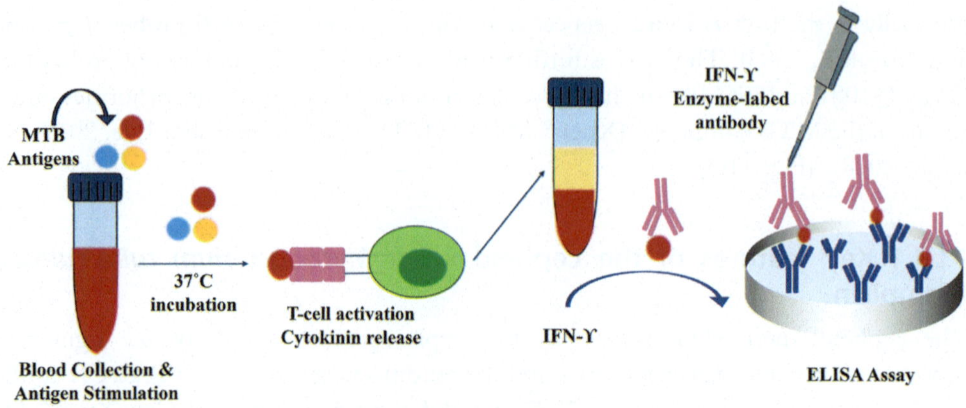

Figure 11.4 Schematic representation of ELISA (quantiFERONVR-TB) working mechanism. This figure explain the methodology and working principle.

using a chemiluminescence immunoassay on the liaison XL analyzer (Kadkhoda et al., 2023).

11.6 Working principle

This technique is basically working on three principle which are based on the reactions between different biological components: (Fig. 11.4)

The working principle of the ELISA used in QFT testing is based on detecting interferon-gamma (IFN-γ) released by T-cells in a patient's blood following stimulation with *MTB*-specific antigens. Whole blood is drawn into a set of specialized tubes: one containing TB-specific peptides such as ESAT-6 and CFP-10, a negative control, and a mitogen, positive control. After thorough mixing and incubation for 16–24 hours at 37°C, sensitized T-cells from individuals previously exposed to MTB secrete IFN-γ in response to the specific antigens (English_QFT_ELISA_R04_082016, n.d). The assay core mechanism involves antigen-antibody binding, where specific antigens interact with corresponding antibodies to form stable immune complexes (English_QFT_ELISA_R04_082016, n.d; ELISA_ What It Is, Purpose, Procedure & Results, n.d). This reaction exhibits high specificity, ensuring the detection of only the intended targets. Subsequently, an enzyme conjugated to one of the antibodies catalyzes the conversion of a colorless substrate into a colored product (English_QFT_ELISA_R04_082016, n.d). The appearance of color confirms antigen-antibody binding, and the resulting signal is measured spectrophotometrically. The intensity of the color directly affects the degree of interaction between the antigens and antibodies and therefore, the intensity of the color can be accurately measured. Immune reaction responses as well as giving sensitive and precise diagnostic output (Cohen et al., 2019; Torati et al., 2016; English_QFT_ELISA_R04_082016, n.d).

11.7 Key features of the technology

The high specificity of QFT ELISA assays is that they are only sensitive to immune responses in cases of immune reactions towards MTB only, unusual in relation to the vaccine strains of BCG and the majority of the no-tuberculosis mycobacteria (For Health Care Providers, 2024). This targeted approach chances of receiving a false-posit and also expands the reliability of the assay among the wide range of populations. The QFT-Plus version has one more sophisticated option, where two antigen tubes are used: TB1, which triggers CD4+ T-cells, and TB 2, which triggers CD4+ and CD8+ T-cells. This dual-tube system has increased test sensitivity and provides a more in-depth explanation on whether recent or recent MTB infections or not (Xpert® MTB/RIF Ultra Instructions For Use, 2019; How Are QFT-Plus Test Results Interpreted?, 2013). The main benefit of QFT testing is that it is quantitative in nature and gives exact measurements of interferon-gamma (IFN-3) concentration in response to a particular antigens of MTB (Lombardi et al., 2018). This standardized test is also quantitative and therefore results in objectivity thus making the diagnosis accurate and therefore facilitating better clinical decision making. Together, these qualities make QFT assays a powerful means of screening and risk stratification as far as tuberculosis is concerned and lead to better prediction and effective prevention in certain conditions, both in the context of population health and in clinical practice (Mahmoudi and Nourazar, 2024; Vo et al., 2023).

11.8 Conclusion

MTB is a major global health concern, necessitating prompt and accurate diagnosis for effective treatment and disease prevention. Successful therapy and prevention of the diseases. Recent molecular diagnostics, including the geneXpert MTB/RIF and Xpert MTB/RIF Ultra tests, are ones that have changed TB diagnosis. These direct detection of the presence of can be done by semi-nested real-time PCR-based technologies. Tuberculosis complex resistance to mycobacterium tuberculosis and RIF, which is an important marker. TB which is multidrug-resistant in clinical samples. The results of these systems are received within two hours and they are very sensitive, very specific with self-contained cartridges, which enhances security and alleviates pollution. They were widely applied especially in resource-restricted and intense-burden environments, have led to significant decreases in the diagnostic delays and achieved better patient outcomes. The QFT test release assay of interferon-gamma that is highly an adjunctive strategy that is efficient in detecting latent infections of TB. Unlike the traditional method of using the tuberculin skin test, QFT is able to give results within a day and is highly specific yet it has not been influenced by prior BCG vaccination. T-cell responses the antigens of TB in whole blood are measured, therefore being a great screening tool to TB. This is useful in ultra-vulcanized high-risk and immunocompromised. Of current-day technologies include geneXpert and the quantiFERON-TB, are transforming the TB control strategies with the ease of offering rapid, accurate, and quantitative results that provide productive information that must be used in global eradication efforts.

11.9 Advancements to be made for devices to diagnose *Mycobacterium tuberculosis*

Various advances in medical diagnostics play a vital role in overcoming the current problem of medical diagnostics of TB, especially in resource-constrained locations and the diagnosis of those difficult to diagnose, such as extrapulmonary TB and TB in children. The developments should be developed so that they are sensitive, specific, quick, mobile, economical, and also those that can detect drug resistance. A number of major developments in diagnostic tools are needed in order to enhance the detection of MTB. The first approach would be to improve the accessibility and affordability of some molecular platforms such as geneXpert so that rural or underserved areas would gain access. Testing in the rural areas can be guaranteed by development of battery-operated or solar-powered machines with minimum infrastructure requirements. Moreover, the integrated ability to test multiplex, i.e. to identify at the same time both MTB and other respiratory pathogenic bacteria would facilitate diagnosis and increase the discrimination diagnosis, especially in individuals with similar symptoms. Still, there is a need to enhance sensitivity of diagnosis of cases of paucibacillary and extrapulmonary TB. Such innovation as CRISPR-based diagnostic platforms, nanotechnology-enhanced detection, or ultra-sensitive tests based on biomarkers may reduce their detection limit significantly. It will also be enhanced with integration with digital health tools, including cloud data sharing, artificial intelligence-aided analysis, and result interpretation through smartphone to provide more efficient real-time surveillance, reduce human error, and assist with timely and efficient clinical decision-making. Accomplished by cheap and noninvasive methods of sample analysis, such as breath- or urine-based diagnostic, scalable and point of care TB screening as well as monitoring would ensure higher patient safety as well as compliance.

References

About Roche Diagnostics; 2025.

Albert H, Nathavitharana RR, Isaacs C, Pai M, Denkinger CM, Boehme CC. Development, roll-out and impact of Xpert MTB/RIF for tuberculosis: what lessons have we learnt and how can we do better? European Respiratory Journal. European Respiratory Society 2016:516–25. Available from https://doi.org/10.1183/13993003.00543-2016.

Bark CM, Okwera A, Joloba ML, Thiel BA, Nakibali JG, Debanne SM, et al. Time to detection of Mycobacterium tuberculosis as an alternative to quantitative cultures. Tuberculosis 2011;91(3):257–9. Available from https://doi.org/10.1016/j.tube.2011.01.004.

Bisognin F, Lombardi G, Re MC, Dal Monte P. QuantiFERON-TB gold plus with chemiluminescence immunoassay: do we need a higher cutoff? Journal of Clinical Microbiology 2020;58(10). Available from https://doi.org/10.1128/JCM.00780-20.

Boehme CC, Nabeta P, Hillemann D, Nicol MP, Shenai S, Krapp F, et al. Rapid molecular detection of tuberculosis and rifampin resistance. New England Journal of Medicine 2010;363(11):1005–15. Available from https://doi.org/10.1056/nejmoa0907847.

Chakravorty S, Simmons AM, Rowneki M, Parmar H, Cao Y, Ryan J, et al. The New Xpert MTB/RIF Ultra: Improving Detection of Mycobacterium tuberculosis and Resistance to Rifampin in an Assay Suitable for Point-of-Care Testing. mBio 2017;8(4):e00812–7. Available from 10.1128/mBio.00812-17.

Cohen A, Mathiasen VD, Schön T, Wejse C. The global prevalence of latent tuberculosis: a systematic review and meta-analysis. European Respiratory Journal 2019. Available from https://doi.org/10.1183/13993003.00655-2019. European Respiratory Society.

ELISA_ What It Is, Purpose, Procedure & Results. n.d.

English_QFT_ELISA_R04_082016. n.d.

For Health Care Providers. Clinical Testing Guidance for Tuberculosis: Interferon Gamma Release Assay. Centers for Disease Control and Prevention; 2024. Available from https://www.cdc.gov/tb/hcp/testing-diagnosis/interferon-gamma-release-assay.html?CDC_AAref_Val=https://www.cdc.gov/tb/publications/factsheets/testing/IGRA.htm.

Frieden, T.R.; Director Harold Jaffe, M.W.; Stephens, J.W.; Thacker, S.B.; Shaw, F.E.; Spriggs Terraye M., et al. Department of Health and Human Services Centers for Disease Control and Prevention Morbidity and Mortality Weekly Report Updated Guidelines for Using Interferon Gamma Release Assays to Detect Mycobacterium Tuberculosis Infection-United States, 2010 Editorial and Production Staff; 2009; Vol. 59. <www.cdc.gov/mmwr>.

General Information. M. Tuberculosis Complex DNA Detection. n.d. Available from https://testguide.labmed.uw.edu/view/TBCDNA?

GeneXpert® Edge Technical Training; 2022.

Global Tuberculosis Report, 2023; World Health Organization, 2023.

Global Tuberculosis Report. 2021; World Health Organization, 2021.

Golichenari B, Velonia K, Nosrati R, Nezami A, Farokhi-Fard A, Abnous K, et al. Label-free nano-biosensing on the road to tuberculosis detection. Biosensors and Bioelectronics 2018:124–35. Available from https://doi.org/10.1016/j.bios.2018.04.059. Elsevier Ltd.

Haas AL, Ma A, Pham J, Verma P, Malhotra U, Church EC, et al. Limitations of the MTB/RIF assay: an xpert review of 4 clinical cases. Open Forum Infectious Diseases 2025;12(4). Available from https://doi.org/10.1093/ofid/ofaf132.

Hopmeier D, Lampejo T, Rycroft J, Tiberi S, Melzer M. The limitations of the Cepheid GeneXpert® Mtb/Rif assay for the diagnosis and management of polyresistant pulmonary tuberculosis. Clinical Infection in Practice 2020:7–8. Available from https://doi.org/10.1016/j.clinpr.2020.100038.

Horne DJ, Kohli M, Zifodya JS, Schiller I, Dendukuri N, Tollefson D, et al. Xpert MTB/RIF and Xpert MTB/RIF ultra for pulmonary tuberculosis and rifampicin resistance in adults. Cochrane Database of Systematic Reviews 2019. Available from https://doi.org/10.1002/14651858.cd009593.pub4.

How Are QFT-Plus Test Results Interpreted?; 2013. <https://www.qiagen.com/us/tb-testing/what-is-quantiferon/how-does-qft-work/quantiferon-tb-test-result-interpretation>.

Interpreting Xpert MTB/RIF Assay Results. n.d. <https://www.cdc.gov/tb/php/laboratory-information/xpert-mtb-rif-assay.html>.

Ioannidis P, Papaventsis D, Karabela S, Nikolaou S, Panagi M, Raftopoulou E, et al. Cepheid geneXpert MTB/RIF assay for Mycobacterium tuberculosis detection and rifampin resistance identification in patients with substantial clinical indications of tuberculosis and smear-negative microscopy results. Journal of Clinical Microbiology 2011;49(8):3068–70. Available from https://doi.org/10.1128/JCM.00718-11.

Jing W, Jiang X, Zhao W, Liu S, Cheng X, Sui G. Microfluidic platform for direct capture and analysis of airborne mycobacterium tuberculosis. Analytical Chemistry 2014;86(12):5815–21. Available from https://doi.org/10.1021/ac500578h.

Kadkhoda K, Dee M, Simeone J. Comparison of LIAISON quantiFERON-TB gold plus with quantiFERON-TB gold plus. Microbiology Spectrum 2023;11(1). Available from https://doi.org/10.1128/spectrum.04754-22.

Lewinsohn DM, Leonard MK, Lobue PA, Cohn DL, Daley CL, Desmond E, et al. Official American thoracic society/Infectious diseases society of America/Centers for disease control and prevention clinical practice guidelines: diagnosis of tuberculosis in adults and children. Clinical Infectious Diseases 2017;64(2):111–5. Available from https://doi.org/10.1093/cid/ciw778.

Lombardi G, Petrucci R, Corsini I, Reggiani MLB, Visciotti F, Bernardi F, et al. Quantitative analysis of gamma interferon release assay response in children with latent and active tuberculosis. Journal of Clinical Microbiology 2018;56(2). Available from https://doi.org/10.1128/JCM.01360-17.

Made-in-India TB Diagnostic Tests Validated by ICMR; Offer Cheaper, Faster Detection. n.d. Available from https://www.business-standard.com/health/made-in-india-tb-tests-validated-by-icmr-offer-cheaper-faster-detection-125100500186_1.html

Mahmoudi S, Nourazar S. Evaluating the diagnostic accuracy of QIAreach QuantiFERON-TB compared to QuantiFERON-TB gold plus for tuberculosis: a systematic review and meta-analysis. Scientific Reports 2024;14(1). Available from https://doi.org/10.1038/s41598-024-65663-4.

Ng KCS, van Deun A, Meehan CJ, Torrea G, Driesen M, Gabriëls S, et al. Xpert ultra can unambiguously identify specific rifampin resistance-conferring mutations. Journal of Clinical Microbiology 2018;56(9):e00686-18. Available from 10.1128/JCM.00686-18.

Ong DLT, Abdullah DA, Loh DTC, Raju DCS, Chong MJ, Atiya DN. Diagnostic utility of auramine O smear microscopy for the detection of non-tuberculous mycobacteria vs Mycobacterium tuberculosis complex in adult clinical samples: a 3-year retrospective cohort study (2019–2021). International Journal of Infectious Diseases 2025;152:107551. Available from https://doi.org/10.1016/j.ijid.2024.107551.

Phd A, Albert H, Nicol P, Boehme CC, Nicol MP, Nabeta P, et al. Feasibility, diagnostic accuracy, and eff ectiveness of decentralised use of the Xpert MTB/RIF test for diagnosis of tuberculosis and multidrug resistance: a multicentre implementation study. The Lancet 2011;377:1495–505. Available from https://doi.org/10.1016/S0140.

QuantiFERON-TB Gold Plus (QFT-Plus). n.d.

Saeed M, Ahmad M, Iram S, Riaz S, Akhtar M, Aslam M. GeneXpert technology: a breakthrough for the diagnosis of tuberculous pericarditis and pleuritis in less than 2 hours. Saudi Medical Journal 2017;38(7):699–705. Available from https://doi.org/10.15537/smj.2017.7.17694.

Salimiyan Rizi K, Aryan E, Meshkat Z, Ranjbar G, Sankian M, Ghazvini K, et al. The overview and perspectives of biosensors and Mycobacterium tuberculosis: a systematic review. Journal of Cellular Physiology 2021;236(3):1730–50. Available from https://doi.org/10.1002/jcp.30007.

Sample to Insight; 2023. <www.qiagen.com>.

Sasikumar C, Utpat K, Desai U, Joshi J. The role of genexpert in the diagnosis of Mycobacterium tuberculosis. Advances in Respiratory Medicine 2020;88(3):184–8. Available from https://doi.org/10.5603/ARM.2020.0102.

Sd Biosensor_ Products. n.d.

Stevens WS, Scott L, Noble L, Gous N, Dheda K. Impact of the geneXpert MTB/RIF technology on tuberculosis control. Microbiology Spectrum 2017;5(1). Available from https://doi.org/10.1128/microbiolspec.tbtb2-0040-2016.

The New GeneXpert® System. n.d.

Torati SR, Reddy V, Yoon SS, Kim CG. Electrochemical biosensor for Mycobacterium tuberculosis DNA detection based on gold nanotubes array electrode platform. Biosensors and Bioelectronics 2016;78:483–8. Available from https://doi.org/10.1016/j.bios.2015.11.098.

Trademark, Patents and Copyright Statements; 2020.

Truenat M.T.B., n.d, MTB Plus and MTB-RIF Dx assays_ TB Knowledge Sharing.

Uppe A, Sharma A, Sawant S, Gupta D, Nair G. Comparison studies of GENXPERT versus MGIT culture of bronchial lavage in sputum smear negative suspected pulmonary tuberculosis. American Journal of Infectious Diseases and Microbiology 2018;6(2):57–61. Available from https://doi.org/10.12691/ajidm-6-2-4.

Vitro, I. GeneXpert® Dx Reference Guide; 2017.

Vo LNQ, Tran TTP, Pham HQ, Nguyen HT, Doan HT, Truong HT, et al. Comparative performance evaluation of QIAreach QuantiFERON-TB and tuberculin skin test for diagnosis of tuberculosis infection in Viet Nam. Scientific Reports 2023;13(1). Available from https://doi.org/10.1038/s41598-023-42515-1.

World Health Organization. Automated Real-Time Nucleic Acid Amplification Technology for Rapid and Simultaneous Detection of Tuberculosis and Rifampicin Resistance: Xpert MTB/RIF Assay for the Diagnosis of Pulmonary and Extrapulmonary TB in Adults and Children. Geneva: World Health Organization; 2013. Available from https://www.ncbi.nlm.nih.gov/books/NBK258608/.

Xpert® MTB/RIF Ultra Instructions For Use; 2019.

Xpert MTB/RIF Implementation Manual: Technical and Operational "How-to": Practical Considerations; World Health Organization, Global TB Programme, 2014.

Xpert MTB_RIF and Xpert MTB_RIF Ultra assays_ TB Knowledge Sharing. n.d.

Xpert® MTB/RIF Instructions for Use For Use with GeneXpert® System with Touchscreen; 2013.

Yancey, J.R. QuantiFERON-TB Gold+ for the Diagnosis of Mycobacterium Tuberculosis Infection; 2021; Vol. 103. <www.aafp.org/afpAmericanFamilyPhysician177>.

Zeka AN, Tasbakan S, Cavusoglu C. Evaluation of the GeneXpert MTB/RIF assay for rapid diagnosis of tuberculosis and detection of rifampin resistance in pulmonary and extrapulmonary specimens. Journal of Clinical Microbiology 2011;49(12):4138–41. Available from https://doi.org/10.1128/JCM.05434-11.

Further reading

510(k) DECISION SUMMARY A. 510(k) Number: K131706 B. Purpose for Submission: De Novo Request for Evaluation of Automatic Class III Designation for the Xpert® MTB/RIF Assay for Use with the GeneXpert® Instrument Systems, Including the GeneXpert® Diagnostic (Dx) Systems and the GeneXpert® Infinity Systems. n.d.

CHAPTER 12

Emerging technologies in advancing *Mycobacterium tuberculosis* diagnostics

Sneh Nema and Neeraj Dwivedi
Academic of Scientific and Innovative Research (AcSIR), CSIR-Advanced Materials and Processes Research Institute, Bhopal, Madhya Pradesh, India

12.1 Introduction

Emerging technologies in tuberculosis (TB) diagnosis are mitigating some of the persistent challenges associated with conventional methods such as poor sensitivity, prolonged turnaround times, and limited accessibility in resource-constrained settings (Steingart et al., 2006; Mandalakas et al., 2011; Yang et al., 2025). Recent breakthroughs in computational sciences, digital health, biosensing platforms, and high-throughput biological data acquisition have altogether accelerated next-generation diagnostic strategies toward improved early detection, disease monitoring, and treatment outcomes (Yadav et al., 2024). Among these advancements, artificial intelligence (AI) has gained prominent attention primarily owing to the fact that this tool has the capabilities to work on complex high-dimensional datasets and extract clinically relevant patterns with minimal human intervention (Hansun et al., 2025). AI-driven approaches, such as machine learning (ML) and deep learning (DL), have immense potential for overcoming the subjectivity, inter-observer variability, as well as scalability issues that are inherent to traditional methods for the diagnosis of TB. These methods also facilitate the automated interpretation of diverse diagnostic data streams with faster, more accurate, and detailed detection and prognosis tools (Liang et al., 2022).

The term AI was coined for the first time by John McCarthy during the Dartmouth Conference in 1956 (Muthukrishnan et al., 2020). It is defined as a broad domain of computer science that enables computers and machines to simulate human cognitive functions such as learning, decision-making, problem-solving, pattern recognition, and language understanding capabilities (Yang et al., 2019). It has proven to be an effective approach and has created a profound impact in areas such as image analysis, natural language processing, large-scale data processing, computer vision, as well as within the medical field (Yu et al., 2018). In this arena, ML has played a major role because it has enabled models to learn patterns from biomedical data and refine predictive models without explicit programming. DL, a subset of ML, relies on multilayer neural

Advances in Biosensor Technology for Mycobacterium Tuberculosis Detection
DOI: https://doi.org/10.1016/B978-0-443-44589-7.00016-9

networks, takes it one step further by automatically discovering rich, hierarchical features from high-dimensional datasets, especially medical images, with minimum human interference (Hansun et al., 2023). Fig. 12.1 depicts the relationship of AI, ML, and DL, along with their short definitions.

With the advent of advanced computer algorithms, these technologies are now leveraging diverse data types, including chest imaging, cough acoustics, genomic and transcriptomic signatures, proteomic, and metabolomics biomarkers, and novel biosensor outputs to deliver more precise, automated, and rapid diagnosis for TB (Hansun et al., 2025; Menon and Koura, 2025). For example, AIRIS-TB, a model evaluated on more than one million CXRs, achieved an area under curve (AUC) of 98.51% and outcompeted radiologist performance, hence proving scalability and the effectiveness of AI at a population level for TB screening (Munjal et al., 2025). AI-enabled technologies have reinforced image interpretation, automated sputum smear microscopy, enhanced cough sound processing, and accelerated molecular and multiomics-based biomarker discovery. Particularly, integrated datasets such as mycobacterial computed tomography (CT) datasets comprising over 1300 scans with clinical metadata, have been developed, reflecting ongoing efforts to train more robust models that can distinguish TB from other pulmonary conditions (Han et al., 2025b). The development of AI-integrated point-of-care biosensors, on the other hand, works in parallel and enables complex physiological signals to be preprocessed, filtered, and interpreted in near real-time conditions (Flynn and Chang, 2024). Along with telemedicine platforms, such

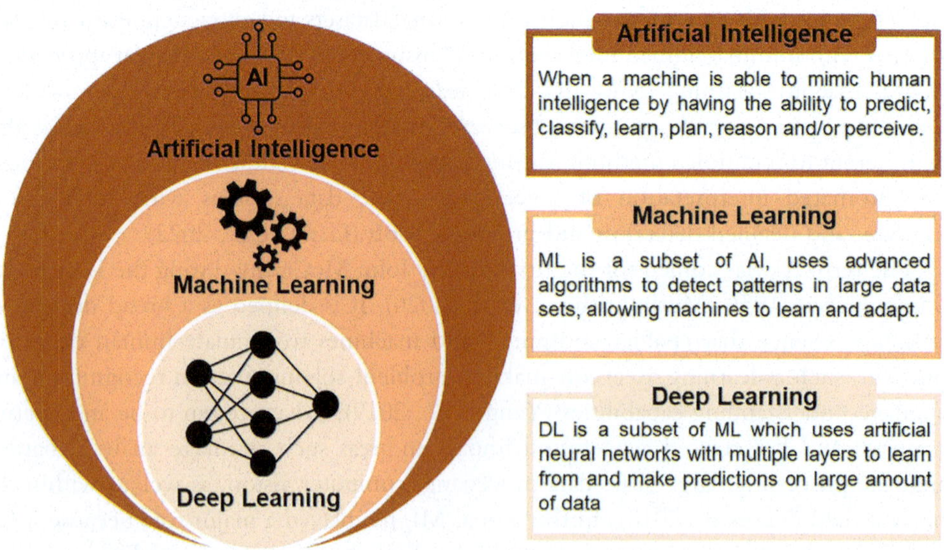

Figure 12.1 Relationship between AI, ML, and DL, along with their brief definitions. *AI*, artificial intelligence; *ML*, machine learning; *DL*, deep learning.

technologies extend diagnostic access in underserved and remote communities. Such technologies are thus in line with global TB elimination strategies and represent a route to more inclusive healthcare delivery (Guseva et al., 2023).

This chapter covers a holistic overview of the fundamentals and applications of AI in TB detection, from imaging to acoustic epidemiology and from multiomics approaches to biosensor integration and telemedicine, and discusses associated challenges and future prospects for global TB management.

12.2 Historical overview of artificial intelligence integration in tuberculosis diagnosis

Many researchers have employed AI in the domain of TB diagnosis in order to overcome the inherent challenges in predicting and accurately assessing the disease. The advent of neural networks (such as the Perceptron (Rosenblatt, 1958) and its advanced variants) brought the concept of pattern recognition; hence, they are applied to identify structural patterns in CXR images for diagnosis purpose (Abe et al., 1976; Maliwan et al., 1991). Artificial neural networks (ANNs) began to influence TB diagnosis during the early 1990s through their pattern-recognition application. One of the earliest models was built to distinguish between nine types of interstitial lung diseases, including TB. The training data comprised ten cases of each of the nine categories of lung disease, the model showed noteworthy diagnostic capability, indicating that ANNs held significant promise for computer-aided lung disease diagnosis (Asada et al., 1990). The earliest automated neural method for detecting TB bacillus in auramine-stained sputum smears was developed in 1998. With a sensitivity of 93.5%, it was postulated to provide rapid and accurate TB diagnosis while simultaneously reducing health hazards for laboratory personnel handling smear slides (Veropoulos et al., 2013). Similarly, in 1999, the first neural network (NN) was introduced using a general regression neural network (GRNN) for active TB diagnosis. This model employed 21 clinical parameters as input features and demonstrated a sensitivity of 100% with a specificity of 72% for detecting active TB (El-Solh et al., 1999). Other than GRNN, some other models such as support vector model (SVM) (Hearst et al., 1998), decision tree (DT) (El-Solh et al., 1997), and random forest (RF) (Ho, 1995), were also designed and applied for TB diagnosis.

One of the major limitations of these early models was the difficulty in finding and exploiting diagnostic features with enough accuracy to make reliable predictions. Beyond these constraints, much research has been dedicated to DL-driven approaches. DL gained prominence around the beginning of the 2010s as the integration of NNs with methods such as genetic algorithms (Osman et al., 2010), and fuzzy logic (Semogan et al., 2011) came into use for TB diagnosis. The ability of DL models to analyze, predict, and interpret complex, heterogeneous datasets has enabled new opportunities for addressing several challenges related to TB, including drug resistance. For example, in 2017, a DL model was designed to differentiate multidrug-resistant TB from drug-sensitive TB using CT scan images, with an accuracy of 91.11% (Gao and Qian, 2018). Other applications of DL in the diagnosis of TB covered the prediction of the severity of TB infection from pulmonary CT images, enabling fast screening with the analysis of chest radiographs and supporting the development of automated diagnostic platforms such as CAD4TB, qXR, inferRead DR, and lunit INSIGHT chest for effective and scalable TB diagnosis.

12.3 Significance of artificial intelligence in early detection of tuberculosis

AI has substantially revolutionized the diagnosis of TB by facilitating advancements in early identification and pattern recognition in clinical data and medical imaging (Srivastava et al., 2025). AI-powered systems provide rapid, consistent, and scalable screening and triage solutions in high-burden areas with limited access to experienced radiologists and laboratory equipment. DL-based computer-aided detection (CAD) systems have shown diagnostic accuracy matching, and sometimes surpassing, that of experienced clinicians in interpreting CXRs, hence aiding prompt case detection and reducing diagnostic delays (Lakhani and Sundaram, 2017). Further, AI algorithms can integrate heterogeneous datasets ranging from radiographic images and electronic health records to cough sound analysis for the identification of subtle early-stage illness patterns that may not be noticeable in conventional methods. Such methods increase sensitivity for asymptomatic or paucibacillary cases and help prioritize patients for confirmatory molecular testing, economizing costs and easing resource demands in resource-constrained settings (Ahmad et al., 2019). Fig. 12.2 shows the pipeline of AI-assisted TB detection and management from multimodal data to clinical outcomes. AI-powered CAD solutions recommended by the WHO have proven particularly effective in mass screening contexts such as community health campaigns, migrant health checks, and prison screening programs

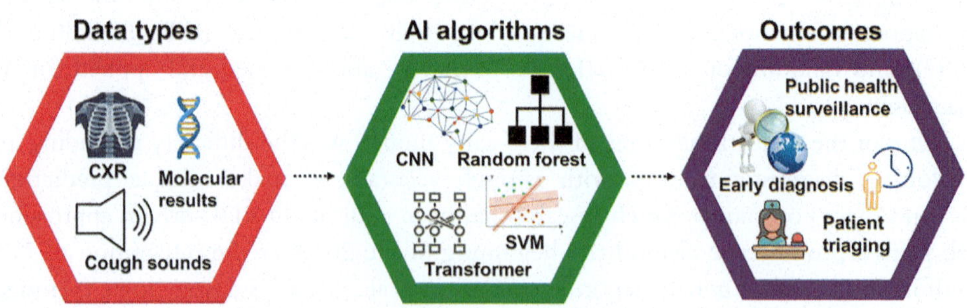

Figure 12.2 Overview of an AI-based framework for TB diagnosis and management. *AI*, artificial intelligence; *TB*, tuberculosis

(World Health Organization, 2021). CAD4TB (Delft Imaging) and qXR (Qure.ai) serve as primary screening methods for TB in community and clinical settings in national TB programs, including extensive deployments in South Africa, Pakistan, and India. (Creswell et al., 2023; Kik et al., 2022). AI-driven strategies therefore immediately contribute to the goals of the WHO end TB strategy to cut the TB incidence and death through early and equitable access to diagnosis, minimizing diagnostic delays and optimizing resource allocation (World Health Organization, 2024).

12.4 Machine learning and deep learning foundations in tuberculosis detection

In TB diagnostics, AI heavily relies on the principles of ML and DL, which allow computational systems to automatically identify patterns in complex biomedical data (Singh et al., 2022). ML, broadly viewed as the most pivotal subdomain of AI, has acquired remarkable momentum within the last couple of decades, although its roots trace back to the conceptualization of ANNs in the 1940s (McCulloch and Pitts, 1943) and Rosenblatt's Perceptron in 1958 (Rosenblatt, 1958). Fundamentally, ML consists of a collection of computational methods that are capable of learning from input data, building a model, and iteratively refining performance to support informed decision-making (Helm et al., 2020; Hosny et al., 2018). ML deploys a variety of algorithms, including SVM, RF, and gradient boosting, which have been used in classifying patient data or imaging details based on identifying and applying learned decision boundaries (Varoquaux and Cheplygina, 2022). Liu et al., for example, developed an NN-based system for TB diagnosis using chest radiographs and achieved accuracy scores ranging between 89.0% to 96.1% for three datasets (Liu et al., 2018). Similarly, Khan et al. presented an NN model that can differentiate between positive TB and negative TB cases for over 12,600 patient records with an accuracy of more than 94% (Khan et al., 2019). Comparative studies by Ghanshala et al. revealed that NN outperformed SVM, k-nearest neighbors (kNN), and RF models, with an accuracy of 80.45% in TB detection (Ghanshala et al., 2020). Chandra et al. further extended this area through the development of a computerized technique to detect CXR abnormalities, including pleural effusion, infiltration, dense consolidation, fibrosis, hila enlargement, etc. associated with TB, using SVM with hierarchical feature extraction, with accuracies ranging between 95.6% and 99.4% on two public datasets (Chandra et al., 2020). Collectively, these efforts have shown ways through which ML has advanced from conceptual foundations to clinically relevant applications and thus formed the basis toward the development of more advanced DL methods in TB diagnostics.

Building on these advances, DL, a specialized branch of ML, has gained prominence because of powerful computational resources and the abundance of large-scale data. It is modeled on the structure and function of the human brain and utilizes multilayered

ANNs capable of automatic learning and optimization of features from complex, high-dimensional data, with minimum preprocessing and manual feature engineering, while offering higher accuracy compared to most traditional ML methods (Shickel et al., 2018; Ho et al., 2019). For example, convolutional neural networks (CNNs) are an efficient architecture for CXR image interpretation, able to extract spatial features ranging from simple edges to complex patterns of lung abnormalities indicative of TB infection (Liu et al., 2017). Recurrent neural networks (RNNs) and their advanced architectures, such as long short-term memory (LSTM) networks are appropriate for analyzing sequential or time-varying data, such as cough sound recordings, because of their ability to effectively capture temporal variability and minute acoustic features relevant to TB diagnosis (Botha et al., 2018). Audio-based LSTM and image-based CNN outputs can be combined using hybrid frameworks that have become one of the effective approaches to improve the diagnostic performance (Rajasekar et al., 2024).

Empirical studies confirm the utility of DL in TB diagnostics. Lakhani and Sundaram evaluated deep convolutional neural networks (DCNNs) for TB detection using an ensemble of AlexNet and GoogLeNet, achieving an accuracy of 98.7% (Lakhani and Sundaram, 2017). Similarly, Hooda et al. presented an ensemble DL-based system based on AlexNet, GoogLeNet, ResNet, and four different datasets, which attained an accuracy of 88.24% (Hooda et al., 2019). Heo et al. further extended the application of CNNs by evaluating five CNN architectures (VGG19, InceptionV3, ResNet50, DenseNet121, and InceptionResNetV2) on chest radiographs from workers' health examinations and found that combining VGC-19 with demographic variables (age, weight, height, and gender) improved performance compared with CXR data alone (Heo et al., 2019). Similarly, Sathitratanacheewin et al. developed a deep CNN based on InceptionV3 for TB CXR classification, reporting AUC value between 0.7054 to 0.8502 across two datasets (Sathitratanacheewin et al., 2020).

The strength of ML and DL lies in their adaptability and scalability. These algorithms, while being trained on large, diverse, well-chosen datasets, can be used for real-time screening, in order to offer quick, accurate, and affordable diagnostic support (Hansun et al., 2023). Moreover, through continuous learning, new data over time gets integrated, enhancing their robustness across different demographics and geographies. These technologies work together to enable imaging, acoustic epidemiology, and multimodal integration applications, while together they form a core on which AI-driven TB diagnosis is based (Xu et al., 2024).

12.5 Learning paradigms in artificial intelligence-based tuberculosis diagnostics

Supervised, unsupervised, and semisupervised learning represent the principal approaches that aid the development of AI models for TB detection. The choice of

Table 12.1 The advantages and disadvantages of the different machine learning approaches applied in tuberculosis diagnosis.

Sr. No.	Learning techniques	Advantages	Disadvantages	References
1.	Supervised learning	• High precision • Aids in problem solving by leveraging previously learned patterns	• There must be labeled training data • Well-curated and sufficiently large datasets are required	Raymond and Medina (2018)
2.	Unsupervised learning	This approach works best with raw or unknown data.	• Less precise. • The output data cannot be specified, in contrast to the supervised approach.	Weber et al. (2000)
3.	Semisupervised learning	Both labeled and unlabeled data can be used.	Cannot handle unseen data.	van Engelen and Hoos (2020)

approach depends on the diagnostic task, with each method offering distinct advantages and limitations (Table 12.1). Supervised learning relies on labeled datasets to train models in identifying particular disease-related features. It remains the most widely used approach in TB diagnostics, which uses labeled datasets with an input (such as a CXR image or cough audio file) and an output label (e.g., "TB positive" or "TB negative") (Gharamaleki et al., 2024). Classification and regression represent two fundamental tasks within supervised learning. In regression, the model is trained using datasets with continuous output variables, aiming to learn the relationship between inputs and predict numeric values. In contrast, in classification, the model is trained on datasets where the output is categorical, and the goal is mapping inputs to discrete classes or categories for prediction (Li, 2019). Specific techniques that are included are logistic regression (LR), NN, DT, RF and DL. Notably, the majority of ML algorithms possess variants tailored for both classification and regression tasks, making the selection of an appropriate algorithm dependent on the specific nature of the problem being addressed.

Unsupervised learning is applied when only input data are available, with no corresponding output labels (Weber et al., 2000). In these types of systems, the algorithm discovers patterns or inherent structures in the data itself without prior labeling of the data (Meier et al., 2021). The primary distinction between unsupervised and supervised learning is that the unsupervised learning lacks a feedback signal to guide signal evaluation, which typically results in decreased accuracy and reduced computational complexity than the supervised approach can offer (Karmani et al., 2020). In recent years, supervised and unsupervised learning have been utilized in Bogota, DC, Colombia, to develop two ANN-based models, one for disease diagnosis and another for data clustering. The dataset was constructed based on variables such as sex, age, AIDS status, and other relevant medical conditions. The model for disease diagnosis used supervised learning, whereas the model used for data clustering applied unsupervised learning based on available information. The models demonstrated a sensitivity of 97% and a specificity of 71% for TB diagnosis, offering the benefits of rapid and low computational cost over traditional approaches (Orjuela-Cañón et al., 2018).

Semisupervised learning falls between the two preceding approaches (Padhy et al., 2019). This AI type uses vast datasets where only a limited amount of data has been labeled. Therefore, in many developing countries, it is highly challenging and time-consuming to gather and label TB patient data, however unlabeled data, such as unconfirmed CXRs, are abundant. These methods take advantage of this imbalance by using a limited set of labeled data combined with larger set of unlabeled data to train models, improving their generalization without the need to exhaustively annotate (Oliver et al., 2018). Active learning is a specific type of semisupervised learning designed to address the challenge of insufficient data through various strategies (Reker et al., 2017). For instance, the model can request the user or another source of information to label unlabeled data for which the input data is uncertain. In this regard, an active learning-based CAD system for identifying TB was proposed, which enhanced model accuracy and reduced labeling efforts by minimizing uncertainty during training (Melendez et al., 2016).

Reinforcement learning (RL) represents yet another important paradigm in AI, extending the powers of both supervised and unsupervised methods by allowing learning through interaction and feedback. It involves the training of a model through direct interaction with an environment, wherein it learns a policy that maximizes total cumulative rewards based on received feedback in terms of rewards or penalties for taken actions (Morales and Escalante, 2022). These different learning paradigms comprise the essential structure for implementing AI in TB diagnostics. Application of all these within workflows related to TB research will enhance diagnostic precision and deepen our knowledge of TB.

12.6 Artificial intelligence applications in tuberculosis diagnostics

AI in TB diagnostics has progressed from a theoretical idea to practical implementation across multiple diagnostic fields. At both the clinical and field levels, the integration of DL, ML, and multimodal data fusion significantly has improved diagnostic accuracy, effectiveness, and accessibility. The uniqueness in these emerging fields consists in their applications based on various modalities, including imaging-based diagnostics, cough sound analysis, AI-based microscopy, multiomics integration, biosensing, and telemedicine platforms. Overall, these innovations tend to mark a clear shift toward data-driven and automated TB diagnostics and management. Schematic illustrations of different applications of AI in TB detection are shown in Fig. 12.3.

12.6.1 Artificial intelligence in imaging-based tuberculosis detection

Imaging has long been a cornerstone in TB diagnosis, with CXRs and CT scans being the most widely used modalities. While CXR gives rapid and low-cost screening, CT provides high-resolution images, which are crucial for a more informed assessment of pulmonary lesions and the extent of disease (Skoura et al., 2015). CT remains one of the most efficient methods for the detection of lung pathologies (Li and Xia, 2020), owing to its very strong sensitivity in the visualization of pulmonary structures such as lobes, trachea, and bronchi. Its ability to capture detailed images rapidly enables efficient identification of abnormalities and quick evaluation of structural alterations within the lungs. Recent studies have adapted DL approaches over multislice CT datasets using 3D-CNNs and hybrid segmentation-classification frameworks for automatic lesion identification, quantification of disease burden, and differentiation of active TB from inactive or healed disease, and from other pulmonary disorders (Li et al., 2024). For example, Han et al. developed and validated a 3D-CNN model using chest CT for

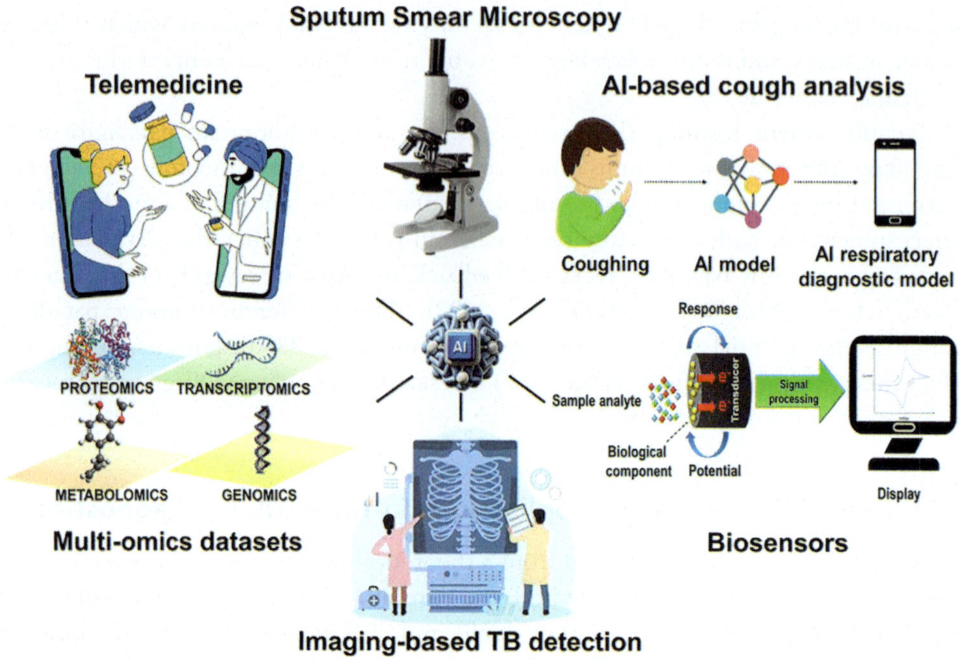

Figure 12.3 Overview of AI integration in TB detection and diagnosis. *AI*, artificial intelligence; *TB*, tuberculosis.

distinguishing active pulmonary TB from community-acquired pneumonia with excellent diagnostic precision (Han et al., 2023). Similarly, Ma et al. developed a U-Net-based AI system for active TB detection from multislice CT images of 846 patients, reporting an AUC of 0.98 with sensitivity, specificity and accuracy of 0.964, 0.971, and 0.968 respectively (Ma et al., 2020).

Complementing these DL approaches, radiomics has emerged as a powerful noninvasive technique that extracts large-scale quantitative features from medical images, thereby converting into mineable data representations (Mayerhoefer et al., 2020). Li et al. developed CT-based radiomics models based on the tree-in-bud sign and nodules to predict multidrug-resistant TB (Li et al., 2023b). Moreover, the combination of radiomics with DL features and clinical or laboratory data within multimodal platforms further developed diagnostic precision and predictability. A positron emission tomography (PET)/CT-based radiomics nanogram, coupled with quantitative and semantic features, successfully differentiated active TB from lung cancer and thus became an accurate, noninvasive diagnostic tool (Du et al., 2021).

WHO recommends CXRs as the first-line screening and triage tool for TB. But CXRs also exhibit significant pitfalls with regard to interobserver variability, dependence on reader expertise, and scarcity of radiologists, especially within TB-epidemic and resource-constrained settings (Samandari et al., 2011). These challenges have resulted in investigations on automated solutions that laid the foundation for AI- and ML-based imaging diagnosis.

AI methods, essentially DL models based on CNNs, have revealed phenomenal capability regarding automatic learning of complex image features and classifying TB into classes with high accuracy (Kotei and Thirunavukarasu, 2024). Early CAD systems relied on handcrafted imaging features combined with traditional ML classifiers; still such methods often failed to generalize among different populations and differences in imaging conditions, leading to inconsistencies in accuracy (Qin et al., 2018). In contrast, use of DCNNs architectures such as ResNet, VGG, and AlexNet has consistently produced significantly better diagnostic outcomes. For example, a 2017 study evaluated two DCNNs models, namely AlexNet and GoogLeNet, in both pretrained and untrained conditions for the detection of TB using CXRs. The study reported the capability of the DL technique using DCNNs for accurate differentiation of TB patients from healthy individuals, with an AUC value of 0.99 (Lakhani and Sundaram, 2017). Another large-scale retrospective study conducted using almost 10,000 CXRs collected from China reported accuracies above 94%, and ResNet was shown to have the best diagnostic performance, as evidenced by its 96.73% accuracy, 95.50% sensitivity, 98.05% specificity, and AUC value of 0.9944 (Nijiati et al., 2022).

There have been a number of DL-based CAD tools that evolved from research prototypes to commercial products, supported by regulatory clearances for clinical use (Table 12.2). In the trial on 23,954 CXRs, five commercial AI algorithms substantially outperformed radiologists, of which qXR and CAD4TB met the criteria of WHO for specificity at 90% sensitivity. Notably, these systems halved the need for confirmatory tests while maintaining high sensitivity (Qin et al., 2021). To overcome infrastructure issues, researchers have developed computationally efficient lightweight CNN models such as LightTBNet, which have been optimized specifically for mobile X-ray devices, reflecting excellent diagnostic accuracy even on low-resource hardware (Raziq et al., 2022). Systematic reviews and meta-analyses strengthened these facts by validating that

Table 12.2 Commercially available AI-based CAD solutions for TB detection.

AI system	Developer	Methodology	Commercial availability
CAD4TB version 6	DIAG, Radbound University, The Netherlands	DL	Yes
qXR	Qure.AI, India	DL	Yes
Lunit INSIGHT CXR	Lunit, South Korea	DL	Yes

AI, artificial intelligence; *CAD*, computer-aided detection; *TB*, tuberculosis; *DL*, deep learning.

CNN-based approaches (e.g., ResNet, VGG, AlexNet, GoogLeNet) consistently enhanced the diagnostic performance in CXR-based TB identification (Harris et al., 2019). A multicountry evaluation across seven countries (India, the Philippines, Madagascar, South Africa, Vietnam, Uganda, and Tanzania) validated the robustness of five CAD systems involving nearly 4000 participants and reached target accuracy rates with specificity as high as 73.8% at 90% sensitivity. Alhough performance was generally varying by country and patient subgroups, but setting-specific thresholds can optimize the results (Worodria et al., 2024).

Beyond the established CAD4TB and qXR commercial systems, recent research has proposed advanced AI architectures that achieve near-perfect performance in experimental settings. For example, the TB-Net model, a tailored self-attention DCNN, reached an excellent accuracy of 99.86%, with sensitivity and specificity approaching 100% and 99.71%, respectively. The decision-making process of the model closely mimicked the radiologists' interpretation, which was validated by the experts' explainability assessments (Wong et al., 2022). Similarly, segmentation-assisted CNNs that target the lung region have boosted diagnostic accuracy as high as 99.9% and proved the value of region-targeted analysis in TB diagnosis (Nafisah and Muhammad, 2024). Ensemble approaches, combining multiple models such as EfficientNetB0 and DenseNet201, further reinforce the diagnostic robustness through leveraging complementary feature extraction capabilities (Oloko-Oba and Viriri, 2021; Thirunavukkarasu et al., 2024).

Several large-scale public health programs have started deploying AI-driven technologies for TB detection in real-world settings in India. In Nagpur, Maharashtra, mobile diagnostic vans equipped with AI-powered CXR and molecular testing have screened over 100,000 high risk individuals. The diagnostic outputs were linked to confirmatory molecular facilities and national TB support schemes such as Nikshay Poshan Yojana to ensure rapid diagnosis, treatment initiation, and patient benefit linkage. Independent implementation studies showed that AI-assisted CXR interpretation (notably with qXR by Qure.ai) specifically enhanced TB case detection by almost 16%, including cases missed by radiologists (Vijayan et al., 2023). Similarly, outreach efforts in urban Delhi have deployed portable AI-enabled X-ray systems, screening thousands so far and aspiring to cover millions in less privileged urban settings. These initiatives together illustrate how AI can accelerate early TB detection, enhance care pathways, and support national TB elimination strategies when embedded within public health infrastructure and linked to national programs.

12.6.2 Artificial intelligence-powered cough sound analysis for tuberculosis diagnosis

Cough remains one of the major clinical symptoms of TB, but traditionally its diagnostic potential has been limited by the difficulty of objectively quantifying subtle acoustic features. AI has dramatically changed the landscape by extracting and analyzing various complex temporal and spectral features related to cough sounds with great accuracy, thereby facilitating noninvasive and portable solutions for TB screening. The early studies, although suffering from relatively small sample sizes and only internal validation, results were nevertheless promising, with sensitivity and specificity levels surpassing 90% for distinguishing TB-related coughs from other respiratory diseases (Alqudaihi et al., 2021). A systemic review and meta-analysis of AI-based cough sound analysis for diagnosis of TB also reported a pooled sensitivity of 91% and a specificity of 89%, and overall an AUC of 0.95, demonstrating its diagnostic potential (Sahoo et al., 2025).

With increased data availability, large-scale projects accelerated progress in the domain. Recent success stories include initiatives such as the CODA TB DREAM Challenge and its related dataset, which collected 733,756 cough audio recordings of 2143 individuals across seven countries and annotated them with detailed microbiological, clinical, and demographic diagnostic information (Huddart et al., 2024). This resource has helped establish training and benchmarking high-performance AI models for TB screening, as well as supporting the potential of developing generalizable algorithms across diverse populations. Based on this dataset, ensemble models combining Mel-spectrogram features, 2D-CNNs, and XGBoost achieved an AUC of 0.88, that adheres to guidelines set by the WHO for proper screening (Suda, 2023).

Several AI tools, including HeAR, Swaasa AI, TBscreen, and Timbre, have been tested through internal validations and clinical pilots conducted in resource-constrained regions of India, South Africa, Kenya, and Zambia. Table 12.3 is a compilation of AI models for the early diagnosis of TB through cough sounds. The Swaasa AI platform reported an overall accuracy of 86.82% for pulmonary TB detection, with 84.67% specificity and 90.36% sensitivity in clinical validation studies (Yellapu et al., 2023). Similarly, TBscreen, a ResNet18-based classifier trained on over 43,000 passive cough recordings, distinguished TB from nonTB with moderate sensitivity (~70%) but with much better performance in patients with high bacterial loads or cavitary disease, which may indicate that passive cough features can capture both disease presence and severity (Sharma et al., 2024).

Table 12.3 Summary of AI models for TB detection using cough sounds, showing algorithm types, validation methods, and diagnostic performance across studies.

AI algorithm	Validation type	Performance	References
HeAR	Internal	AUC: 0.652	Baur et al. (2024)
TBscreen	Internal	Sensitivity: 70%	Sharma et al. (2024)
Swaasa	External	Sensitivity: 90.36%	Yellapu et al. (2023)
Timbre	External	Sensitivity: 80-83%	Pathri et al. (2022)

AI, artificial intelligence; *TB*, tuberculosis.

Most of these systems record cough sounds using microphones or smartphones, followed by preprocessing steps such as denoising, feature extraction, and signal enhancement. The output from the final model is then validated against established microbiological or clinical gold standards (Pahar et al., 2021). Although performance has been variable between systems, these data support the feasibility of AI-driven cough analysis as a scalable, low-cost triage approach in resource-limited settings. However, future work should be focused on standardized datasets, better methods for robust handling of noise in real-world conditions, and multicountry prospective evaluations to support routine integration into clinical practice (Hegde et al., 2024).

12.6.3 Artificial intelligence-assisted sputum smear microscopy for tuberculosis detection

Computer vision and DL-based sputum smear microscopy automate the detection and quantification of acid-fast bacilli (AFB) in Ziehl–Neelsen (ZN) or fluorescent-stained slides, overcoming limitations of manual reading such as reader fatigue, inter-operator variability, and reduced sensitivity at low bacillary counts while retaining microscopy's inherent speed and affordability in resource-constrained settings (Khare et al., 2019).

Recent clinical validation studies have demonstrated the potential of AI-based tools in routine workflows. A prospective, multicentric clinical trial across three hospitals in northern India compared an AI-based microscopy system against three expert microscopists and achieved sensitivity of 89.25%, specificity of 92.15%, and diagnostic accuracy of 91.53%, demonstrating the technology's strong potential as an adjunctive diagnostic tool (Gupta et al., 2023).

Advanced whole-slide and patch-level DL approaches have further enhanced AFB detection. Zurac et al. designed a model for AI-assisted detection system for whole-slide ZN images that is able to identify bacilli with 95.65% sensitivity, 98.33% accuracy, and 100% specificity, outperforming or matching pathologists, reducing errors, and saving one-third of examination time. Active learning with annotated positive slides will further improve robustness across different ZN stains and clinical settings (Zurac et al., 2022). Similarly, a cross-domain study trained on histopathology AFB datasets and directly tested on 1059 ZN-stained sputum smears achieved 100% sensitivity for high-confidence patches and 86.84% specificity. The modest specificity is caused by smear peculiarities such as dust contamination, uneven thickness, and lack of a coverslip, and proposed retraining on smear images to exceed 95% specificity for screening applications (Witarto et al., 2024).

ML has also been applied to quality control. A color-based ML algorithm using K-Means clustering was developed to screen sputum smears for adequacy based on Bartlett's criteria. Analysis of smartphone images of satisfactory and unsatisfactory smears showed that the model's cell counts closely matched manual counts and both were within acceptable limits (Sirohi et al., 2022). Furthermore, Aulia et al. have addressed the challenges of standardized grading by introducing microscopic TB detection through the development of a digitized whole-slide imaging database of ZN-stained sputum samples from Indonesian patients, called MIDTI, following WHO regulations. Using this dataset, a CAD system based on the YOLOv7 algorithm was developed to classify sputum smears into International Union Against Tuberculosis and Lung Disease (IUATLD) grades, providing a novel, automated approach for accurate and reliable TB diagnosis (Aulia et al., 2024).

Across the literature, the field has evolved from traditional image-processing and handcrafted feature-based ML models to advanced CNN techniques integrated with digital microscopy platforms and human-in-the-loop validation. These advancements have increased throughput, ensured standardized grading consistent with guidelines by the WHO/IUATLD, and delivered clinically reliable performance when coupled with culture or nucleic acid amplification test (NAAT) confirmation during routine practice (Huang et al., 2022a; Fu et al., 2022).

Digitization hardware also has a central role in enabling AI-assisted microscopy. High-throughput scanners and dedicated fluorescence AFB modules, including MetaSystems Metafer with NEON data management, produce consistent, analysis-ready images and have demonstrated strong concordance with manual fluorescence microscopy in retrospective studies, especially for assisted modes that preserve sensitivity with only slight decrements in specificity (Desruisseaux et al., 2024). At the point of care, low-cost alternatives, including smartphone based fluorescence adapters and compact digital microscopes such as CellScope, extend access to AI- enabled smear reading in peripheral laboratories and community settings (Tapley et al., 2013; Chang et al., 2012). However, to maintain high sensitivity with these low-cost solutions, requires precise control of focus, illumination, and mechanical alignment.

Recent studies using advanced DL architectures, such as Faster R-CNN detectors, U-Net segmenters, attention-residual networks, and multitask frameworks, have all shown very good high internal performance, although results vary with staining method, dataset characteristics, magnification, and evaluation protocols, pointing to a need for standardized benchmarking and large-scale external validation (Tamura et al., 2024). Overall, AI-assisted sputum smear microscopy is gradually shifting from early prototypes to clinical application, with future progress depends on large, standardized datasets, robust external validation, and seamless integration with existing laboratory infrastructure.

12.6.4 Artificial intelligence in multiomics data for tuberculosis detection

Beyond imaging and acoustic diagnostics, AI is also powerful in analyzing complex multiomics data for the identification of robust biomarkers and enhancements in diagnostics of TB infection. Integration of multiomics data such as genomics, transcriptomics, proteomics, and metabolomics has transformed biomedical science and provides profound integrated insights into disease mechanisms and potential therapeutic interventions (Ali, 2023). AI-assisted approaches open new avenues toward a systems-level understanding of TB biology by evaluating host and pathogen molecular profiles simultaneously to support the development of diagnostic tools, against such as latency, drug resistance, and clinical heterogeneity.

Genomics plays a significant role in identifying mutations in the *Mycobacterium tuberculosis* (MTB) genome that drive antimicrobial resistance (Cohen et al., 2019; Nimmo et al., 2022). ML models trained on whole-genome sequencing (WGS) data can rapidly predict resistance to first- and second-line drugs, providing genotypic susceptibility predictions significantly faster than traditional phenotypic testing workflows (Deelder et al., 2019). CNNs and gradient boosting algorithms have successfully classified multidrug-resistance and extensively drug-resistance TB strains with high accuracy (Green et al., 2022; Deelder et al., 2019). Based on these advances, recent multi-country evaluations of WGS combined with ML resistance prediction pipelines show consistent performance across different settings, thus serving as a great potential for clinical decision support and programmatic surveillance (Dixit et al., 2024).

Transcriptomic profiling of host immune responses has been equally instrumental in TB biomarker discovery (Singhania et al., 2018). Analysis of blood RNA expression signatures with AI techniques such as SVM, NNs, and ensemble classifiers can differentiate active TB from latent infection as well as from other respiratory diseases with high precision (Mann et al., 2024). Early studies indicated that whole-blood transcriptomic patterns could accurately distinguish TB cases from other diseases, especially among individuals with HIV coinfection (Kaforou et al., 2013). A multi-cohort study with 14 datasets comprising 2572 samples from 10 countries identified a concise three-gene blood signature (GBP5, DUSP3, and KLF2) distinguished active TB from latent infection, healthy individuals, and other diseases with global AUCs of approximately 0.84–0.90 across independent validations, and showed a decline during therapy, supporting its potential utility for both diagnosis and treatment monitoring (Sweeney et al., 2016). The inclusion of DL models, which incorporate multiple gene expression profiles, further boosted predictive performance, making it easier to utilize for early detection, risk stratification, and monitoring of treatment outcomes.

Further, by understanding the dynamics of changes in proteins and small molecules in host or bacterial models, proteomics and metabolomics approaches have not only helped to broaden the spectrum of biomarker discovery in TB but have also facilitated a comprehensive understanding of TB in a more holistic manner (Yu et al., 2023). ML-driven proteomic analysis was able to identify host immune proteins serving as disease markers and indicators of treatment response (Schiff et al., 2024; Zhou et al., 2025). Similarly, metabolomic profiling has identified diagnostic biomarkers that can reliably differentiate TB-infected individuals from healthy individuals when combined with AI algorithms such as RF and LASSO regression (Sun et al., 2025; Anh et al., 2024). More recently, the integration of metabolomics data with clinical markers through ML tools has also aided in the identification of key diagnostic metabolites, while urinary metabolomics profiling has shown the capacity to distinguish TB from nontuberculous mycobacteria, thus emphasizing further advancements in developing nonsputum based diagnostic tools (Anh et al., 2024; Hu et al., 2022). Altogether, AI-enabled integration of the genomic, transcriptomic, proteomic, and metabolomic layers presents a more holistic understanding of TB biology, whereas combined analysis increases the accuracy and specificity of candidate biomarkers for better diagnosis.

12.6.5 Artificial intelligence-integrated biosensor for tuberculosis detection

Biosensors represent a new generation of analytical devices that combine a biological sensing component with transducers, which convert biochemical recognition events into a readable signal. They have applications ranging from food quality control and environmental monitoring to diagnostics and drug screening (Mehrotra, 2016). Nonetheless, conventional biosensors often suffer from signal drift, matrix effects induced by complex clinical samples, nonspecific interactions, and loss of sensitivity caused by biofouling and interfering agents (Campuzano et al., 2020). In fact, the current advances in the biosensing techniques have been propelled extensively through the integration of nanotechnology, microfluidics, and AI, allowing the simultaneous detection of multiple analytes with high sensitivity and accuracy (Noor et al., 2023; Sajjan et al., 2025). Table 12.4 compares conventional and AI-assisted biosensors.

Advances in hardware, algorithms, and computational power have improved the uses of biosensors immensely. Currently, trends in applications based on biosensors include the use of intelligent, interconnected systems that integrate big data analytics, the Internet of Things (IoT), as well as AI (Jin et al., 2020; Cui et al., 2020). The present advanced approaches in ML and DL provide tremendous opportunities to enhance the real-time predictive capability of biosensors by rapidly processing large datasets, uncovering complex patterns and anomalies (Tafadzwa Mpofu and Mthunzi-Kufa, 2025; Kılıç et al., 2022). AI-driven systems are highly promising for the development of versatile monitoring platforms that are capable of making real-time decisions in dynamic environments (Bhaiyya et al., 2024). Further, AI also has strong potential to markedly enhance the specificity, sensitivity, and stability characteristics of biosensors through the elimination of unwanted noise and signal artifacts, thereby providing more precise and reliable measurements (Zhang et al., 2023; Şen et al., 2024).

By enabling sophisticated analytical capabilities, DL algorithms, particularly CNNs and RNNs, have revolutionized diverse fields. These include early-stage TB diagnosis

Table 12.4 Comparative analysis of traditional and artificial intelligence (AI)-integrated biosensors.

Feature	Traditional biosensors	AI-integrated biosensors
Calibration	Manual and static calibration	Performs dynamic recalibration based on continuously monitored data
Sensitivity	Constrained by sensor design and hardware	Improved through AI-driven optimization and predictive modeling
Data processing	Based on a calibration curve	Employs sophisticated analytical methods and pattern recognition for multiplexed data interpretation
Error handling	Manual	Automatic error identification and rectification
Decision support	Dependent on expert interpretation	Provides automated decision- making assistance
Result interpretability	Straightforward and easily interpretable outputs	Reduced transparency and complex to interpret
Robustness/ adaptability	Limited tolerance to environmental fluctuations	Capable of learning and adapting to complex and dynamic conditions
Personalization	One-size-fits-all	Enables personalized diagnostics through data driven learning
Cybersecurity	Minimal risk	Susceptible to software attacks and data breaches

in medical imaging, such as X-ray and CT scans and identifying environmental hazards or pathogens through smart sensors (Mahmoudi et al., 2022), and efficient feature extraction. Over the past decade, several CNN architectures have been developed to strengthen the analytical performance of a wide range of applications such as disease detection using wearable sensors (Huang et al., 2022b), analysis of the activity of the brain obtained from Electroencephalogram signals (Li et al., 2020), and real-time glucose monitoring during diabetes management (Ellahham, 2020). Emerging architectures such as 3D CNNs, Transformers, and deep image priors have considerably broadened the computational scope of AI-integrated biosensing platforms. Specifically, 3D CNNs, which represent an important extension of 2D CNNs designed to capture spatiotemporal information (Akkaş et al., 2025), and operate simultaneously across three spatial dimensions (Cho and Hwang, 2020). This feature makes them highly effective for analyzing volumetric medical data analysis, such as CT scans (Niyas et al., 2022) and continuous biosensor streams such as video recordings. In biosensing applications, this capability enables the modeling of complex temporal-spatial relationships, for example, monitoring metabolic variations over time (Hu et al., 2024), thus allowing for more precise and context-aware classifications, for example, distinguishing pathological changes across different organs (Singh et al., 2021).

In view of such advances, CRISPR-Cas technology presents promising avenues for the rapid and precise detection of infectious diseases such as TB. CRISPR-Cas13a systems, characterized by their collateral cleavage activity, have been adapted into biosensing systems that enable robust and target-specific identification of MTB nucleic acid sequences (Razavi et al., 2024; Kumaran et al., 2023). These approaches usually combine isothermal amplification methods such as recombinase polymerase amplification (RPA) or loop-mediated isothermal amplification (LAMP) with Cas-based detection modules such as SHERLOCK or DETECTR for rapid point-of-care (PoC) workflows (Gootenberg et al., 2018). ML approaches add to the capabilities of CRISPR diagnostics through the processing and interpretation of high-throughput outputs such as fluorescence signals, cleavage kinetics, or microfluidic images and classify the outcomes using algorithms such as SVMs, RFs, or CNNs (Ren et al., 2023; Huang et al., 2023). These models improve the overall robustness of the assay by ensuring optimal threshold settings and separating true positives from background noise. CRISPR-Cas technology has also been researched for drug susceptibility testing in TB, wherein ML-driven integration of genomic, phenotypic, and clinical data improves the analysis of resistance-associated mutations (Kaminski et al., 2021).

The application of ML in point-of-care-testing (PoCT) biosensors has enhanced the quality of predictive analytics, interpretation of results, and tailored healthcare, so that instant diagnosis is now possible along with a personalized treatment plan (Han et al., 2025a). The future prospects of ML are extremely bright for the PoCT biosensors to carry out automatic device calibration and maintenance, critical statistical analysis for early detection of diseases, and enhanced accuracy in determining complex biomarkers. In addition, it would promote ease of data transfer for innovation and thereby enhance the quality of healthcare services (Bhaiyya et al., 2024).

In field-specific applications, such as biosensing technologies, electrochemical biosensors have become particularly promising for TB diagnostics in conjunction with AI that can effectively distinguish true binding or hybridization events from background signals in voltammetric and impedimetric measurements. ML applied to voltammetry and electrochemical impedance spectroscopy has enhanced the classification of binding states and limits of detection in nucleic acid-based genosensors and immunosensors (Seele et al., 2023; Ribeiro et al., 2024). This has positively impacted the level of accuracy of detection at low target concentrations in complex matrices across both laboratory and pilot-scale conditions (Bocan et al., 2025). For protein biomarkers, portable electrochemical immunosensors for the detection of TB antigens CFP-10 and ESAT-6, specifically in its heterodimer form, were proven to be more specific and applicable for sputum testing (Mohd Azmi et al., 2021). Incorporation of AI-assisted feature extraction further improves threshold stability and reduces the need for expertise in these assays.

Noninvasive methods of biosensing are also emerging as a topic of interest. Electronic nose (e-nose) platforms that have the ability to sense volatile organic compounds in exhaled breath, using AI algorithms such as SVMs and DL, have shown their potential to effectively differentiate between TB from other respiratory diseases (Li et al., 2023a). These benefit from AI's ability to deal with complex high-dimensional sensor arrays and detection of subtle biochemical changes that are difficult to access by usual means.

Critically, combining biosensor outputs with multiparametric data such as patient demographics, symptoms, radiography triage scores, and treatment history makes it possible to develop diagnostic pathways ideal for PoC decision-making, triage, and longitudinal monitoring (Wasilewski et al., 2024). Hence, such systems would be able to underpin rapid drug-resistance risk stratification when combined with molecular assays and curated genomic or phenotypic datasets, reducing dependency on centralized laboratories and expanding diagnostic access in areas with limited resources. Nevertheless, this will be accompanied by extensive prospective, multisite validation before these approaches can reach clinical practice (Schwab et al., 2024; Zhao et al., 2023).

12.6.6 Telemedicine, remote screening, and mobile applications

Emerging telemedicine and mobile health technologies are critical to expanding the reach of TB screening, easy diagnosis, and even prolonged surveillance, particularly for underserved and hard-to-reach communities (Olowoyo et al., 2025). These tools, when integrated with AI, expand service availability, improve diagnostic throughput and accuracy, and reduce care delays across low-resource geographies (Parthasarathi et al., 2024). Telemedicine technology allows the sharing of health professionals' input with specialists through remote consultations, digital referrals, and cloud-based analysis of imaging or biosensor information. These solutions provide centralized interpretation and AI-assisted diagnostic support that mitigate human-resource gaps and reduces turnaround times (Unanah and Mbanugo, 2025). A successful use of the CAD4TB system has been seen in the context of mobile-based screening initiatives, particularly for South Africa and Lesotho, illustrating the utility of AI-driven radiographic interpretation within larger-scale screening settings (Nzimande et al., 2025). Table 12.5 summarizes key studies regarding the application of AI-based telemedicine technology, mobile health technology, or even distant health screening for the diagnosis of TB.

A recent review has pointed out that telemedicine has emerged as a potential solution for the management of multidrug-resistant TB (MDR-TB), reduces delays in treatment as well as improves patient outcomes in resource-constrained settings. It facilitates early diagnosis of disease, treatment initiation, sustained adherence through immediate reminders and patient education, and accelerated follow-up with continuous monitoring, together with professional assistance. It further increases access to high-quality healthcare for rural and under-resourced communities by reducing geographical

Table 12.5 Key studies on AI-enabled telemedicine, mobile applications, and remote screening for tuberculosis detection.

Study/report	Key findings	Relevance to telemedicine and mobile AI	References
Feasibility and effectiveness of mobile app for active case finding for tuberculosis in India (Zhang et al., 2020)	Mobile survey app was launched in Bhopal and Indore slums, encompassing approximately 828,020 individuals, detecting large numbers of presumptive pulmonary and extrapulmonary TB cases through symptom checklists and follow-up.	Demonstrates the scalability of mobile apps for active case findings in marginalized urban populations	Zhang et al. (2020)
Performance of CAD4TB artificial intelligence technology in TB screening programmes among the adult population in South Africa and Lesotho (N. Nzimande et al., 2025)	Included 6524 chest X-ray images (288 TB cases and 6236 controls). CAD4TB v7 demonstrated an overall enhancement in performance accuracy as compared to v6.	Highlights the use of AI-based CAD in extensive screening initiatives, underscores the importance of calibrating systems for local contexts.	Nzimande et al. (2025)
Active case finding using mobile vans equipped with artificial intelligence aided radiology tests and sputum collection for rapid diagnostic tests to reduce tuberculosis prevalence in rural China: protocol for a pragmatic trial	Ongoing study in Guangxi, plans to deploy mobile vans equipped with AI-enabled radiology and molecular diagnostic testing to serve rural areas.	Illustrates the future convergence of AI-driven imaging and diagnostics within mobile platforms, highlighting the advancement of telemedicine solutions	Wei et al. (2024)

(*Continued*)

Table 12.5 (Continued)

Study/report	Key findings	Relevance to telemedicine and mobile AI	References
An intelligent mobile-enabled expert system for tuberculosis disease diagnosis in real time (A. M. Shabut et al., 2018)	This proposed intelligent system enhanced with real-time processing capabilities and data portability, enables predictions without requiring any opto-mechanical components and has demonstrated accuracy of 98.4% in detecting TB antigen-specific antibody on a mobile platform.	This study demonstrate a mobile AI-based system for real time TB antibody detection.	Shabut et al. (2018)

AI, artificial intelligence.

constraints. When coupled with electronic health records, telemedicine further enhances data collection, enhances coordination of care, and data-driven clinical decision support, while also strengthening surveillance, identifying disease trends, and supporting targeted public health interventions (Olawade et al., 2024).

Mobile applications further extend this paradigm by providing facilities for large-scale active case finding and community-based triage. In Indore and Bhopal's slum areas, a mobile application screened over 8.28 lakh individuals, identified several thousand probable pulmonary and extrapulmonary TB cases, and thus showcased how digital platforms can support community-level surveillance at scale (Zhang et al., 2020). Similarly, in Nigeria, portable digital X-ray units along with CAD4TB screened over 86,000 people in hard-to-reach regions, getting strong presumptive TB yields and further emphasizing the role of AI-guided imaging in detection in remote areas (Useni et al., 2025).

Another major effort, the Chennai mobile screening program, using the "Genki," an AI-based CADe software, screened 25,598 individuals and showed outstanding diagnostic performance with a specificity of 96.9%, sensitivity of 98%, and accuracy of 96.9% in detecting TB from CXR in a decentralized mobile setting. It deployed across four mobile diagnostic units, the CADe system facilitated rapid triage from radiographs while maintaining high accuracy across sexes and all age groups during real-world community outreach (Jayaraman et al., 2025). BITScreen PTB is another telemedicine platform that was designed to improve pediatric TB diagnosis by using digital CXR and clinical information in resource-limited settings (Gómez-Valverde et al., 2024). A mobile application prototype, TB Kids, was developed using AI to predict and support the diagnosis of pulmonary TB in children. It has features for nutritional assessment, risk assessment, investigation of antibiotic therapy and contacts, the tuberculin skin test, and AI-based CXR interpretation with risk graph and decision-making, complementary guidance and recording of the clinical picture (Picoli et al., 2025). Similarly, another innovative study introduced an intelligent mobile-enabled expert system for real-time detection of TB-specific antibodies using images from plasmonic ELISA assays. The system uses supervised ML; it extracts color features from assay images and classifies results by using an RF ensemble model, which gave an accuracy of 98.4% (Shabut et al., 2018). Furthermore, DeepGB-TB is also a noninvasive system, that uses cough audio and basic demographic data to generate TB risk scores via lightweight DL models. This offers offline and real-time inference directly on common mobile devices, hence making this model particularly suitable for deployment in resource-constrained settings (Lu et al., 2026). Table 12.6 summarizes AI-integrated mobile and digital platforms for TB detection and management. Together, AI-enabled telemedicine and mobile platforms help close diagnostic gaps by connecting frontline providers with specialist expertise, scaling community screening, and extending services into underserved areas.

12.7 Challenges in artificial intelligence-driven tuberculosis diagnosis

12.7.1 Technical challenges

Although AI has enormous potential for the modernization of medical practice, there are still many technical challenges to be addressed. Since ML approaches are essentially bound by the availability of substantial volumes of high-quality training data, it is imperative to ensure that data collected captures the relevant population of interest. For instance, datasets from diverse healthcare settings can have varying degrees of bias and noise, and thus a model trained on data from one hospital may not generalize well to another (Obermeyer and Emanuel, 2016). Another critical issue is interpretability. The outputs of many high-performing ML models are difficult for humans to decipher without assistance. Even when such algorithms are able to perform better than humans, it remains challenging to provide intuitive explanations for their results, finding faults in them, or obtaining additional biological insights from these black-box models. However, applications of DL have resulted in outstanding performance in fields such as image classification, language translation, voice recognition (Lecun et al., 2015), and even automated neural network design (Zoph and Le, 2016). Yet, clinical diagnostics and treatment tasks might demand more complex context information than the narrowly defined tasks that DL has mastered. Moreover, the application of transfer

Table 12.6 Summary of AI-integrated mobile and digital platforms for TB detection and management.

Application/ platform	Key features	Performance	References
qXR, qTrack	qXR and qTrack, set the gold standard in TB care, transforming detection and management at a global scale.	According to Qure.ai FY 23-24 Impact Report, these advanced AI-driven screening solutions positively impact over 20 million lives across approximately 2700 healthcare facilities in more than 90 countries.	https://www.qure.ai/impact_stories/qure-ai-fy-23-24-impact-report-global-ai-healthcare-innovations
Genki	It analyze X-ray scans and classifies each scan as either "TB suggestive" or "TB not suggestive."	The system exhibits a combined sensitivity of 98%, accuracy of 96.9%, and specificity of 96.9%, in diagnosing TB from chest X-ray scans within the screened population.	Jayaraman et al. (2025)
TBScreen	It is a ResNet18-based classifier designed to distinguish TB from nonTB cases using RGB scalogram representations.	This model achieved a fivefold cross-validation sensitivity of 0.70 and showed better performance in patients with higher bacterial load or the presence of lung cavities.	Sharma et al. (2024)
DeepGB-TB	Its key innovation lies in the Cross-Modal Bidirectional Cross-Attention (CM-BCA) module, which facilitates the exchange of important information between different modalities.	This model achieved AUC-0.903 and F1-score of 0.851	Lu et al. (2025)

learning techniques to leverage insights learned from extensive nonmedical datasets for the analysis of multimodal clinical data remains uncertain. There is a pressing need for large-scale efforts in data collection and annotation for the complete integration of AI systems within the clinical domain.

12.7.2 Socioeconomic and legal concerns

Nevertheless, new social, economic, as well as legal issues are bound to arise in the further clinical adoption of AI technologies (Dignum, 2018; Price II, 2017). From a social standpoint, AI can facilitate accuracy by eliminating human errors and easing some of the burdens on clinicians by automation of routine clinical tasks, thus allowing health professionals to focus on complex decision making and patient care. However, this will not necessarily lower the overall workload because guidelines may require more frequent testing of high-risk patients, and automation of routine tasks may lead to workforce restructuring. Economically, proper integration of AI into the clinical workflow is essential but nontrivial. Poorly designed systems may introduce unintended consequences, such as alert fatigue, increased clinician workload, disruptions in doctor-patient communication, or diagnostic errors owing to over-reliance on AI outputs (Carspecken et al., 2013; Ash et al., 2004). Hence, the efficient integration of the workflow process and comprehensive provider training are crucial. From a legal perspective, regulatory bodies are still building their foundations for adaptive ML models, which are constantly updated with new data. Initiatives such as the FDA's "pre-certification approach" emphasize the technology developer rather than primarily at the product and bring out the need to define clear standards for validation, model updates, and accountability (U.S. Food & Drug Administration, 2019). In parallel, legal frameworks must clearly articulate liability in cases of medical malpractice where decisions are supported by AI, while insurance policies will have to adapt to cover such scenarios. A robust, end-to-end policy framework is also required in terms of privacy and consent for ubiquitous health data collection, particularly from mobile and wearable devices (Shortliffe, 1987). To address such challenges, for safe, effective, and equitable integration of AI into clinical practice calls for close collaboration among AI developers, clinicians, hospital administrators, insurers, regulators, and policymakers.

12.8 Conclusion and future prospects

AI has proven to be an important driver within the realm of TB diagnosis, bridging the gap between clinical expertise and data-driven accuracy. This chapter has traced the progress of AI in TB detection, from early computer-aided imaging systems to modern DL frameworks that make use of multiomics, biosensing, and mobile-based technologies. AI has demonstrated remarkable potential in enhancing TB detection across both imaging-based and nonimaging diagnostic modalities. The breadth of innovation within this domain has ranged from deep CNNs analyzing chest radiographs to capsule networks identifying TB through cough sounds. AI adoption across different verticals in TB diagnosis, from imaging to acoustic epidemiology, sputum smear microscopy, multiomics analysis, biosensors, and mobile-enabled telemedicine, emphasizes the revolutionary capability of AI not only for improved and early diagnosis but also for disease management. In resource-constrained environments, AI-driven algorithms have consistently outperformed conventional approaches with quick, reproducible, and scalable solutions. AI applications have proven to be most beneficial in improving the reach of diagnostics and offering continuous surveillance for people at risk and even community-level screenings on large scales using telemedicine applications. Systems such as Genki, TBScreen, DeepGB-TB, and qXR/qTrack exemplify AI real-world implementations where AI integrates imaging, clinical, and acoustic inputs with predictive analytics to support public health initiatives and connect patients to timely care. Multimodal integration, federated learning to use distributed datasets without sacrificing privacy, and lightweight architectures designed for deployment in rural or resource-constrained contexts will form important thrust areas of future work. As global efforts to eliminate TB gain further momentum, AI comes across as a disruptive technology in augmenting human capabilities, expanding diagnostic reach, and thereby accelerating progress toward eradication. To attain full value from such a prospect, continued investment in rich, diverse, and high-quality datasets, strong validation frameworks, and ethical deployment practices will be essential. In conclusion, the functionality of AI techniques to learn and interpret complex patterns is a promising potential for future advancements in the detection of TB, although considerable challenges remain, to work on and further technological refinement is necessary to increase their predictive potential.

References

Abe H, Kumazawa S, Taji T, Sasaki S. Applications of computerized pattern recognition: a survey of correlations between pharmacological activities and mass spectra. Biological Mass Spectrometry 1976;3(4):151–4. Available from https://doi.org/10.1002/bms.1200030402.

Ahmad R, Xie L, Pyle M, Suarez MF, Broger T, Steinberg D, et al. A rapid triage test for active pulmonary tuberculosis in adult patients with persistent cough. Science Translational Medicine 2019;11(515). Available from https://doi.org/10.1126/scitranslmed.aaw8287.

Akkaş T, Reshadsedghi M, Şen M, Kılıç V, Horzum N. The role of artificial intelligence in advancing biosensor technology: past, present, and future perspectives. Advanced Materials 2025;37(34). Available from https://doi.org/10.1002/adma.202504796.

Ali H. Artificial intelligence in multiomics data integration: advancing precision medicine, biomarker discovery and genomic-driven disease interventions. International Journal of Science and Research Archive 2023;8(1):1012–30. Available from https://doi.org/10.30574/ijsra.2023.8.1.0189.

Alqudaihi KS, Aslam N, Khan IU, Almuhaideb AM, Alsunaidi SJ, Ibrahim NMAR, et al. Cough sound detection and diagnosis using artificial intelligence techniques: challenges and opportunities. IEEE Access 2021;9:102327–44. Available from https://doi.org/10.1109/ACCESS.2021.3097559.

Anh NK, Phat NK, Thu NQ, Tien NTN, Eunsu C, Kim HS, et al. Discovery of urinary biosignatures for tuberculosis and nontuberculous mycobacteria classification using metabolomics and machine learning. Scientific Reports 2024;14(1). Available from https://doi.org/10.1038/s41598-024-66113-x.

Asada N, Doi K, MacMahon H, Montner SM, Giger ML, Abe C, et al. Potential usefulness of an artificial neural network for differential diagnosis of interstitial lung diseases: pilot study. Radiology 1990;177(3):857–60. Available from https://doi.org/10.1148/radiology.177.3.2244001.

Ash JS, Berg M, Coiera E. Some unintended consequences of information technology in health care: the nature of patient care information system-related errors. Journal of the American Medical Informatics Association 2004;11(2):104–12. Available from https://doi.org/10.1197/jamia.M1471.

Aulia S, Suksmono AB, Mengko TR, Alisjahbana B. A novel digitized microscopic images of ZN-stained sputum smear and its classification based on IUATLD grades. IEEE Access 2024;12:51364–80. Available from https://doi.org/10.1109/ACCESS.2024.3386208.

Baur S, Nabulsi Z, Weng W-H, Garrison J, Blankemeier L, Fishman S, et al. HeAR – health acoustic representations. arXiv:2403.02522 2024. Available from https://doi.org/10.48550/arXiv.2403.02522.

Bhaiyya M, Panigrahi D, Rewatkar P, Haick H. Role of machine learning assisted biosensors in point-of-care-testing for clinical decisions. ACS Sensors 2024;9(9):4495–519. Available from https://doi.org/10.1021/acssensors.4c01582.

Bocan A, Siavash Moakhar R, del Real Mata C, Petkun M, De Iure-Grimmel T, Yedire SG, et al. Machine-learning-aided advanced electrochemical biosensors. Advanced Materials 2025;37(33). Available from https://doi.org/10.1002/adma.202417520.

Botha GHR, Theron G, Warren RM, Klopper M, Dheda K, van Helden PD, et al. Detection of tuberculosis by automatic cough sound analysis. Physiological Measurement 2018;39(4):045005. Available from https://doi.org/10.1088/1361-6579/aab6d0.

Campuzano S, Pedrero M, Gamella M, Serafín V, Yáñez-Sedeño P, Pingarrón JM. Beyond sensitive and selective electrochemical biosensors: toward continuous, real-time, antibiofouling and calibration-free devices. Sensors 2020;20(12):3376. Available from https://doi.org/10.3390/s20123376.

Carspecken CW, Sharek PJ, Longhurst C, Pageler NM. A clinical case of electronic health record drug alert fatigue: consequences for patient outcome. Pediatrics 2013;131(6):e1970. Available from https://doi.org/10.1542/peds.2012-3252.

Chandra TB, Verma K, Singh BK, Jain D, Netam SS. Automatic detection of tuberculosis related abnormalities in Chest X-ray images using hierarchical feature extraction scheme. Expert Systems with Applications 2020;158. Available from https://doi.org/10.1016/j.eswa.2020.113514.

Chang J, Arbeláez P, Switz N, Reber C, Tapley A, Davis JL, et al. Automated tuberculosis diagnosis using fluorescence images from a mobile microscope. Medical Image Computing and Computer-Assisted Intervention – MICCAI 2012. MICCAI 2012. Lecture Notes in Computer Science, vol 7512. Berlin, Heidelberg: Springer; 2012. p. 345–52. Available from https://doi.org/10.1007/978-3-642-33454-2_43.

Cho J, Hwang H. Spatio-temporal representation of an electoencephalogram for emotion recognition using a three-dimensional convolutional neural network. Sensors 2020;20(12):3491. Available from https://doi.org/10.3390/s20123491.

Cohen KA, Manson AL, Desjardins CA, Abeel T, Earl AM. Deciphering drug resistance in Mycobacterium tuberculosis using whole-genome sequencing: progress, promise, and challenges. Genome Medicine 2019;11(1). Available from https://doi.org/10.1186/s13073-019-0660-8.

Creswell J, Vo LNQ, Qin ZZ, Muyoyeta M, Tovar M, Wong EB, et al. Early user perspectives on using computer-aided detection software for interpreting chest X-ray images to enhance access and quality of care for persons with tuberculosis. BMC Global and Public Health 2023;1(1). Available from https://doi.org/10.1186/s44263-023-00033-2.

Cui F, Yue Y, Zhang Y, Zhang Z, Zhou HS. Advancing biosensors with machine learning. ACS Sensors 2020;5(11):3346–64. Available from https://doi.org/10.1021/acssensors.0c01424.

Deelder W, Christakoudi S, Phelan J, Benavente ED, Campino S, McNerney R, et al. Machine learning predicts accurately mycobacterium tuberculosis drug resistance from whole genome sequencing data. Frontiers in Genetics 2019;10. Available from https://doi.org/10.3389/fgene.2019.00922.

Desruisseaux C, Broderick C, Lavergne V, Sy K, Garcia DJ, Barot G, et al. Retrospective validation of MetaSystems' deep-learning-based digital microscopy platform with assistance compared to manual fluorescence microscopy for detection of mycobacteria. Journal of Clinical Microbiology 2024;62(3). Available from https://doi.org/10.1128/jcm.01069-23.

Dignum V. Ethics in artificial intelligence: introduction to the special issue. Ethics and Information Technology 2018;20(1):1–3. Available from https://doi.org/10.1007/s10676-018-9450-z.

Dixit A, Freschi L, Vargas R, Gröschel MI, Nakhoul M, Tahseen S, et al. Estimation of country-specific tuberculosis resistance antibiograms using pathogen genomics and machine learning. BMJ Global Health 2024;9(3). Available from https://doi.org/10.1136/bmjgh-2023-013532.

Du D, Gu J, Chen X, Lv W, Feng Q, Rahmim A, et al. Integration of PET/CT radiomics and semantic features for differentiation between active pulmonary tuberculosis and lung cancer. Molecular Imaging and Biology 2021;23(2):287–98. Available from https://doi.org/10.1007/s11307-020-01550-4.

Ellahham S. Artificial intelligence: the future for diabetes care. The American Journal of Medicine 2020;133(8):895–900. Available from https://doi.org/10.1016/j.amjmed.2020.03.033.

El-Solh A, Mylotte J, Sherif S, Serghani J, Grant BJ. Validity of a decision tree for predicting active pulmonary tuberculosis. American Journal of Respiratory and Critical Care Medicine 1997;155(5):1711–6. Available from https://doi.org/10.1164/ajrccm.155.5.9154881.

El-Solh AA, Hsiao CB, Goodnough S, Serghani J, Grant BJB. Predicting active pulmonary tuberculosis using an artificial neural network. Chest 1999;116(4):968–73. Available from https://doi.org/10.1378/chest.116.4.968.

Flynn CD, Chang D. Artificial intelligence in point-of-care biosensing: challenges and opportunities. Diagnostics 2024;14(11):1100. Available from https://doi.org/10.3390/diagnostics14111100.

Fu HT, Tu HZ, Lee HS, Lin YE, Lin CW. Evaluation of an AI-based TB AFB smear screening system for laboratory diagnosis on routine practice. Sensors 2022;22(21). Available from https://doi.org/10.3390/s22218497.

Gao XW, Qian Y. Prediction of multidrug-resistant TB from CT pulmonary images based on deep learning techniques. Molecular Pharmaceutics 2018;15(10):4326–35. Available from https://doi.org/10.1021/acs.molpharmaceut.7b00875.

Ghanshala T, Tripathi V, Pant B. An effective vision based framework for the identification of tuberculosis in chest X-ray images. In: Singh M, Gupta P, Tyagi V, Flusser J, Ören T, Valentino G, editors. Advances in Computing and Data Sciences. ICACDS 2020. Communications in Computer and Information Science, vol 1244. Singapore: Springer; 2020. p. 36–45. Available from https://doi.org/10.1007/978-981-15-6634-9_4.

Gharamaleki OG, Colijn C, Sekirov I, Johnston JC, Sobkowiak B. Early prediction of Mycobacterium tuberculosis transmission clusters using supervised learning models. Scientific Reports 2024;14(1). Available from https://doi.org/10.1038/s41598-024-78247-z.

Gómez-Valverde JJ, Sánchez-Jacob R, Ribó JL, Schaaf HS, García Delgado L, Hernanz-Lobo A, et al. Chest X-ray–based telemedicine platform for pediatric tuberculosis diagnosis in low-resource settings: development and validation study. JMIR Pediatrics and Parenting 2024;7. Available from https://doi.org/10.2196/51743.

Gootenberg JS, Abudayyeh OO, Kellner MJ, Joung J, Collins JJ, Zhang F. Multiplexed and portable nucleic acid detection platform with Cas13, Cas12a and Csm6. Science (New York, N.Y.) 2018;360(6387):439–44. Available from https://doi.org/10.1126/science.aaq0179.

Green AG, Yoon CH, Chen ML, Ektefaie Y, Fina M, Freschi L, et al. A convolutional neural network highlights mutations relevant to antimicrobial resistance in Mycobacterium tuberculosis. Nature Communications 2022;13(1). Available from https://doi.org/10.1038/s41467-022-31236-0.

Gupta P, Khare V, Srivastava A, Agarwal J, Mittal V, Sonkar V, et al. A prospective observational multicentric clinical trial to evaluate microscopic examination of acid-fast bacilli in sputum by artificial intelligence-based microscopy system. Journal of Investigative Medicine 2023;71(7):716–21. Available from https://doi.org/10.1177/10815589231171402.

Guseva V, Testov V, Parolina L, Otpushchennikova O, Lizinfeld I, Nazarova L, et al. Using telemedicine to increase access to treatment for tuberculosis patients. International Journal of Infectious Diseases 2023;130:S123. Available from https://doi.org/10.1016/j.ijid.2023.04.304.

Han D, Chen Y, Li X, Li W, Zhang X, He T, et al. Development and validation of a 3D-convolutional neural network model based on chest CT for differentiating active pulmonary tuberculosis from community-acquired pneumonia. Radiologia Medica 2023;128(1):68–80. Available from https://doi.org/10.1007/s11547-022-01580-8.

Han GR, Goncharov A, Eryilmaz M, Ye S, Palanisamy B, Ghosh R, et al. Machine learning in point-of-care testing: innovations, challenges, and opportunities. Nature Communications 2025;16(1). Available from https://doi.org/10.1038/s41467-025-58527-6.

Han Z, Zhang Y, Ding W, Zhao X, Jia B, Liu T, et al. An integrated mycobacterial CT imaging dataset with multispecies information. Scientific Data 2025;12(1). Available from https://doi.org/10.1038/s41597-025-04838-8.

Hansun S, Argha A, Liaw ST, Celler BG, Marks GB. Machine and deep learning for tuberculosis detection on chest X-rays: systematic literature review. Journal of Medical Internet Research 2023;25. Available from https://doi.org/10.2196/43154.

Hansun S, Argha A, Bakhshayeshi I, Wicaksana A, Alinejad-Rokny H, Fox GJ, et al. Diagnostic performance of artificial intelligence-based methods for tuberculosis detection: systematic review. Journal of Medical Internet Research 2025;27. Available from https://doi.org/10.2196/69068.

Harris M, Qi A, Jeagal L, Torabi N, Menzies D, Korobitsyn A, et al. A systematic review of the diagnostic accuracy of artificial intelligence-based computer programs to analyze chest x-rays for pulmonary tuberculosis. PLoS One 2019;14(9):e0221339. Available from https://doi.org/10.1371/journal.pone.0221339.

Hearst MA, Dumais ST, Osuna E, Platt J, Scholkopf B. Support vector machines. IEEE Intelligent Systems and their Applications 1998;13(4):18–28. Available from https://doi.org/10.1109/5254.708428.

Hegde S, Sreeram S, Alter IL, Shor C, Valdez TA, Meister KD, et al. Cough sounds in screening and diagnostics: a scoping review. The Laryngoscope 2024;134(3):1023–31. Available from https://doi.org/10.1002/lary.31042.

Helm JM, Swiergosz AM, Haeberle HS, Karnuta JM, Schaffer JL, Krebs VE, et al. Machine learning and artificial intelligence: definitions, applications, and future directions. Current Reviews in Musculoskeletal Medicine 2020;13(1):69–76. Available from https://doi.org/10.1007/s12178-020-09600-8.

Heo SJ, Kim Y, Yun S, Lim SS, Kim J, Nam CM, et al. Deep learning algorithms with demographic information help to detect tuberculosis in chest radiographs in annual workers' health examination data. International Journal of Environmental Research and Public Health 2019;16(2). Available from https://doi.org/10.3390/ijerph16020250.

Ho TK. Random decision forests. Proceedings of the International Conference on Document Analysis and Recognition, ICDAR 1995;1:278–82. Available from https://doi.org/10.1109/ICDAR.1995.598994.

Ho TKK, Gwak J, Prakash O, Song JI, Park CM. Utilizing pretrained deep learning models for automated pulmonary tuberculosis detection using chest radiography. In: Nguyen N, Gaol F, Hong TP, Trawiński B, editors. Intelligent Information and Database Systems. ACIIDS 2019. Lecture Notes in Computer Science, vol 11432. Cham: Springer; 2019. p. 395–403. Available from https://doi.org/10.1007/978-3-030-14802-7_34.

Hooda R, Mittal A, Sofat S. Automated TB classification using ensemble of deep architectures. Multimedia Tools and Applications 2019;78(22):31515–32. Available from https://doi.org/10.1007/s11042-019-07984-5.

Hosny A, Parmar C, Quackenbush J, Schwartz LH, Aerts HJWL. Artificial intelligence in radiology. Nature Reviews. Cancer 2018;18(8):500–10. Available from https://doi.org/10.1038/s41568-018-0016-5.

Hu L, De Hoyos D, Lei Y, West AP, Walsh AJ. 3D convolutional neural networks predict cellular metabolic pathway use from fluorescence lifetime decay data. APL Bioengineering 2024;8(1). Available from https://doi.org/10.1063/5.0188476.

Hu X, Wang J, Ju Y, Zhang X, Qimanguli W, Li C, et al. Combining metabolome and clinical indicators with machine learning provides some promising diagnostic markers to precisely detect smear-positive/negative pulmonary tuberculosis. BMC Infectious Diseases 2022;22(1). Available from https://doi.org/10.1186/s12879-022-07694-8.

Huang HC, Kuo KL, Lo MH, Chou HY, Lin YE. Novel TB smear microscopy automation system in detecting acid-fast bacilli for tuberculosis- a multi-center double blind study. Tuberculosis 2022;135. Available from https://doi.org/10.1016/j.tube.2022.102212.

Huang JD, Wang J, Ramsey E, Leavey G, Chico TJA, Condell J. Applying artificial intelligence to wearable sensor data to diagnose and predict cardiovascular disease: a review. Sensors 2022;22(20). Available from https://doi.org/10.3390/s22208002.

Huang Z, Zhang G, Lyon CJ, Hu TY, Lu S. Outlook for CRISPR-based tuberculosis assays now in their infancy. Frontiers in Immunology 2023;14. Available from https://doi.org/10.3389/fimmu.2023.1172035.

Huddart S, Yadav V, Sieberts SK, Omberg L, Raberahona M, Rakotoarivelo R, et al. A dataset of solicited cough sound for tuberculosis triage testing. Scientific Data 2024;11(1). Available from https://doi.org/10.1038/s41597-024-03972-z.

Jayaraman P, Saumit Paul SS, Pant R, Gupte T, Kulkarni V, Kharat A. Artificial intelligence as a proficient tool in detecting pulmonary tuberculosis in massive population screening programs: a case study in Chennai, India. Journal of Rural Medicine 2025;20(1):13–9. Available from https://doi.org/10.2185/jrm.2024-015.

Jin X, Liu C, Xu T, Su L, Zhang X. Artificial intelligence biosensors: challenges and prospects. Biosensors and Bioelectronics 2020;165:112412. Available from https://doi.org/10.1016/j.bios.2020.112412.

Kaforou M, Wright VJ, Oni T, French N, Anderson ST, Bangani N, et al. Detection of tuberculosis in HIV-infected and uninfected african adults using whole blood RNA expression signatures: a case-control study. PLoS Medicine 2013;10(10). Available from https://doi.org/10.1371/journal.pmed.1001538.

Kaminski MM, Abudayyeh OO, Gootenberg JS, Zhang F, Collins JJ. CRISPR-based diagnostics. Nature Biomedical Engineering 2021;5(7):643–56. Available from https://doi.org/10.1038/s41551-021-00760-7.

Karmani P, Chandio AA, Karmani V, Soomro JA, Korejo IA, Chandio MS. Taxonomy on healthcare system based on machine learning approaches: tuberculosis disease diagnosis. International Journal of Computing and Digital Systems 2020;9(6):1199–212. Available from https://doi.org/10.12785/ijcds/0906017.

Khan MT, Kaushik AC, Ji L, Malik SI, Ali S, Wei DQ. Artificial neural networks for prediction of tuberculosis disease. Frontiers in Microbiology 2019;10. Available from https://doi.org/10.3389/fmicb.2019.00395.

Khare V, Agrawal A, Gupta P, Saxena S. Artificial Intelligence based AFB microscopy for pulmonary tuberculosis in North India: a pilot study. International Journal of Scientific and Research Publications (IJSRP) 2019;9(12):p9669. Available from https://doi.org/10.29322/ijsrp.9.12.2019.p9669.

Kik SV, Gelaw SM, Ruhwald M, Song R, Khan FA, van Hest R, et al. Diagnostic accuracy of chest X-ray interpretation for tuberculosis by three artificial intelligence-based software in a screening use-case: An individual patient meta-analysis of global data. medRxiv 2022. Available from https://doi.org/10.1101/2022.01.24.22269730.

Kılıç V, Mercan ÖB, Tetik M, Kap Ö, Horzum N. Nonenzymatic colorimetric glucose detection based on Au/Ag nanoparticles using smartphone and machine learning. Analytical Sciences 2022;38(2):347–58. Available from https://doi.org/10.2116/analsci.21P253.

Kotei E, Thirunavukarasu R. A comprehensive review on advancement in deep learning techniques for automatic detection of tuberculosis from chest x-ray images. Archives of Computational Methods in Engineering 2024;31(1):455–74. Available from https://doi.org/10.1007/s11831-023-09987-w.

Kumaran A, Jude Serpes N, Gupta T, James A, Sharma A, Kumar D, et al. Advancements in CRISPR-based biosensing for next-gen point of care diagnostic application. Biosensors 2023;13(2):202. Available from https://doi.org/10.3390/bios13020202.

Lakhani P, Sundaram B. Deep learning at chest radiography: automated classification of pulmonary tuberculosis by using convolutional neural networks. Radiology 2017;284(2):574–82. Available from https://doi.org/10.1148/radiol.2017162326.

Lecun Y, Bengio Y, Hinton G. Deep learning. Nature 2015;521(7553):436–44. Available from https://doi.org/10.1038/nature14539.

Li G, Lee CH, Jung JJ, Youn YC, Camacho D. Deep learning for EEG data analytics: a survey. Concurrency and Computation: Practice and Experience 2020;32(18). Available from https://doi.org/10.1002/cpe.5199.

Li J. Regression and classification in supervised learning. ICCBD '19: Proceedings of the 2nd International Conference on Computing and Big Data. New York, NY: Association for Computing Machinery; 2019. p. 99–104. Available from https://doi.org/10.1145/3366650.3366675.

Li P, Wang J, Tang M, Li M, Han R, Zhou S, et al. A CT-based radiomics predictive nomogram to identify pulmonary tuberculosis from community-acquired pneumonia: a multicenter cohort study. Frontiers in Cellular and Infection Microbiology 2024;14. Available from https://doi.org/10.3389/fcimb.2024.1388991.

Li Y, Xia L. Coronavirus disease 2019 (COVID-19): role of chest CT in diagnosis and management. American Journal of Roentgenology 2020;214(6):1280–6. Available from https://doi.org/10.2214/AJR.20.22954.

Li Y, Wei X, Zhou Y, Wang J, You R. Research progress of electronic nose technology in exhaled breath disease analysis. Microsystems and Nanoengineering 2023;9(1). Available from https://doi.org/10.1038/s41378-023-00594-0.

Li Y, Xu Z, Lv X, Li C, He W, Lv Y, et al. Radiomics analysis of lung CT for multidrug resistance prediction in active tuberculosis: a multicentre study. European Radiology 2023;33(9):6308–17. Available from https://doi.org/10.1007/s00330-023-09589-x.

Liang S, Ma J, Wang G, Shao J, Li J, Deng H, et al. The Application of artificial intelligence in the diagnosis and drug resistance prediction of pulmonary tuberculosis. Frontiers in Medicine 2022;9. Available from https://doi.org/10.3389/fmed.2022.935080.

Liu C, Cao Y, Alcantara M, Liu B, Brunette M, Peinado J, et al. TX-CNN: Detecting tuberculosis in chest X-ray images using convolutional neural network. 2017 IEEE International Conference on Image Processing, ICIP 2017 - Proceedings. (Proceedings - International Conference on Image Processing, ICIP; Vol. 2017-September). IEEE Computer Society; 2017. p. 2314–8. Available from https://doi.org/10.1109/ICIP.2017.8296695.

Liu J, Liu Y, Wang C, Li A, Meng B, Chai X, et al. An original neural network for pulmonary tuberculosis diagnosis in radiographs. In: Kůrková V, Manolopoulos Y, Hammer B, Iliadis L, Maglogiannis I, editors. Artificial Neural Networks and Machine Learning – ICANN 2018. ICANN 2018. Lecture Notes in Computer Science, vol 11140. Cham: Springer; 2018. p. 158–66. Available from https://doi.org/10.1007/978-3-030-01421-6_16.

Lu Z, Li Y, Tang F, Jiang Z, Li C, Zhou M, et al. DeepGB-TB: A Risk-Balanced Cross-Attention Gradient-Boosted Convolutional Network for Rapid, Interpretable Tuberculosis Screening. arXiv:2508.02741 2026. Available from https://doi.org/10.48550/arXiv.2508.02741.

Ma L, Wang Y, Guo L, Zhang Y, Wang P, Pei X, et al. Developing and verifying automatic detection of active pulmonary tuberculosis from multi-slice spiral CT images based on deep learning. Journal of X-ray Science and Technology 2020;28(5):939–51. Available from https://doi.org/10.3233/XST-200662.

Mahmoudi T, Naghdi T, Morales-Narváez E, Golmohammadi H. Toward smart diagnosis of pandemic infectious diseases using wastewater-based epidemiology. TrAC Trends in Analytical Chemistry 2022;153:116635. Available from https://doi.org/10.1016/j.trac.2022.116635.

Maliwan N, Reid RW, Pliska SR, Bird TJ, Zvetina JR. Direct diagnosis of tuberculosis by computer assisted pattern recognition gas chromatographic analysis of sputum. Biomedical Chromatography 1991;5(4):165–70. Available from https://doi.org/10.1002/bmc.1130050406.

Mandalakas AM, Detjen AK, Hesseling AC, Benedetti A, Menzies D. Interferon-gamma release assays and childhood tuberculosis: systematic review and meta-analysis. International Journal of Tuberculosis and Lung Disease 2011;15(8):1018–32. Available from https://doi.org/10.5588/ijtld.10.0631.

Mann T, Gupta RK, Reeve BWP, Ndlangalavu G, Chandran A, Krishna AP, et al. Blood RNA biomarkers for tuberculosis screening in people living with HIV before antiretroviral therapy initiation: a diagnostic accuracy study. The Lancet Global Health 2024;12(5):e783. Available from https://doi.org/10.1016/S2214-109X(24)00029-9.

Mayerhoefer ME, Materka A, Langs G, Häggström I, Szczypiński P, Gibbs P, et al. Introduction to radiomics. Journal of Nuclear Medicine 2020;61(4):488–95. Available from https://doi.org/10.2967/JNUMED.118.222893.

McCulloch WS, Pitts W. A logical calculus of the ideas immanent in nervous activity. The Bulletin of Mathematical Biophysics 1943;5(4):115–33. Available from https://doi.org/10.1007/BF02478259.

Mehrotra P. Biosensors and their applications - a review. Journal of Oral Biology and Craniofacial Research 2016;6(2):153–9. Available from https://doi.org/10.1016/j.jobcr.2015.12.002.

Meier NR, Sutter TM, Jacobsen M, Ottenhoff THM, Vogt JE, Ritz N. Machine learning algorithms evaluate immune response to novel mycobacterium tuberculosis antigens for diagnosis of tuberculosis. Frontiers in Cellular and Infection Microbiology 2021;10. Available from https://doi.org/10.3389/fcimb.2020.594030.

Melendez J, van Ginneken B, Maduskar P, Philipsen RHHM, Ayles H, Sanchez CI. On combining multiple-instance learning and active learning for computer-aided detection of tuberculosis. IEEE Transactions on Medical Imaging 2016;35(4):1013–24. Available from https://doi.org/10.1109/tmi.2015.2505672.

Menon S, Koura KG. Artificial intelligence for tuberculosis control: a scoping review of applications in public health. Journal of Global Health 2025;15. Available from https://doi.org/10.7189/jogh.15.04192.

Mohd Azmi UZ, Yusof NA, Abdullah J, Alang Ahmad SA, Mohd Faudzi FN, Ahmad Raston NH, et al. Portable electrochemical immunosensor for detection of Mycobacterium tuberculosis secreted protein CFP10-ESAT6 in clinical sputum samples. Microchimica Acta 2021;188(1). Available from https://doi.org/10.1007/s00604-020-04669-x.

Morales EF, Escalante HJ. Chapter 6 - A brief introduction to supervised, unsupervised, and reinforcement learning. In: Torres-García AA, Reyes-García CA, Villaseñor-Pineda L, Mendoza-Montoya O, editors. Biosignal processing and classification using computational learning and intelligence. Academic Press; 2022. p. 111–29.

Munjal P, Mahrooqi AA, Rajan R, Jeremijenko A, Ahmad I, Akhtar MI, et al. Population-scale cross-sectional observational study for AI-powered TB screening on one million CXRs. npj Digital Medicine 2025;8(1). Available from https://doi.org/10.1038/s41746-025-01832-7.

Muthukrishnan N, Maleki F, Ovens K, Reinhold C, Forghani B, Forghani R. Brief history of artificial intelligence. Neuroimaging cClinics of North America 2020;30(4):393–9. Available from https://doi.org/10.1016/j.nic.2020.07.004.

Nafisah SI, Muhammad G. Tuberculosis detection in chest radiograph using convolutional neural network architecture and explainable artificial intelligence. Neural Computing and Applications 2024;36(1):111–31. Available from https://doi.org/10.1007/s00521-022-07258-6.

Nijiati M, Ma J, Hu C, Tuersun A, Abulizi A, Kelimu A, et al. Artificial intelligence assisting the early detection of active pulmonary tuberculosis from chest x-rays: a population-based study. Frontiers in Molecular Biosciences 2022;9. Available from https://doi.org/10.3389/fmolb.2022.874475.

Nimmo C, Millard J, Faulkner V, Monteserin J, Pugh H, Johnson EO. Evolution of Mycobacterium tuberculosis drug resistance in the genomic era. Frontiers in Cellular and Infection Microbiology 2022;12. Available from https://doi.org/10.3389/fcimb.2022.954074.

Niyas S, Pawan SJ, Anand Kumar M, Rajan J. Medical image segmentation with 3D convolutional neural networks: a survey. Neurocomputing 2022;493:397–413. Available from https://doi.org/10.1016/j.neucom.2022.04.065.

Noor J, Chaudhry A, Batool S. Microfluidic technology, artificial intelligence, and biosensors as advanced technologies in cancer screening: a review article. Cureus 2023;15(5). Available from https://doi.org/10.7759/cureus.39634.

Nzimande N, Murphy K, Reither K, Bosman S, Ayakaka I, Glass TR, et al. Performance of CAD4TB artificial intelligence technology in TB screening programmes among the adult population in South

Africa and Lesotho. Journal of Clinical Tuberculosis and Other Mycobacterial Diseases 2025;40. Available from https://doi.org/10.1016/j.jctube.2025.100540.

Obermeyer Z, Emanuel EJ. Predicting the future—big data, machine learning, and clinical medicine. The New England Journal of Medicine 2016;375(13):1216.

Olawade DB, Eberhardt J, David-Olawade AC, Balogun MA, Bolarinwa OA, Esan DT. Transforming multidrug-resistant tuberculosis care: the potentials of telemedicine in resource-limited settings. Health Sciences Review 2024;12. Available from https://doi.org/10.1016/j.hsr.2024.100185.

Oliver A, Odena A, Raffel C, Cubuk ED, Goodfellow IJ. Realistic evaluation of deep semi-supervised learning algorithms. Advances in Neural Information Processing Systems. United States: Neural Information Processing Systems Foundation; 2018. p. 3235–46. Available from https://proceedings.neurips.cc/paper_files/paper/2018/file/c1fea270c48e8079d8ddf7d06d26ab52-Paper.pdf.

Oloko-Oba M, Viriri S. Ensemble of efficientNets for the diagnosis of tuberculosis. Computational Intelligence and Neuroscience 2021;2021. Available from https://doi.org/10.1155/2021/9790894.

Olowoyo KS, Esan DT, Olowoyo P, Oyinloye BE, Fawole IO, Aderibigbe S, et al. Treatment adherence and outcomes in patients with tuberculosis treated with telemedicine: a scoping review. Tropical Medicine and Infectious Disease 2025;10(3). Available from https://doi.org/10.3390/tropicalmed10030078.

Orjuela-Cañón AD, Camargo Mendoza JE, Awad García CE, Vergara Vela EP. Tuberculosis diagnosis support analysis for precarious health information systems. Computer Methods and Programs in Biomedicine 2018;157:11–7. Available from https://doi.org/10.1016/j.cmpb.2018.01.009.

Osman MK, Ahmad F, Saad Z, Mashor MY, Jaafar H. A genetic algorithm-neural network approach for Mycobacterium tuberculosis detection in Ziehl-Neelsen stained tissue slide images. 2010 10th International Conference on Intelligent Systems Design and Applications, Cairo, Egypt. IEEE; 2010. p. 1229–34. Available from 10.1109/ISDA.2010.5687018.

Padhy S, Takkar B, Chawla R, Kumar A. Artificial intelligence in diabetic retinopathy: a natural step to the future. Indian Journal of Ophthalmology 2019;67(7):1004–9. Available from https://doi.org/10.4103/ijo.IJO_1989_18.

Pahar M, Klopper M, Reeve B, Warren R, Theron G, Niesler T. Automatic cough classification for tuberculosis screening in a real-world environment. Physiological Measurement 2021;42(10). Available from https://doi.org/10.1088/1361-6579/ac2fb8.

Parthasarathi A, George T, Kalimuth MB, Jayasimha S, Kaleem Ullah M, Patil R, et al. Exploring the potential of telemedicine for improved primary healthcare in India: a comprehensive review. The Lancet Regional Health - Southeast Asia 2024;27. Available from https://doi.org/10.1016/j.lansea.2024.100431.

Pathri R, Jha S, Tandon S, GangaShetty S. Acoustic epidemiology of pulmonary tuberculosis (TB) & COVID19 leveraging AI/ML. medRxiv 2022. Available from https://doi.org/10.1101/2022.02.05.22269707.

Picoli KS, Ramos FRS, da Silva DMGV, Thurner BDV, Sacramento DS, Antunes IA, et al. Mobile application prototyping using Artificial Intelligence to support childhood tuberculosis diagnosis. Revista Brasileira de Enfermagem 2025;78(3). Available from https://doi.org/10.1590/0034-7167-2024-0398.

Price II WN. Artificial intelligence in health care: applications and legal implications. The SciTech Lawyer 2017;14(1). Available from https://repository.law.umich.edu/facarticles/1932/.

Qin C, Yao D, Shi Y, Song Z. Computer-aided detection in chest radiography based on artificial intelligence: a survey. Biomedical Engineering Online 2018;17(1). Available from https://doi.org/10.1186/s12938-018-0544-y.

Qin ZZ, Ahmed S, Sarker MS, Paul K, Adel ASS, Naheyan T, et al. Tuberculosis detection from chest x-rays for triaging in a high tuberculosis-burden setting: an evaluation of five artificial intelligence algorithms. The Lancet Digital Health 2021;3(9):e543. Available from https://doi.org/10.1016/S2589-7500(21)00116-3.

Rajasekar SJS, Balaraman AR, Balaraman DV, Mohamed Ali S, Narasimhan K, Krishnasamy N, et al. Detection of tuberculosis using cough audio analysis: a deep learning approach with capsule networks.

Discover Artificial Intelligence 2024;4(1). Available from https://doi.org/10.1007/s44163-024-00179-4.

Raymond JL, Medina JF. Computational principles of supervised learning in the cerebellum. Annual Review of Neuroscience 2018;41:233–53. Available from https://doi.org/10.1146/annurev-neuro-080317-061948.

Razavi ZS, Soltani M, Souri M, Pazoki-Toroudi H. CRISPR-driven biosensors: a new frontier in rapid and accurate disease detection. Critical Reviews in Analytical Chemistry 2024:1–25. Available from https://doi.org/10.1080/10408347.2024.2400267.

Raziq A, Ahmed N, Khan S, Bizanjo M, Uddin N, Baloch R. Development of light-weight convolutional neural network model to diagnose tuberculosis. VFAST Transactions on Software Engineering 2022;10(3):43–50. Available from https://doi.org/10.21015/vtse.v10i3.1148.

Reker D, Schneider P, Schneider G, Brown JB. Active learning for computational chemogenomics. Future Medicinal Chemistry 2017;9(4):381–402. Available from https://doi.org/10.4155/fmc-2016-0197.

Ren W, Zhou Y, Li H, Shang Y, Zhang X, Yuan J, et al. Development and clinical evaluation of a CRISPR/Cas13a-based diagnostic test to detect Mycobacterium tuberculosis in clinical specimens. Frontiers in Microbiology 2023;14. Available from https://doi.org/10.3389/fmicb.2023.1117085.

Ribeiro LV, Cancino-Bernardi J, Razzino CDA, Machado TR, Tuesta MAM, Zucolotto V. Exploring electrochemical impedance spectroscopy for the early diagnosis of Mycobacterium tuberculosis using CFP10:ESAT6 protein detection. Frontiers in Sensors 2024;5. Available from https://doi.org/10.3389/fsens.2024.1512936.

Rosenblatt F. The perceptron: a probabilistic model for information storage and organization in the brain. Psychological Review 1958;65(6):386–408. Available from https://doi.org/10.1037/h0042519.

Sahoo RK, Sinha A, Mishra M, Bhattacharya D, Pati S, Sahoo KC. A systematic review and meta-analysis of the diagnostic accuracy of artificial intelligence in detecting tuberculosis using cough sounds. The Lancet 2025. Available from 10.2139/ssrn.5242653.

Sajjan AC, Jha H, Borulkar T, Regalla S, Yadav R. Systematic review: biosensors for early detection of infectious diseases. Journal of Contemporary Clinical Practice 2025;11:195–202.

Samandari T, Bishai D, Luteijn M, Mosimaneotsile B, Motsamai O, Postma M, et al. Costs and consequences of additional chest X-ray in a tuberculosis prevention program in Botswana. American Journal of Respiratory and Critical Care Medicine 2011;183(8):1103–11. Available from https://doi.org/10.1164/rccm.201004-0620OC.

Sathitratanacheewin S, Sunanta P, Pongpirul K. Deep learning for automated classification of tuberculosis-related chest X-ray: dataset distribution shift limits diagnostic performance generalizability. Heliyon 2020;6(8). Available from https://doi.org/10.1016/j.heliyon.2020.e04614.

Schiff HF, Walker NF, Ugarte-Gil C, Tebruegge M, Manousopoulou A, Garbis SD, et al. Integrated plasma proteomics identifies tuberculosis-specific diagnostic biomarkers. JCI Insight 2024;9(8). Available from https://doi.org/10.1172/jci.insight.173273.

Schwab TC, Perrig L, Göller PC, Guebely De la Hoz FF, Lahousse AP, Minder B, et al. Targeted next-generation sequencing to diagnose drug-resistant tuberculosis: a systematic review and meta-analysis. The Lancet Infectious Diseases 2024;24(10):1162–76. Available from https://doi.org/10.1016/S1473-3099(24)00263-9.

Seele PP, Dyan B, Skepu A, Maserumule C, Sibuyi NRS. Development of gold-nanoparticle-based lateral flow immunoassays for rapid detection of TB ESAT-6 and CFP-10. Biosensors 2023;13(3). Available from https://doi.org/10.3390/bios13030354.

Semogan ARC, Gerardo BD, Tanguilig III BT, de Castro JT, Cervantes LF. A rule-based fuzzy diagnostics decision support system for tuberculosis. 2011 Ninth International Conference on Software Engineering Research, Management and Applications, Baltimore, MD, USA. IEEE; 2011. p. 60–3. Available from 10.1109/SERA.2011.30.

Şen M, Oğuz M, Avcı İ. Nontoxic flexible screen-printed MWCNT-based electrodes for noninvasive biomedical applications. Talanta 2024;268(Pt 2):125341. Available from https://doi.org/10.1016/j.talanta.2023.125341.

Shabut AM, Hoque Tania M, Lwin KT, Evans BA, Yusof NA, Abu-Hassan KJ, et al. An intelligent mobile-enabled expert system for tuberculosis disease diagnosis in real time. Expert Systems with Applications 2018;114:65–77. Available from https://doi.org/10.1016/j.eswa.2018.07.014.

Sharma M, Nduba V, Njagi LN, Murithi W, Mwongera Z, Hawn TR, et al. TBscreen: a passive cough classifier for tuberculosis screening with a controlled dataset. Science Advances 2024;10(1). Available from https://doi.org/10.1126/sciadv.adi0282.

Shickel B, Tighe PJ, Bihorac A, Rashidi P. Deep EHR: a survey of recent advances in deep learning techniques for electronic health record (EHR) analysis. IEEE Journal of Biomedical and Health Informatics 2018;22(5):1589–604. Available from https://doi.org/10.1109/jbhi.2017.2767063.

Shortliffe EH. Computer programs to support clinical decision making. JAMA: The Journal of the American Medical Association 1987;258(1):61–6. Available from https://doi.org/10.1001/jama.258.1.61.

Singh M, Pujar GV, Kumar SA, Bhagyalalitha M, Akshatha HS, Abuhaija B, et al. Evolution of machine learning in tuberculosis diagnosis: a review of deep learning-based medical applications. Electronics (Switzerland) 2022;11(17). Available from https://doi.org/10.3390/electronics11172634.

Singh SP, Wang L, Gupta S, Gulyas B, Padmanabhan P. Shallow 3D CNN for detecting acute brain hemorrhage from medical imaging sensors. IEEE Sensors Journal 2021;21(13):14290–9. Available from https://doi.org/10.1109/JSEN.2020.3023471.

Singhania A, Wilkinson RJ, Rodrigue M, Haldar P, O'Garra A. The value of transcriptomics in advancing knowledge of the immune response and diagnosis in tuberculosis. Nature Immunology 2018;19(11):1159–68. Available from https://doi.org/10.1038/s41590-018-0225-9.

Sirohi M, Lall M, Yenishetti S, Panat L, Kumar A. Development of a machine learning image segmentation-based algorithm for the determination of the adequacy of gram-stained sputum smear images. Medical Journal Armed Forces India 2022;78(3):339–44. Available from https://doi.org/10.1016/j.mjafi.2021.09.012.

Skoura E, Zumla A, Bomanji J. Imaging in tuberculosis. International Journal of Infectious Diseases 2015;32:87–93. Available from https://doi.org/10.1016/j.ijid.2014.12.007.

Srivastava V, Kumar R, Wani MY, Robinson K, Ahmad A. Role of artificial intelligence in early diagnosis and treatment of infectious diseases. Infectious Diseases 2025;57(1):1–26. Available from https://doi.org/10.1080/23744235.2024.2425712.

Steingart KR, Ng V, Henry M, Hopewell PC, Ramsay A, Cunningham J, et al. Sputum processing methods to improve the sensitivity of smear microscopy for tuberculosis: a systematic review. Lancet Infectious Diseases 2006;6(10):664–74. Available from https://doi.org/10.1016/S1473-3099(06)70602-8.

Suda C. Early detection of tuberculosis with machine learning cough audio analysis: towards more accessible global triaging usage. arXiv:2310.17675 2023. Available from https://doi.org/10.48550/arXiv.2310.17675.

Sun G, Wang Q, Shan X, Kuerbanjiang M, Ma R, Zhou W, et al. Metabolomics and lipidomics of plasma biomarkers for tuberculosis diagnostics using UHPLC-HRMS. Frontiers in Cellular and Infection Microbiology 2025;15. Available from https://doi.org/10.3389/fcimb.2025.1526740.

Sweeney TE, Braviak L, Tato CM, Khatri P. Genome-wide expression for diagnosis of pulmonary tuberculosis: A multicohort analysis. The Lancet Respiratory Medicine 2016;4(3):213–24. Available from https://doi.org/10.1016/S2213-2600(16)00048-5.

Tafadzwa Mpofu K, Mthunzi-Kufa P. Recent advances in artificial intelligence and machine learning based biosensing technologies. In: Karakuş S, editor. Current Developments in Biosensors and Emerging Smart Technologies. IntechOpen; 2025. Available from http://dx.doi.org/10.5772/intechopen.1009613.

Tamura G, Llano G, Aristizábal A, Valencia J, Sua L, Fernandez L. Machine-learning methods for detecting tuberculosis in Ziehl-Neelsen stained slides: systematic literature review. Intelligent Systems with Applications 2024;22:200365. Available from https://doi.org/10.1016/j.iswa.2024.200365.

Tapley A, Switz N, Reber C, Davis JL, Miller C, Matovu JB, et al. Mobile digital fluorescence microscopy for diagnosis of tuberculosis. Journal of Clinical Microbiology 2013;51(6):1774–8. Available from https://doi.org/10.1128/JCM.03432-12.

Thirunavukkarasu J, Charumathi R, Parthiban M, Pavithra V, Kirthana Mohan R, Thiyagarajan J. Optimizing Tuberculosis Detection Through DenseNet Architecture and Ensemble Learning Techniques. 2024 International Conference on Innovative Computing, Intelligent Communication and Smart Electrical Systems (ICSES), Chennai, India. IEEE; 2024. p. 1–6. Available from 10.1109/ICSES63760.2024.10910865.

U.S. Food & Drug Administration. Digital Health Software Precertification (Pre-Cert) Pilot Program. U.S. Food & Drug Administration; 2019. Available from https://www.fda.gov/medical-devices/digital-health-center-excellence/digital-health-software-precertification-pre-cert-program (accessed on 7 May 2022).

Unanah OV, Mbanugo OJ. Telemedicine and mobile health imaging technologies: Business models for expanding U.S. Healthcare Access. International Journal of Science and Research Archive 2025;14(02):479–89. Available from https://doi.org/10.30574/ijsra.2025.14.2.0398.

Useni S, Odume B, Tukur M, Chukwu E, Nwokoye N, Ogbudebe C, et al. Comparative analysis of the impact of portable digital X-ray on TB screening in hard-to-reach areas in Nigeria. Public Health Action 2025;15(1):17–20. Available from https://doi.org/10.5588/pha.24.0040.

van Engelen JE, Hoos HH. A survey on semi-supervised learning. Machine Learning 2020;109(2):373–440. Available from https://doi.org/10.1007/s10994-019-05855-6.

Varoquaux G, Cheplygina V. Machine learning for medical imaging: methodological failures and recommendations for the future. npj Digital Medicine 2022;5(1). Available from https://doi.org/10.1038/s41746-022-00592-y.

Veropoulos K, Campbell C, Learmonth G, Knight B, Simpson J. The Automated Identification of Tubercle Bacilli using Image Processing and Neural Computing Techniques. In: Niklasson L, Bodén M, Ziemke T, editors. ICANN 98. ICANN 1998. Perspectives in Neural Computing. London: Springer; 2013. p. 797–802. Available from https://doi.org/10.1007/978-1-4471-1599-1_123.

Vijayan S, Jondhale V, Pande T, Khan A, Brouwer M, Hegde A, et al. Implementing a chest X-ray artificial intelligence tool to enhance tuberculosis screening in India: lessons learned. PLOS Digital Health 2023;2(12):e0000404. Available from https://doi.org/10.1371/journal.pdig.0000404.

Wasilewski T, Kamysz W, Gębicki J. AI-Assisted detection of biomarkers by sensors and biosensors for early diagnosis and monitoring. Biosensors 2024;14(7):356. Available from https://doi.org/10.3390/bios14070356.

Weber M, Welling M, Perona P. Unsupervised learning of models for recognition. Computer Vision - ECCV 2000. Berlin, Heidelberg: Springer; 2000. p. 18–32.

Wei X, Liang D, Zhang Z, Thorpe K, Zhou L, Zhao J, et al. Active case finding using mobile vans equipped with artificial intelligence aided radiology tests and sputum collection for rapid diagnostic tests to reduce tuberculosis prevalence in rural China: protocol for a pragmatic trial. medRxiv 2024. Available from https://doi.org/10.1101/2024.12.08.24318678.

Witarto AB, Ceachi B, Popp C, Zurac S, Daha IC, Sari FE, et al. AI-based analysis of Ziehl-Neelsen-stained sputum smears for Mycobacterium tuberculosis as a screening method for active tuberculosis. Life (Basel, Switzerland) 2024;14(11):1418. Available from 10.3390/life14111418.

Wong A, Lee JRH, Rahmat-Khah H, Sabri A, Alaref A, Liu H. TB-net: a tailored, self-attention deep convolutional neural network design for detection of tuberculosis cases From chest x-ray images. Frontiers in Artificial Intelligence 2022;5. Available from https://doi.org/10.3389/frai.2022.827299.

World Health Organization. WHO consolidated guidelines on tuberculosis: Module 2: screening – systematic screening for tuberculosis disease [Internet]. Geneva: World Health Organization; 2021. Available from https://www.who.int/publications/i/item/9789240022676.

World Health Organization. WHO consolidated guidelines on tuberculosis (Module 1 – Prevention): Tuberculosis preventive treatment, second edition. Geneva: World Health Organization; 2024. Available from https://www.who.int/publications/i/item/9789240097773.

Worodria W, Castro R, Kik SV, Dalay V, Derendinger B, Festo C, et al. An independent, multi-country head-to-head accuracy comparison of automated chest x-ray algorithms for the triage of pulmonary tuberculosis. medRxiv [Preprint] 2024. Available from 10.1101/2024.06.19.24309061.

Xu X, Li J, Zhu Z, Zhao L, Wang H, Song C, et al. A comprehensive review on synergy of multi-modal data and AI technologies in medical diagnosis. Bioengineering 2024;11(3):219. Available from https://doi.org/10.3390/bioengineering11030219.

Yadav S, Jeyaraman N, Jeyaraman M, Rawal G. Artificial intelligence in tuberculosis diagnosis: revolutionizing detection and treatment. IP Indian Journal of Immunology and Respiratory Medicine 2024;9(2):85–7. Available from https://doi.org/10.18231/j.ijirm.2024.017.

Yang X, Wang Y, Byrne R, Schneider G, Yang S. Concepts of artificial intelligence for computer-assisted drug discovery. Chemical Reviews 2019;119(18):10520–94. Available from https://doi.org/10.1021/acs.chemrev.8b00728.

Yang Z, Li J, Shen J, Cao H, Wang Y, Hu S, et al. Recent progress in tuberculosis diagnosis: insights into blood-based biomarkers and emerging technologies. Frontiers in Cellular and Infection Microbiology 2025;15. Available from https://doi.org/10.3389/fcimb.2025.1567592.

Yellapu GD, Rudraraju G, Sripada NR, Mamidgi B, Jalukuru C, Firmal P, et al. Development and clinical validation of Swaasa AI platform for screening and prioritization of pulmonary TB. Scientific Reports 2023;13(1). Available from https://doi.org/10.1038/s41598-023-31772-9.

Yu KH, Beam AL, Kohane IS. Artificial intelligence in healthcare. Nature Biomedical Engineering 2018;2(10):719–31. Available from https://doi.org/10.1038/s41551-018-0305-z.

Yu Y, Jiang XX, Li JC. Biomarker discovery for tuberculosis using metabolomics. Frontiers in Molecular Biosciences 2023;10. Available from https://doi.org/10.3389/fmolb.2023.1099654.

Zhang W, Dogar ME, Jain M, Rodriges E, Pathak S, Bhargava S, et al. Feasibility and effectiveness of mobile app for active case finding for tuberculosis in India. Telehealth and Medicine Today 2020;5(3). Available from https://doi.org/10.30953/tmt.v5.177.

Zhang Y, Hu Y, Jiang N, Yetisen AK. Wearable artificial intelligence biosensor networks. Biosensors and Bioelectronics 2023;219:114825. Available from https://doi.org/10.1016/j.bios.2022.114825.

Zhao Y, Wang X, Sun T, Shan P, Zhan Z, Zhao Z, et al. Artificial intelligence-driven electrochemical immunosensing biochips in multi-component detection. Biomicrofluidics 2023;17(4). Available from https://doi.org/10.1063/5.0160808.

Zhou Z, Tian J, He Y, Xiong H, Wang W. Proteomic analysis of plasma unravels dynamic pathways and potential biomarkers indicating disease stages following MTB infection. mSystems 2025;10(8):e00616–25. Available from 10.1128/msystems.00616-25.

Zoph B, Le QV. Neural architecture search with reinforcement learning. arXiv:1611.01578 2016. Available from https://doi.org/10.48550/arXiv.1611.01578.

Zurac S, Mogodici C, Poncu T, Trăscău M, Popp C, Nichita L, et al. A new artificial intelligence-based method for identifying Mycobacterium tuberculosis in Ziehl–Neelsen stain on tissue. Diagnostics 2022;12(6):1484. Available from https://doi.org/10.3390/diagnostics12061484.

CHAPTER 13

Ethical and safety consideration for *Mycobacterium tuberculosis* diagnostics

Kanishka Bhatia[1], Maumita Das Mukherjee[1], and Chetna Dhand[2]
[1]Amity Institute of Applied Sciences, Amity University, Uttar Pradesh, India,
[2]Academic of Scientific and Innovative Research (AcSIR), CSIR-Advanced Materials and Processes Research Institute, Bhopal, Madhya Pradesh, India

13.1 Introduction

The diagnostic scene of *Mycobacterium tuberculosis* (MTB), the etiological agent of tuberculosis (TB), has continued to experience an incredible transformation over the last decades and has shifted its diagnostic focus to traditional phenotypic approaches of the conventional ones to advanced molecular techniques and innovative biosensor systems (Mishra et al., 2025; Joshi et al., 2022). Although conventional diagnostic methods such as sputum smear microscopy and culturebased diagnostic methods are considered the mainstay of TB control programs in most countries, their drawbacks that include long turnaround time, low sensitivity in cases of paucibacillary infections and high infrastructure needs have necessitated the invention and use of more sophisticated diagnostic tools (Mishra et al., 2025; Mousavi-Sagharchi et al., 2024).

The modern diagnostic spectrum includes quick nucleic acid amplification assays (NAATs) like Xpert MTB/rifampicin (RIF), isothermal amplification techniques: TB-LAMP, and extensive genomic methods such as whole genome sequencing (WGS) to profile drug resistance and epidemiological monitoring (Chaturvedi et al., 2024; Joshi et al., 2022). These molecular technologies have redefined the situation where a week-long diagnostic process was done to a same-day identification of results that would help initiate treatment earlier and have better patient outcomes (Joshi et al., 2022). At the same time, with the advent of nanotechnology-based biosensor platforms based on the principles of electrochemical, optical, piezoelectric, and magnetic detection, even more democratization of TB diagnostics is expected through the introduction of portable, inexpensive, and easytouse point of care devices based on the latter (Hasan et al., 2024; Thangarasu et al., 2023). However, this rapid technological advancement in MTB diagnostics brings critical ethical and safety considerations that demand rigorous examination and proactive management (Shivakumar and Shettigar, 2022). The deployment of novel diagnostic technologies raises fundamental questions about

Advances in Biosensor Technology for Mycobacterium Tuberculosis Detection
DOI: https://doi.org/10.1016/B978-0-443-44589-7.00002-9

biosafety protocols, particularly regarding specimen handling and containment to protect health care workers and prevent cross-contamination (Shivakumar and Shettigar, 2022). Equally significant are concerns related to diagnostic accuracy and its clinical consequences, where false negative results may perpetuate disease transmission and false positive outcomes can lead to unnecessary treatment and psychological distress (Mishra et al., 2025; Mousavi-Sagharchi et al., 2024). The challenge of ensuring equitable access to advanced diagnostic technologies remains paramount, as cost barriers, infrastructure limitations, and supply chain constraints often restrict the availability of lifesaving diagnostics in high-burden, resource-limited settings where they are most urgently required (World Health Organization, 2023; Joshi et al., 2022).

Nonetheless, such a fast technological rise in the field of MTB diagnostics is associated with urgency in ethical and safety issues that require strict analysis and active monitoring (Shivakumar and Shettigar, 2022). The introduction of new diagnostics technologies lays some critical doubts about the biosafety measures that need to be taken, especially when it comes to the handling and safe seclusion of the specimen so that the health care systems stay safe and the cross-contamination can be avoided (Shivakumar and Shettigar, 2022). Essentially, It is the issue of the diagnostic accuracy and its clinical implications whereby, false-negative outcomes can infuse disease and its spread, and false-positive outcomes can cause unnecessary treatment and psychological trauma (Mishra et al., 2025; Mousavi-Sagharchi et al., 2024). An equitable access to highlevel diagnostic technologies is the main challenge because the supply barriers, infrastructure, and constraints on supply chains make the introduction of life saving diagnostics in high-burden, resource constrained environments, where it is needed most of all, often rather challenging (World Health Organization, 2023; Joshi et al., 2022). Moreover, the validation and regulatory process of novel biosensor technologies remain complicated because a number of potentially viable prototypes, have to undergo large-scale clinical validation and standardization processes before they can be safely applied to regular clinical use (Hasan et al., 2024; Thangarasu et al., 2023). Human-centred design is especially important when dealing with situations with limited resources where diagnostic equipment needs to be both highly reliable and requires minimal infrastructure but with userfriendly interfaces reducing the number of errors committed by the operator and safeguarding such health care personnel (Shivakumar and Shettigar, 2022). With the field of the field moving on to more sophisticated and digitally integrated solutions in diagnosis, the concerns of data privacy, digital security and the conscientious use and protection of patient data are becoming a more prominent question when it comes to responsible implementation of the diagnostic solutions of TB (Price and Cohen, 2019).

This covers all the different facets of the diagnostics of MTB, that includes the traditional analysis to highly complicated molecule diagnostics and the latest technological biosensors. It provides a paradigm towards which the tremendous potential of tuberculosis diagnostic tools can be perceived as well as the accountability of the continuous improvement of the tools.

13.1.1 Overview of *Mycobacterium tuberculosis* and its global impact

MTB or the microbe that causes TB is the source of the current epidemic which engulfs millions of people all over the globe. Even though TB is an illness mainly linked to pulmonary sickness, the universal bacterium is capable of penetrating and making critical infections in almost all other areas of the body. Transmission of the TB organism occurs by a method of inhalation of the airborne fluid as the person with active pulmonary TB coughs, sneezes, or even talks that is infected, which makes the TB organism very contagious, especially in well-heated or not wellventilated facilities. A significant part of the world population is infected with latent TB that may develop into active disease especially in immunocompromised persons (Chaturvedi et al., 2024; Hasan et al., 2024). Moreover, the small percentage of infected individuals that may cause the manifestation of the disease is latent TB infection, which is present in a high percentage of the world population, contributes to the significance of the challenge. Although this latent condition is asymptomatic in its onset, it may resurface and progress to the full blown active TB disease especially in individuals who have weakened immune systems. TB can be described as having an enormous worldwide effect, leading to millions of fresh infections and deaths. Low and middle income countries bear this financial burden disproportionately. These areas have interlocked health and socio-economic determinants such as poverty, malnutrition, co-infection with HIV, and poor access to quality health care, which enhance rates of incidence. This issue is worsening due to the alarming rate of drug-resistant TB including multidrug-resistant TB (MDR-TB) and the more intractable extensively drug-resistant TB (XDR-TB). The increased rates of drug-resistant phenotypes among populations of MTB patients present a significant complex challenge into the TB control strategies which requires an urgent paradigm shift to a rapid and high level of reliability in diagnostic techniques. The fast and accurate identification of antimicrobial resistance profiles has been at the core of informing the effective treatment regimens thus maximising patient outcome and reducing treatment time. At the same time, early and precise etiological characterization is essential to timely applying to the public health interventions to break the chain of transmission and reduce the high risk of contagiousness of these hard to treat strains (Shivakumar and Shettigar, 2022).

13.1.2 Evolution of tuberculosis diagnostic technologies

The diagnosis of TB has experienced an impressive revolution and the last century, the methods that were used in diagnosis have gone through metamorphosis to modern day advanced technology involving molecules. Traditional diagnostic tools are sputum smear microscopy, which is the basis of TB diagnosis since the late 19th century and mycobacterial culture, which is the gold standard of TB diagnosis, although it takes a long time to turnaround (Ghebreyesus et al., 2023). These techniques are basic, but they possess severe shortcomings on sensitivity, specificity, and time to results. Introduction of molecular diagnostics in this field has redefined the testing of TB, the NAATs including GeneXpert MTB/RIF, Xpert MTB/RIF Ultra and other quick molecular tests has provided better sensitivity, quicker result and additionally it has the ability to detect drug resistance (Price and Cohen, 2019; Elbehiry et al., 2025). Such technologies have been shown to be more effective than traditional ones, and in particular, in the detection of the paucibacillary TB and the identification of the drug-resistant strains in several hours instead of weeks (Cao et al., 2021). Recently, biosensor technologies have turned out to be a promising frontier in the diagnostic of TB with a potential of providing the point of care testing which is rapid, sensitive and specific in its detection. These platforms combine the aspects of biological recognition with physical or chemical transducers to identify MTB-specific biomarkers, such as cells of bacteria, have sequences of DNA, protein, or metabolites (Sharma et al., 2015; Peeling et al., 2006).

13.1.3 Role of biosensor technology in tuberculosis detection

Biosensor technology is evolving rapidly and can be used to entirely transform the diagnostic aspect of TB. These technological devices are intelligent sensors through an integration of a biological recognition component whereby they are specifically designed to target the components of the TB bacterium with a transduction system that converts that binding reaction into a measurable signal (Mehrotra, 2016). The combination allows determining how fast, sensitive, and extremely specific identification of MTB and biological markers associated with this disease can take place. Biosensors can be tailored to detect a broad range of different targets, including entire bacterial cells, single specific DNA or RNA sequences, signature proteins and unique metabolic products which are indicative of an MTB infection, which is one of their most powerful and empowering features. The latter feature allows the diagnostic services to be offered in centralised labs and be transferred to local clinics or even to the bedside of the patient (Pai et al., 2012). This can allow individuals to obtain a diagnosis and initiate life-saving treatment much earlier and this can be especially useful in an environment with constraints on available resources in which access to specialised institutions is often grossly limited. Moreover, some biosensors platforms are demonstrating a high level of sensitivity and specificity, which may largely impact the accuracy of diagnostic power. There is a great prospect of biosensor technology improving the global fight against TB since this technology is in its fast development.

These methods can significantly enhance patient care and contribute to the global control efforts with the help of faster diagnoses and more well informed treatment choices (Drain et al., 2014). Since the first creation and clinical testing of the technologies to their extensive application, the moral and safety concerns regarding the entire life cycle are to be considered attentively and comprehensively. The second important thing in achieving the full potential of such high-tech solutions to humanity is to ensure that they are fitted correctly and ethically (Peeling and Mcnerney, 2014).

13.1.4 Importance of ethical and safety considerations in tuberculosis diagnostics

There is a need to proactively change established ethical standards as there has been continuous evolution in the diagnostic technology particularly the discovery of complex molecular tests and new biosensor solutions. When it comes to the application of any diagnostic tool, whether highly technical or otherwise, we should ensure that it is rooted in some kind of fundamental respect of the individual since the scientific innovations get integrated into the clinical practice. This will involve placing the welfare of patients and personal autonomy as the primary concern, and offering to everyone equal access to these new innovations, and maintaining the utmost safety of all medical practitioners and other people (Najjari et al., 2022).

The ethical environment of modern diagnostics is also very difficult, and it is necessary to overcome it with caution. Basic notions, such as informed consent, are very fundamental; patients should understand what a test involves, what are its implications, and its limitations to an extent before consenting to a test. Also, the need to secure patient confidentiality and high data privacy to protect the most sensitive genetic and health data is gaining more importance as more and more powerful diagnostic equipment is interconnected with the digital health network (Daftary, 2012). The other urgent ethical imperative that deals with equity and access is ensuring that the most affected populations by the diseases such as TB are able to maximize on the benefits of technological advances without a sense of financial or geographical barriers. To transform the diagnosis into a care giving approach instead of a source of social evil, the diagnostic process should be proactive in the minimization of the prospect of stigma and discrimination that are at times related to favourable outcome (Selgelid and Reichman, 2011). Ethical environment of TB diagnostics is also inherently complicated, and the consideration of individual rights, the principle of social justice, and community implications of testing technologies should be considered generally, which is emphasized by the range of problems as depicted in Fig. 13.1.

The high safety and scientific criteria required with the use of advanced diagnostics are also significant. In the area of safety, following the biosafety precautions on how to

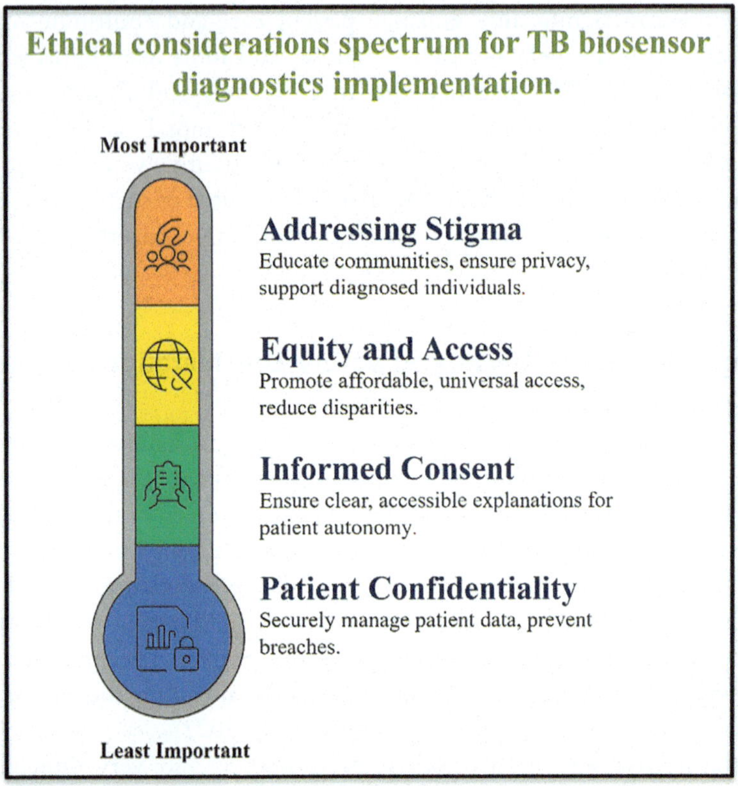

Figure 13.1 Ethical consideration spectrum for TB biosensor diagnostics implementation. *TB*, tuberculosis. illustrates the hierarchical spectrum of ethical considerations essential for the implementation of TB biosensor diagnostics, ranging from foundational patient confidentiality to the overarching priority of addressing social stigma.

handle potentially infectious agents, including the samples of MTB, is necessary to prevent exposure either in a laboratory or in a clinical facility (Das et al., 2011). This relates greatly with the protection of health and safety of all the dedicated medical and laboratory staff that handle these samples on a daily basis. In addition, proper enforcement of proper biological waste disposal protocols is required to protect the environment and the human health in a show of a responsible implementation process (Shickle and Chadwick, 1994). The value of accurate diagnosis cannot be overrated scientifically and in the public health perspective. Since causing unwarranted, possibly dangerous treatment to an infectious disease to spread without being detected, not being able to eliminate false positive or false negative outcomes can be very dizadvantageous. In effect therefore, the scientific integrity and reliability of these assays is more of a moral obligation to patient care and health outcomes in the community (Swaminathan et al., 2021).

13.2 Comparative analysis of tuberculosis diagnostic methods and associated ethical issues

13.2.1 Conventional diagnostic methods

13.2.1.1 Sputum smear microscopy

Microscopy of Sputum smear had been developed during the late 19th century, and is common until today especially in resource deprived setting due to its cheapness and ease of practice. The sensitivity of this method lies between 20% and 80% according to the quantity of bacteria and quality of the specimen, presents significant drawbacks (Steingart et al., 2006). Studies have also revealed that microscopy is sensitive and specific with a sensitivity of approximately 52% relative to molecular methods, specificity of 99.4 % and lack of power to identify paucibacillary specimen (de Vos et al., 2024). To employ the procedure, several sputum samples should be collected over a period of days and this may delay the diagnosis and commencement of treatment.

Ethical considerations:

- **Accessibility:** It is cheap, making it accessible in the resource poor environments which encourages equity.
- **Diagnostic delays:** Poor sensitivity could lead to missed diagnoses and continued transmission.
- **Patient burden:** Repeated specimen collection over several days causes inconvenience to the patient, with a greater risk of loss to follow-up.

13.2.1.2 Mycobacterial culture

The most available tests include culture upon *mycobacterium* associated with greatest sensitivity (approximately 98.8%), and thus the method of TB diagnosis of the gold standard (Steingart et al., 2006). It is, however, the long time required to allow culture development and additional time to perform drug susceptibility testing that greatly delay the proper start of treatment (Cruciani et al., 2004).

Ethical considerations:

- **Diagnostic accuracy:** High sensitivity lowers the rate of missed cases.
- **Treatment delays:** Extended turnaround time may lead to disease progression and ongoing transmission.
- **Resource allocation:** Requires a well equipped advanced level laboratory infrastructure along with skilful personnel.

13.2.2 Modern molecular diagnostic methods

13.2.2.1 Nucleic acid amplification tests

In comparison with conventional methods, NAATs offer faster results as it is based on PCR assays, and are more sensitive (approximately 79.8% when compared to other studies) (Steingart et al., 2006). These tests have even been found to provide results in an hour as opposed to weeks since they can directly identify the presence of the DNA or the RNA of the MTB in clinical samples.

Ethical Considerations:

- **Improved outcomes:** Faster diagnosis enables earlier treatment initiation and reduced transmission.
- **Cost barriers:** Higher per test costs may limit accessibility in resource-poor settings.
- **Technical requirements:** Need for trained personnel and infrastructure may create implementation challenges.

13.2.2.2 GeneXpert Mycobacterium tuberculosis*/rifampicin and xpert ultra*

The GeneXpert technology is also another significant advance in the TB diagnosis in that it offers automated test results based on a cartridge format and the results are available within a maximum of 2 hours. It has been demonstrated that there are high sensitivity and specificity in both TB and the determination of the resistance of rifampicin (Boehme et al., 2011). The Xpert Ultra version is more sensitive and paucibacillary specimens are particularly sensitive to the Xpert Ultra version (Chakravorty et al., 2017).

Ethical considerations:

- **Rapid diagnosis:** Median treatment initiation delay reduced by approximately 2.6 days for drug-sensitive TB (Cruciani et al., 2004).
- **Drug resistance detection:** Simultaneous rifampicin resistance testing enables appropriate treatment selection.
- **Equity concerns:** Higher costs and infrastructure requirements may limit access in high burden, resource limited settings.
- **Invalid results:** Some tests produce invalid or indeterminate results, requiring repeat testing or alternative methods.

13.2.2.3 Line probe assays and advanced molecular platforms

It is now possible to test multiple antiTB resistance simultaneously on more recent molecular platforms, like line probe assays and full resistance test panel (such as Xpert MTB/XDR). Featuring comparatively to the conventional methods, the technologies have demonstrated a substantial shortening of the time to appropriate drug resistant TB therapy, resulting in delays of approximately 45 days (Lee et al., 2022).

Ethical Considerations:

- **Comprehensive resistance detection:** Enables rapid initiation of appropriate regimens for drug-resistant TB.
- **Resource intensity:** High costs and technical requirements may limit implementation.
- **Diagnostic accuracy:** While highly accurate for common resistance mutations, may miss rare or novel resistance mechanisms (Pillay et al., 2022).

13.2.3 Comparative ethical analysis

13.2.3.1 Diagnostic performance and patient outcomes

The newest molecular techniques have an improved sensitivity and speed over the traditional ones and thus they have a direct influence on the outcome of patients by providing an earlier diagnosis and earlier initiation of therapy as shown in Table 13.1. This advancement is consistent with the ethical principle of beneficence by decreasing morbidity, mortality, and risk of transmission (Varkey, 2020). Nonetheless, the benefits of performance should be balanced with the provisions of accessibility and the cost.

13.2.3.2 Equity and access

Although standard approaches continue to be more affordable in a resourcerestrictive environment since they are cheaper and need less infrastructure, their poorer performance can continue to facilitate health inequalities by exposing cases to inadequate treatment. On the other hand, improved molecular applications have high performance, but they can be used to promote inequity when used without offering them to everyone (Daniels, 2007).

13.2.3.3 Resource allocation and sustainability

The increased price of the molecular diagnostics poses significant concerns with regard to allocation of resources and the opportunity cost. Cost-effectiveness studies recommend the combined approach based on the application of both molecular tests and selective culture to be economical in detecting MDR-TB in some settings (Yadav et al., 2022). The policymakers need to find the right balance between the high performances of superior methods on one side and the need of resources and longterm sustainability on the other.

13.2.3.4 Safety and occupational health

Various diagnostic procedures have diverse biosafety hazards. Although some of the handling requirements can be minimized using molecular methods, issues of incomplete inactivation of the specimen in some of these protocols have been expressed with a need of strong biosafety measures despite the mode of diagnosis being used (Luukinen et al., 2024).

Table 13.1 Comparison of TB diagnostic methods and ethical considerations.

Diagnostic method	Sensitivity	Specificity	Time to results	Cost	Key ethical issues	References
Smear microscopy	50%–60%	95–98%	1–2 hours	$1–2	False negatives leading to delayed treatment; inadequate for HIV patients	de Vos et al. (2024)
Culture (Solid)	80%–85%	>99%	4–8 weeks	$5–10	Delayed treatment initiation, prolonged patient anxiety, and biosafety risks	Steingart et al. (2006)
Culture (Liquid)	85%–90%	>99%	2–4 weeks	$10–15	Laboratory-acquired infections; complex waste disposal requirements	Cruciani et al. (2004)
GeneXpert MTB/RIF	67%–89%	99%	2 hours	$10–15	Equity issues due to high cost; potential overreliance on technology	Boehme et al. (2011)
Xpert ultra	78%–95%	96%	2 hours	$12–18	False positives leading to unnecessary treatment; patient anxiety	Chakravorty et al. (2017)
Line probe assays	97%	99%	1–2 days	$20–30	Limited access in low-resource settings; complex interpretation	Lee et al. (2022), Pillay et al. (2022)

MTB, *Mycobacterium tuberculosis*; *RIF*, rifampicin; *TB*, tuberculosis

13.3 Safety considerations in tuberculosis diagnostics

In the worldwide fight against TB, preserving the integrity of diagnostic results and putting the safety of medical personnel, laboratory personnel, and the general public first are essential (Sen, 2018). With a focus on the consequences of incorporating biosensor technology, this section thoroughly covers important safety issues of the full range of MTB diagnostics. Strict biosafety procedures for handling potentially contagious MTB samples, reducing occupational hazards for test takers, guaranteeing the safe disposal of biohazardous material, and putting them into practice.

13.3.1 Biosafety in handling *Mycobacterium tuberculosis* samples for biosensor analysis

With the high transmissibility of MTB, stringent biosafety practices are very much essential to avoid laboratory acquired infection and environmental contamination, which is depicted in Fig. 13.2. When implying biosensor technology for MTB detection, it's crucial to strongly stick to existing biosafety guidelines for handling

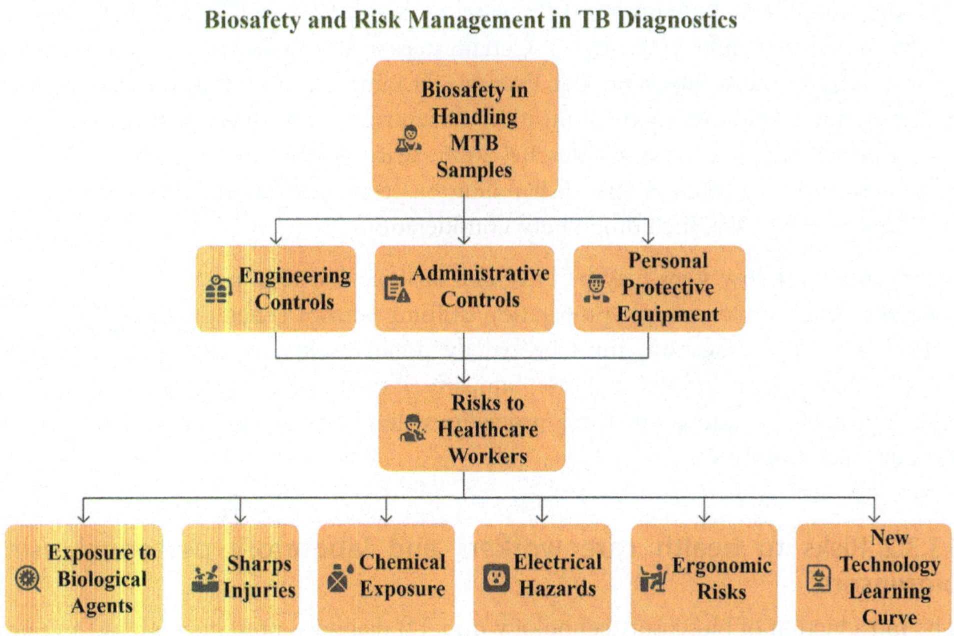

Figure 13.2 Biosafety and risks to health care workers in handling the MTB sample illustrates a hierarchical approach to biosafety management when handling *Mycobacterium tuberculosis* (MTB) samples, highlighting core control measures, Engineering, administrative, and PPE, alongside a comprehensive assessment of clinical risks ranging from biological exposure to technological learning curves.

clinical specimens. These guidelines may require adaptation based on the specific design and operational characteristics of the biosensor platform (Singh, 2009).

Laboratories processing MTB samples should be at least Biosafety level 2 (BSL-2), with recommendations for BSL-3 for procedures that involve high concentrations of the organism or those that produce aerosols (Kruse et al., 1991). Core biosafety practices include:

1. **Engineering controls:** These include laboratory design and infrastructure features with the purpose of risk reduction, including directional airflow, negative pressure environments, and the required use of BSCs for any aerosol generating activity, such as sample processing and, if part of biosensor preparation, sonication (Berche, 2026).
2. **Administrative controls:** These are policies and procedures established to ensure safe work practices, including strong standard operating procedures on the handling of samples, proper utilization of personal protective equipment, effective spill management, and proper waste disposal (Qin et al., 2023).
3. **Personal protective equipment (PPE):** Health care and laboratory workers should always use appropriate PPE, including gloves, lab coats, respiratory protection (e.g., N95 respirators for potentially aerosolised samples), and eye protection. Requirements for PPE vary by procedure and by biosafety level of the laboratory (Miller et al., 2012). Certain steps in a workflow for biosensor analysis may require more biosafety. For example, if sample preparation requires aerosol-generating techniques such s sonication or centrifugation, those operations must be performed only under a BSC (Berche, 2026; Sharp and Elder, 2004). Similarly, if the biosensor device requires special handling or decontamination, such steps must be added to the SOPs, including safety considerations.

Specific workflow steps may be biosensor analysis that requires additional biosafety concerns. As examples, sample preparation utilising aerosol generating activity, such as sonication or centrifugation, must be strictly done inside a biosafety cabinet (BSC) (Berche, 2026; Sharp and Elder, 2004). Similarly, if the biosensor device requires special handling or decontamination, these procedures should be integrated into SOPs with relevant safety measures.

13.3.2 Risks to health care workers and laboratory personnel using biosensors

The introduction of biosensor technology into TB diagnosis presents both old and new hazards for health care and laboratory practitioners (Azyabi et al., 2024). While certain biosensors may simplify certain steps of a certain diagnosis, it is imperative to locate and to mitigate potential hazards. The major considerations to be made include:

1. **Exposure to biological agents:** Although it is true that some biosensors could potentially be used at a point of care, it is still true that many will require initial sample processing in a laboratory environment, concurrent with the inherent danger of exposure to viable MTB if biosafety protocols are not rigidly observed (Salvaraji et al., 2022).
2. **Sharps injuries:** The handling of the needle and other sharp instruments used in the sample collection process still runs a risk. Adherence to safe injection practices and consistent application of designated sharps containers are important.
3. **Chemical exposure:** Some biosensor systems may contain chemical reagents that are used for sample preparation, cleaning or device operation. Proper handling, safe storage and disposal of these chemicals, along with proper use of appropriate PPE, are important to prevent chemical burns, inhalation or other adverse effects. Safety data sheets (SDS) for all the chemicals used must be readily available (Salvaraji et al., 2022).
4. **Electrical hazards:** Most biosensor devices usually need electrical power. Users of the equipment require full training in the proper use of such equipment and must constantly remain aware of electrical safety hazards. The equipment should be properly grounded and regularly maintained (Mengistu et al., 2021).
5. **Ergonomic risks:** Repetitive tasks related to biosensor device operation and/or sample processing could be factors in ergonomic injuries. Careful workstation design and implementation of proper techniques should be considered in order to minimise these risks (Berche, 2026).
6. **New technology learning curve:** Sufficient training on operation, routine maintenance and troubleshooting of biosensor devices is essential to avoid operational mistakes and potential safety hazards. This training needs to specifically include biosafety aspects particular to the technology (Berche, 2026; Papadopoli et al., 2020).

13.3.3 Safe disposal of biological waste from biosensor-based diagnostics

Effective management and disposal of biological waste products produced from TB diagnostics, such as materials which have been in contact with MTB samples (e.g. used biosensor cartridges, sample containers, gloves, swabs), are important to avoid the dissemination of infection (Peters et al., 2013). Waste disposal procedures must strictly adhere to all appropriate regulations on a local and national level (Layre et al., 2022). Generally, biological wastes related to MTB diagnostics should be

1. **Decontaminated:** Before disposal, waste should be thoroughly decontaminated through means of valid methods such as autoclaving, chemical disinfection (i.e., with

appropriate concentrations of bleach or other approved disinfectants), or by incineration. The approach adopted has to be suitable for the particular waste type.(Ministry of Health and Family Welfare, Government of India, 2024; Denkinger et al., 2015).

2. **Segregated:** At the point of creation, biological waste must be separated from other waste flows and collected in any containers marked with biohazard symbols and not leaking.
3. **Transported safely:** Whenever waste needs off-site treatment or disposal, it needs to be transported in safe, properly labelled transport containers in conformity to applicable transportation regulations (Ministry of Health and Family Welfare, Government of India, 2024).

13.3.4 Minimising false positives and false negatives in biosensor-based diagnostics

For biosensor-specific waste, such as disposable cartridges or sensors that have come into contact with MTB samples (Ministry of Health and Family Welfare, Government of India, 2024), the same principles of decontamination and safe disposal apply, as shown below, Fig. 13.3.

Manufacturers guidelines can also include specific instructions for disposing of their products. While biosensors aim to improve the diagnostic accuracy, the risk of false positive and negative results is a major safety issue with serious consequences on patient management and public health issues (Denkinger et al., 2015).

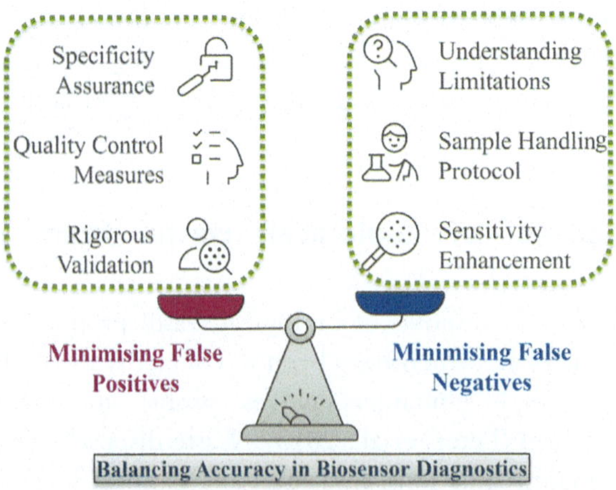

Figure 13.3 Balancing accuracy in biosensor diagnostics with evaluation of false positives and false negatives illustrates the critical balance required in biosensor diagnostics to optimize accuracy by simultaneously minimizing false positives and false negatives. The graphic highlights key technical pillars such as specificity assurance and quality control (left) against sensitivity enhancement and rigorous sample handling protocols (right).

False positives: The result of a false positive can cause excessive patient anxiety, unnecessary treatment using potentially toxic antituberculosis drugs, and misdirection of limited health care resources (Asgharzadeh et al., 2020; Peters et al., 2013; World Health Organization et al., 2022). To minimise false positives:

1. **Rigorous validation**: The rigorous analytical and clinical validation of the biosensor technology is a must, with analysis of well characterised positive and negative samples.
2. **Quality control;** Follow-up of stringent quality control procedures, either in the laboratory or at the point of care, such as the consistent use of internal and external controls, can facilitate the identification and prevention of errors (Matias et al., 2025).
3. **Specificity;** The biosensor should have a high level of specificity for MTB and not cross react with other microorganisms or substances that may be found in clinical samples is of utmost importance (Bashir et al., 2024).
4. **Proper training:** Good training of personnel for proper operation and interpretation of the biosensor test is essential to prevent technical errors that may cause test results to be reported as false positives (Pang et al., 2016).

False negatives: A false negative result can lead to a delay in diagnosis and treatment, which allows the person to continue passing the infection to others and cause disease progression (Parrish and Carroll, 2011). To minimise false negatives:

1. **Sensitivity:** The biosensor should have adequate analytical sensitivity to detect MTB even at low concentrations in clinical samples (Bashir et al., 2024).
2. **Appropriate sample collection and handling:** It is important that the collection and proper handling of high-quality sputum samples are critical for accurate results (Ministry of Health and Family Welfare, Government of India, 2024).
3. **Understanding limitations:** Limitations The health care providers need to be well aware of the limitations associated with the biosensor technology, including potential lowered sensitivity of the biosensor in certain patient populations (e.g., HIV coinfected patients) (Eivazzadeh-Keihan et al., 2024).
4. **Confirmation testing:** In some cases, especially where clinical suspicion for TB is raised despite a negative biosensor result, confirmation testing with a more sensitive reference method (e.g. mycobacterial culture or molecular test with high sensitivity) may be necessary (Das et al., 2010).

13.4 Specific ethical and safety considerations for biosensor technology

Integrating biosensor technology into the diagnostic framework for MTB opens a specific set of ethical and safety issues, which must be analysed carefully. While some of these concerns can be grouped into broader themes related to TB diagnostics, the following characteristics of biosensors, such as the possibility of rapid, point of care implementation, reliance on digital infrastructures and combination of novel materials, present complex challenges and opportunities that must be characterised carefully (Matias et al., 2025; Eivazzadeh-Keihan et al., 2024). This section aims to comprehensively consider these specific ethical and safety issues to foster a responsible and human-centred approach to the development and application of biosensor-based MTB diagnostics.

13.4.1 Ethical implications of rapid, point-of-care biosensor diagnostics

The use of the TB rapid point-of-care biosensor collagen diagnostics raises ethical issues, especially information handling, at the point of prompt. While accelerated diagnosis can certainly result in early treatment initiation and minimised transmission (Layre et al., 2022), it also requires thinking through how people receive and process this important information.

1. **Psychological impact of rapid results:** Receiving an immediate positive TB diagnosis, especially outside a conventional clinical environment, can provoke much anxiety, fear, and distress. Health workers and medical practitioners should be adequately trained to achieve the results in a defective way, offering instant counselling and connecting individuals to the relevant support services (Nahid et al., 2016). The human aspect of care and support which is so vital should not be sacrificed to overall diagnostic speed.
2. **Risk of coercion and forced testing:** The simplicity and broadscale applicability of point of care biosensors can instigate a risk to establish damaging or possibly compulsory testing (including of vulnerable groups or in a location of high prevalence). The most important thing is to make sure that the testing is voluntary and informed consent-based testing (McNerney and Daley, 2011). There should be individual autonomy and the right to natural choice so that one would not be tested without the fear of retaliation.
3. **Connection to care and treatment:** Only a speedy diagnosis can be beneficial in the sense that it needs proper and timely connection with proper care and treatment. This would involve ethical considerations when it comes to implementation because it is imperative to ensure that the infrastructure and resources needed to establish the implementation be available so that those diagnosed with the use of the point of care biosensors can access readily available follow-up services like a clinical evaluation, initiation and adherence to treatment. Lack of the necessary provision of such vital connection may render quick diagnostics ethically challenging (World Health Organization, 2017).

13.4.2 Digital biosensors: data security and privacy threats

Most high-level biosensors have been based on digital systems to manage, transmit and analyse data. This relatedness demonstrates the existence of major data security and privacy challenges, which have to be pre-empted and openly handled in order to maintain patient confidence and ethical standards.

1. **Exposure to cyberattacks and data breaches:** Electronic storage and transference of sensitive diagnostic data make such platforms vulnerable to a cyberattack and data breach. Strong cybersecurity protocols, which comprise encryption, firewalls, and safe authentications measures, ensure patient data is not exposed to any form of unauthorised access, data theft, or manipulation, among other untrusted individuals or agents (Mennella et al., 2024).
2. **Secondary use data and consent:** The data obtained through the biosensor platforms might also be used as a secondary use, including the use of data in the public health surveillance, in research or even commercial interests. Controversial and explicit policies on the use of data, including the system of receiving explicit and informed consent on any secondary use other than the immediate diagnostic focal point, are the key. The patients must have the right to choose how their data can be used.
3. **Data ownership and accessibility:** Transparency on data ownership (e.g. patient, health care provider, platform provider) and patient rights to access, review and possibly correct their diagnostic data is morally significant. Such platforms should be created to make patient access to patient information comfortable and safe in the way that patients can access it easily and effectively (Mennella et al., 2024).

13.4.3 Overdependence on the biosensor technology

The main aspects that make biosensors attractive, its speed, ease of use, and the prospects of being used in the automation context, also create issues of how it can be abused or overutilized, which might be ethically and socially unsafe. Some are as mentioned below:

1. **No adequate clinical evaluation:** The accessibility of rapid point of care testing may promote the use of a test-and-treat strategy in the absence of a comprehensive clinical evaluation that might result in unwarranted treatment of false positives or false negativities of coinfections or other diseases. In TB management, clinical judgment and holistic examination of the patient should be the focus points (Pang et al., 2016; Drain et al., 2014).
2. **Use without a clear public health rationale:** Pressure of Poor Deplorability might result in biosensors being used to implement mass screening programs that do not have a clear public health rationale, and which might be resource-wasting and

cause undue anxiety and possible stigma to individuals with positive screening results that have not been verified as active TB. The ethical standards require that the screening programs need to be evidence-based and well-focused.

3. **Risk of unauthorised or inappropriate use:** The portability and convenience of certain biosensor devices may have an inexpressible risk of unauthorised or inappropriate use outside of the constructed health care premises, which is something that needs to be updated with robust regulations and guidelines in terms of the distribution and usage of these devices.

13.4.4 Ethical challenges in deploying biosensors in low-resource settings

Although biosensors promise an enormous potential to improve TB diagnostics in a low resource context, their usage also creates distinct ethical issues that should be circumvented.

1. **Sustainability and resource allocation:** Biosensor device, consumable and maintenance costs, as well as the requirement to maintain sufficient infrastructure and personnel training may be a major preventive barrier to the sustainability of biosensors in resource constrained settings in the long-term (Kanabalan et al., 2021). The issue of ethics would require that the increase of these technologies should be financially advantageous in the long run without stealing funds of other health services of high importance.
2. **Training and capacity building**: Biosensors require proper training of the health care workers and laboratory staff to ensure proper utilization. Ethical issues are the necessity to offer full training and continuous education to be able to perform competent testing and interpretation of the results. This is especially vital in the environment where only a minimal amount of preexisting expertise exists (Denholm et al., 2012).
3. **Community engagement, cultural awareness:** The new diagnostic technologies should be introduced with the robust community engagement to create the understanding, acceptance, and trust. Attitudes to the testing and result interpretation can be culturally-based and should be carefully addressed not to create unwarranted adverse effects or further promote stigma (Lee et al., 2012).

13.4.5 Concerns related to safety of nanomaterials in biosensor

Other advanced biosensors involve the use of nanomaterial or other emerging materials so that its sensitivity and capability can be enhanced. All these materials are potential but at the same time the materials also raise special safety issues which involve consideration as mentioned below:

1. **Toxicity and biocompatibility:** When nanomaterials are involved, in biosensor device under consideration, they may need to come into contact with a biological sample or the patients, in that case, their toxicity and biocompatibility should be reconsidered. It is also extremely crucial that research on the potential health and environmental impacts of the specific nanomaterials utilized must be done prior (Verma et al., 2004).
2. **Environmental impact of disposing:** Biosensors that are made up of nanomaterials materials may have an environmental impact when they are being disposed of. On ways of disposing such devices in a way that would be safe and more environment friendly, the appropriate guidance should be drafted and followed.
3. **Manufacturing and handling procedures:** Occupational exposure to biosensors that contain these sophisticated materials may pose a risk to the workers who are engaged in the manufacturing or handling of these biosensors. Proper safety measures and PPE should be observed in the production, transportation and use in order to reduce these risks (Singh et al., 2013).

It is in these specific ethical and safety issues concerning the biosensor technology that we can strive to exploit the potential as completely as possible to our best against the devastation that is MTB in the pursuit of achieving the best diagnostics to help people and the society at large in the best possible way.

13.5 Regulatory and policy frameworks

The biosensors used in diagnostics of MTB require a robust regulatory and policy framework landscape to be in a position to have the responsible development, validation, and implementation of such biosensor-based diagnostics. These standards are used to guarantee the safety, effectiveness, quality, and moral utilization of these emerging technologies to build confidence between users, medical professionals, and communities at large (Acharya et al., 2020). This section will discuss the major international guidelines, ethical review procedures as well as compliance regulations at the global and local levels that will influence the course of biosensor based TB diagnostics.

13.5.1 International standards of biosensor based TB-related diagnostics (e.g., WHO, CDC)

The presence of several international organizations as mentioned below, is important in formulating guidelines and recommendations that determine the progress and application of TB diagnostics including the emergent biosensor technologies. These standards tend to give a guideline to the regulatory bodies in a country, and they can give best practices to ensure the quality and effectiveness of diagnostic instruments.

1. **World health organisation (WHO):** The WHO develops evidence-based recommendations on the diagnosis, treatment, prevention and treatment of TB. The recommendation of WHO on how to assess novel diagnostics, point of care testing, and molecular assays is exceptionally relevant despite the fact that TB biosensors have special instructions. In many of the cases, their recommendations emphasize the importance of extensive validation, assessment of clinical utility, and simply taking a look at their application in numerous circumstances, in particular, in low-resource countries, where biosensors can become very powerful pioneers (World Health Organization et al., 2022; Getahun et al., 2015; Timimi et al., 2012).
2. **Centres for disease control and prevention (CDC):** CDC also provides TB diagnostic guidelines and standards in the US, both of which often influence US and international practices. They focus on laboratory practice, the evaluation of the new tests, and the techniques of TB control. They can be used to provide advice concerning the validation and application of the latest diagnostic technology even though this is not only restricted to biosensors (Miller et al., 2012).
3. **Other International Bodies:** Other organizations such as Foundation of innovative new diagnostics (FIND) are also involved in the assessment and recommendation of new TB diagnostics, usually collaborating with WHO and national initiatives. Their reports and assessment may be of great use in the performance and potential of biosensors technologies (World Health Organization, 2014; Timimi et al., 2012).

13.5.2 Review of biosensor technologies acceptance of approval in ethical review boards

The foregoing consideration of ethical standards of research and development, and the application of TB diagnostics, which involve biosensors is the ultimate one. Ethical review boards (ERBs) or Institutional review boards (IRBs) are essential to getting these technologies created and utilized in a way that honours the rights and welfare of individuals and communities as illustrated below, Fig. 13.4.

1. **Research ethics review:** Before any research can commence, an independent committee undertake the rigorous ethical analysis of the research (including clinical trials to determine the performance of a novel biosensor) involving human subjects. All this is considered in this evaluation because the research design, process of informed consent, data privacy, potential risks and benefits, and the overall ethical soundness of the research is considered. Due to its peculiarities, biosensor technologies should be thoroughly scrutinized to manage any possible ethics of data processing, new technologies permission, and potential outcomes of fast results (Selgelid, 2008; Coker et al., 2007; Edginton et al., 2013).

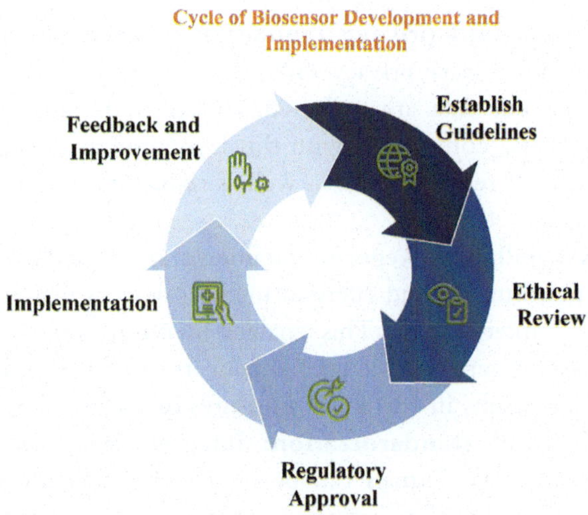

Figure 13.4 Cycle of biosensor development, ethical review boards and approval processes for biosensor technologies outlines the iterative life cycle of biosensor development and implementation, emphasizing the integration of ethical and regulatory oversight. The circular model tracks the progression from establishing guidelines and ethical review to regulatory approval, clinical implementation, and continuous feedback for technological improvement.

2. **Regulatory approval pathways:** Prior to broad applications, TB biosensor tests normally need to undergo regulatory approval at a national level in medical devices agencies. Such organizations make use of preclinical and clinical information to determine the quality, performance and safety of the device. The approval process can be altered depending on the country or region and the classification of a medical device. To deal with the special characteristics of cutting-edge technology such as biosensors, regulators might have to revise existing frameworks or develop unique regulations to cover them (Verma et al., 2004; Scandurra et al., 2020; Enemark, 2013).

13.5.3 Local and global health regulations of biosensors

Local and global health regulations should be respected to ensure the responsible and extensive implementation of the biosensor-based TB diagnostics. The aim of these regulations is to guarantee the quality, safety, and interoperability so that they could be used in various health care environments as mentioned below:

1. **Manufacturing standards and quality control:** The biosensors are manufactured in accordance with the international standards of producing medical devices standards like the ISO 13485, (ISO 13485, n.d.) meaning that they are manufactured under strict quality management systems. This aids in ensuring the reliability as well as uniformity of the devices.

2. **Data management and privacy regulations:** When biosensors platforms are used, local and global data privacy laws (e.g., GDPR, HIPAA) are considered essential when patient data are collected, stored, and transferred. Effective data security protocols and compliance with this legislation are necessary in preserving patient trust and legal requirements, as well as remaining within the bounds of the law (Coleman et al., 2010).
3. **Post-market surveillance:** Regulatory frameworks frequently impose post-market surveillance of performance and safety of medical devices that are already approved and are in the market in use. This enables identification and reporting of any ominous instances or performance related concerns in terms of biosensor based TB diagnostics, which results in corrective measures or recalls.
4. **Interoperability and standardization:** Interoperability and standardization of medical devices and data formats can be used to enhance the implementation of biosensor data in larger health information systems to enhance surveillance and patient care. It is possible to support the usability and the impact of these technologies with the help of compliance with the corresponding standards (Jackson et al., 2019; Spiegel et al., 2021; Ahmad Khan et al., 2017; Wild et al., 2017).

With the successful handling of these regulatory and policy frameworks, stakeholders can be able to create a conducive environment where biosensor-based TB diagnostics can be developed and implemented responsible policies, and finally establish more effective TB control efforts in the international arena.

13.6 Case studies and real-world applications of ethical and safety considerations in *Mycobacterium tuberculosis* diagnostics

As stated in the foregoing analysis, the diagnosis of MTB has not only become an aspect of sophistication as opposed to the traditional usage of a microbiological technique but also an objective of biosensor-based techniques. Both approaches have specific ethical and safety issues, which impact patient outcomes and the strategy of addressing health problems in the population. In this part, it examines actual case studies, significant dilemmas, safety accidents, and best practices in different methods of diagnosing.

13.6.1 Ethical dilemmas in *Mycobacterium tuberculosis* diagnostics

13.6.1.1 Case Study 1: Misdiagnosis of pulmonary tuberculosis and associated factors in peripheral laboratories (Addis Ababa, Ethiopia, 2018)

This case study investigated the prevalence of pulmonary tuberculosis misdiagnosis in Addis Ababa's peripheral laboratories, highlighting significant diagnostic gaps. Their findings emphasize that factors such as laboratory workload and technician experience critically impact the accuracy of tuberculosis identification in resource-constrained urban settings.

- Results Misdiagnosis: Among 9,816 reexamined slides, the discordant (false reading) rate between the peripheral laboratories and between the CRLs was found to be 0.61% (Desalegn et al., 2018).
- False Positive: 2.4% (25 slides) of the slides were positive were found to be false positives, states that patients could be unnecessarily treated.
- False Negative: 0.4% (35 slides) of the slides reported as negative were found to be false negatives, meaning TB cases were missed.
- Contributing Factors: Poor laboratory quality control was common. Some Issues are stated as below:
- Shortage of critical supplies: 63.7% of labs did not have lens cleaning tissue paper.
- Gaps in quality assurance: Only 28.1% of labs ran internal quality control (IQC) slides in parallel with routine tasks.
- Poor preparation of the slides: There were high rates of poor quality of the specimen (35.4%), poor smear thickness (39.2), and poor background cleanness with staining (30.9).

  Ethical Lesson:

- **No-tolerance of transmissibility:** Low falsenegative misdiagnosis rates are ethically unacceptable in infectious disease because such increases directly lead to the social health imperfection, which means that patients may turn out to be vectors of further transmission.
- **Quality assurance is non-negotiable:** Adequate Laboratory infrastructure, consistent staff training as well as strict internal quality control (IQC) are ethical demands of the delivery of the public health services rather than extras.

13.6.1.2 Case study 2: The tuberculin skin test versus interferon-gamma release assay conflict

The presented case study involving a hospital in Germany comparing traditional tuberculin skin test (TST) to the new interferon-gamma release assay (IGRA, in particular, QFT-GIT) on measuring latent TB infection (LTBI) poses a number of different ethical dilemma (Ringshausen et al., 2009):

1. The dilemma of false positives:

The TST is also cheap, however it is highly likely to give a false positive especially with populations who have been vaccinated against BCG.

- The conflict: The participants revealed that TST had a positive result in 28.0% of contacts whereas IGRA was positive only in 9.1%. The TST-positive/IGRA-negative discordant outcome was the most prevalent one, which proved the low specificity of TST.

- The ethical harm: The FP result of TST contributes to unwarranted anxiety, medical referral and prescribing toxic preventive chemotherapy (INH). The fact that a person made a diagnosis that was not accurate justifies treating a healthy person toxicly is a significant ethical issue.

2. The dilemma of treatment refusal:

Preventive chemotherapy (PCT) for LTBI is essential for preventing future active TB cases and protecting the community. However, the drugs carry a risk of severe side effects.

- The conflict: The study offered preventive chemotherapy to the 13 health care workers (HCW) who tested positive by the more accurate IGRA. Remarkably, only one HCW (7.7%) actually took the recommended medication.
- The ethical harm: This highlights a conflict between Patient Autonomy and the Public Health obligation to minimise TB transmission risks in the hospital environment. If HCW refuse treatment, the risk of nosocomial TB persists.

3. The dilemma of test choice:

The case forces a decision on whether to prioritise low cost or high accuracy. The IGRA is clearly diagnostically superior in the BCG-vaccinated population of HCW, as it is much more specific. However, it is also more expensive than the TST.

13.6.2 Case study 3: The ethical and policy challenge of artificial intelligence in tuberculosis diagnostics ()

This editorial discusses the rise of Computer-Aided Detection (CAD) software for reading chest radiography (CXR) images to screen for TB Ahmad Khan et al., 2017 Khan et al., 2017, especially in lowresource settings.

- Medical CAD systems identify abnormalities in the score of TB by analysing the chest Xrays through artificial intelligence (AI), which is potentially a fast, cheap, and reliable diagnostic method.
- CAD may radically reduce and speed up TB screening in regions where the human radiologist is limited in number or microscopy fails to give accurate results (such as HIV positive patients). It is also a useful triage device and can be used in quick identification of persons that require more testing that is more complex (such as GeneXpert).
- The editorial is contradictory, despite the promise, to the fact that significant gaps in research that are necessary to make CAD widely accepted and financed by health

policies of the whole world are missing. These loopholes pose major both ethical and policy challenges.

13.6.2.1 Ethical and policy dilemmas

The editorial poses three fundamental dilemmas during the implementation of CAD:

1. Diagnostic error and accountability
   - CAD systems have wide variation in their accuracy and are in most cases only tested in highly controlled conditions and not in the high burden conditions they are likely to be used in the real world.
   - In a scenario when AI is providing a false-negative, what is to blame the stunted treatment of the patient and the subsequent disease propagation? The vendor of the software or the medical professional or the policy-approving government? There must be policy in place to determine diagnostic criteria and responsibility in AI diagnoses.
2. Issue of equity and cost-effectiveness dilemma
   - Although CAD may prove cheaper than the work of a radiologist, it should be demonstrated that the original cost will have to be balanced with its practical value. The technology itself may not be costly in rich countries but a big risk in terms of investment to the low and middle income countries (LMICs).
   - Policymakers confront a justice issue: Do they spend insufficient government health resources in developing a state of the art CAD platform or rather use the resources enhancing the current, less sophisticated diagnostic systems, such as, educating more human microscopists? The moral objective is to make sure that the new technology really provides equal access, and not improved access to the minority.

13.6.2.2 Case study 4: Ethical dilemmas of occupational risk in health workers

da Silva et al., 2022 The assembled results prove that HCWs comprise an occupational risk group that is at risk of being infected with TB many times more than the general population. The general global findings regarding the prevalence of LTBI in HCWs overall were 28% with the rates in single studies being highly varied ranging up to 62.0% in some high-burden areas. The identified risk poses critical ethical considerations to health care facility and policy in the area of health care. The dilemma that fulfills the main ethical contradiction is the principle of non-maleficence and the duty of care of the institution. The infection rate is very high, which indicates a systemic failure to safeguard the employees by applying sufficient infection control (IC) measures. The lack of facilities (particularly in resourcelimited regions) to offer even the most basic protective measures (such as proper ventilation, N95 respirators that have been fit-tested, and patient triaging as a mandatory condition) is frequent. The ethical issue that the institutions may face is during the time when they compromise the safety of their employees through cost or logistics challenge, in other words, compelling HCWs to sell their health and their lives to earn a living. Moreover, this case is characterized by a root problem of Justice in the distribution of resources. The policy-makers are faced with the challenge of making an ethical decision on how to allocate a scarce budget between the two important priorities. These alarming infection rates of HCWs are ethically obligatory that there should be a great utterance in the provision of occupational health programs to respect the rights and safety of the personnel that offers indispensable diagnostic and care services.

Ethical lessons:

- **Non-maleficence:** The institutions should put the HCW protection first. The 28% prevalence rate of LTBI worldwide indicates that there has been failure to institute the required and moral measures of infection control leaving the workers at the risk of receiving unnecessary harm.
- **Autonomy:** HCWs should be full and freely informed of the high, intermittent TB risk in its work setting (up to 62) in order to make an informed and independent decision about their work and preventive care.
- **Justice:** There should be a moral decision between the use of budget on patient diagnostics and that on safety measures by HCWs. Resource justice means the protection of the workforce as a key to laying out the maintenance of the public health services.

13.6.3 Best practices for ethical and safe *Mycobacterium tuberculosis* diagnostics

As these case studies and additional literature analysis reveal, there is a set of best practices related to ethical and safe implementation of TB diagnostics on all technological platforms:

1. Equity-focused implementation strategies
   - Prioritise deployment of advanced diagnostics in highburden, resourcelimited settings
   - Develop sustainable financing mechanisms that do not perpetuate health inequities
   - Implement hub and spoke models to extend access to rural and remote populations
   - Monitor and address geographic and demographic disparities in diagnostic access
2. Comprehensive data protection frameworks
   - Implement end to end encryption for all digital diagnostic platforms
   - Conduct regular cybersecurity assessments and penetration testing
   - Develop incident response plans for data breaches
   - Ensure patient consent processes address digital privacy risks

3. Robust quality assurance programs
   - Establish comprehensive validation requirements for novel diagnostic technologies
   - Implement continuous quality monitoring with real-time feedback systems
   - Conduct regular external quality assessments
   - Develop rapid response protocols for performance issues
4. Enhanced safety protocols
   - Maintain current biosafety training and certification programs
   - Conduct regular risk assessments and safety audits
   - Implement redundant safety measures for highrisk procedures
   - Establish clear protocols for incident reporting and investigation
5. Community centered approaches
   - Engage communities in diagnostic technology selection and implementation
   - Develop culturally appropriate education and counseling materials
   - Address stigma and discrimination through targeted interventions
   - Establish patient advocacy and support systems
6. Integrated care delivery models
   - Ensure diagnostic services are linked to comprehensive treatment and support
   - Develop patient navigation systems to facilitate care transitions
   - Implement case management approaches for complex cases
   - Monitor treatment outcomes and patient satisfaction

These case studies show that ethical and safety issues related to the TB diagnostics are not associated only with the technical performance but also with the overall considerations of the health system, social and policy. To ensure the successful deployment of any diagnostic technology, both traditional approaches and hitech biosensors, it is necessary to pay full attention to such multidimensional challenges and systematize the evidencebased best practices.

13.7 Future directions

The method of biosensor based diagnostics of MTB is the dynamic area that rapidly evolves. Ethical implication of the emerging trends and potential future directions is explored in this area as well as the necessity to enhance global health systems to justify their safe and ethical application (Chopra and Singh, 2020) and the critical role that patient centred approaches would play in their development.

13.7.1 Emerging biosensor technologies and their ethical implications

Emerging technologies in biosensors to diagnose TB have had a highly progressive effect on the horizon of the field with each offering distinct benefits, yet presenting new ethical challenges.

Next-generation sequencing (NGS) based biosensors: This type of biosensor would integrate NGS functionality capabilities directly into the biosensor systems to enable identification of drug resistance and strain of the strains of MTB at the point of care.

Ethical implications: This provokes complicated questions regarding the areas of information produced, such as scrutinized incidental discoveries, the impacts on patient privacy within highly genetic facts and the necessity of proper counselling and interpretation bundles.

Volatile organic compound (VOC) detection biosensors: The technology is a newly developed instrument that detect volatile organic compounds in the breath samples that identify an active TB infection (Lim et al., 2016), and this is a possible non-invasive means of diagnosis.

Ethical implications: Non-invasive, but using breath samples, the reliability and interpretation of findings, the possibility of environmental modulation in the VOC profiles (Pathak et al., 2023) and explicit validation in the established diagnostic approaches are debatable. It is important to be open with the limitations of the technology and keep trust (Chew et al., 2025).

13.7.2 Strengthening global health systems for ethical and safe biosensor use

To ensure the successful and responsible implementation of biosensor technologies in TB diagnostics, a collective action was required to reinforce the global health systems in numerous respects.

Infrastructure development: Biosensor systems require the efficient infrastructure and integration such as reliable electricity, internet connectivity and transportation to deploy and maintain the system, particularly in resource constrained environments (Misera et al., 2024). Ethical matters require that these infrastructural projects must be fair and access the least served groups of people.

Workforce training and capacity building: To guarantee proper and safe usability of biosensors and ethical reporting of the results and follow-up care, detailed training programs of health care workers and laboratory personnel are necessary (Davoudian et al., 2025).

Regulatory harmonization: By harmonizing the regulatory standards of medical devices, including the biosensors, in various countries, it would be easier to increase their use in the different countries in line with their quality and safety. International partnerships can be important on this process (Pawnikar and Patel, 2025).

13.7.3 Advocacy for patientcentred approaches in biosensor development

As mentioned below, the key to the moral and efficient use of the biosensor technologies is the insistence that the creation and application of these technologies would be user-centered as concerns patient needs, preferences and values.

User-friendly design and accessibility: Biosensors are supposed to be enduser friendly where ease of use, mobility, and cost are viewed as some of the factors to be considered in designing these biosensors. One of the design principles should be accessibility to the various populations including the disabled (Srinivasan and Tung, 2015).

Respect of patient autonomy and preferences: Informed consent and Shared decision making and respect of patient preferences on the location of testing, provision of results and sharing of patient data should be developed and implemented (Rubeis et al., 2018).

Stigmatizing and supporting: The development and implementation of biosensors should be supported by the attention that the stigma surrounding TB and other related diseases may cause and the underprivileged individuals with TB found by these technologies must be subjected to effective psychosocial support and connection with care (McLamore and Datta, 2023).

The successful and responsible integration of biosensor technologies into TB diagnostics necessitates a concerted effort to strengthen global health systems across multiple dimensions.

13.8 Call to action on stakeholders in tuberculosis diagnostics and biosensor technology

Fulfilment of the biosensor technology in the promise of revolution in the TB diagnostics can just be made possible through concerted and collaborative effort by all the stakeholders as illustrated below in Fig. 13.5.

- **Researchers and developers:** To prioritise ethical considerations and safety by design in the development of new biosensor technologies, engaging with ethicists, health care providers, and patients throughout the innovation process (Ogwu and Izah, 2025).
- **Policymakers and regulatory agencies:** To establish clear, adaptive, and harmonised regulatory frameworks that facilitate the safe and ethical implementation of biosensors while fostering innovation (Engel and Wolffs, 2020).
- **Health care providers and public health programs:** To ensure adequate training, infrastructure, and support systems are in place for the responsible use of biosensors, with a strong emphasis on patient-centred care, confidentiality, and linkage to treatment (Davoudian et al., 2025).
- **Global health organisations and funding agencies:** To support research, development, and equitable access to biosensor technologies, particularly in high burden, low resource settings, while prioritising ethical and safety safeguards (Ogwu and Izah, 2025).

Figure 13.5 Ethical biosensor implementation for TB. *TB*, tuberculosis. presents a collaborative stakeholder framework for the ethical implementation of TB biosensors. The archway structure identifies the critical roles of researchers, policymakers, health care providers, funding agencies, and patient communities in bridging the gap between diagnostic limitations and accelerated TB elimination.

- **Communities and patients:** To engage actively in discussions about the ethical and social implications of new diagnostic technologies and to advocate for their rights and preferences (Engel and Wolffs, 2020).

With a positive and considerate attitude to the ethical and safety issues discussed in this chapter, the world community can adopt the potential transforming ability of biosensor technology to speed up the process of TB eradication so that technological progress can benefit humanity safely.

13.9 Conclusion

The critical moral and safety concerns that should be considered to effectively and successfully introduce the biosensor technologies in MTB diagnostics have been discussed in detail in this chapter. We have explained why patient privacy and confidentiality matter so much in the digital era, why the highly developed technologies demand the use of valid informed consent, why it is so vital to ensure equity and access to such tools, and why it is quite challenging to decimate stigma and discrimination surrounding the diagnosis of TB. We have also highlighted the importance of reducing diagnostic errors, associated risks to medical staff, safe waste disposal and the importance of the biosafety protocols that are essential when dealing with infectious samples. Specific ethical and safety concerns involving biosensor technology in particular (such as the concerns relating to rapid point of care testing, data security, potential abuse, application in low-resource settings, and the utilization of new materials) have also been widely discussed. The opportunities and complexities in this emerging field have also been illuminated by the international and local regulatory and policy issue and the lessons learned or the further possibilities in practical use.

References

Acharya B, charya A A, Gautam S, Ghimire SP, Mishra G, Parajuli N, Sapkota B, et al. Advances in diagnosis of Tuberculosis: an update into molecular diagnosis of Mycobacterium tuberculosis. Mol Biol Rep 2020;47(5):4065–75. Available from https://doi.org/10.1007/s11033-020-05413-7.

Ahmad Khan F, Salim MAH, du Cros P, Casas EC, Khamraev A, Sikhondze W, Benedetti A, Bastos M, Lan Z, Jaramillo E, Falzon D, Menzies D. Effectiveness and safety of standardised shorter regimens for multidrug-resistant tuberculosis: individual patient data and aggregate data meta-analyses. Eur Respir J 2017;50(1). Available from https://doi.org/10.1183/13993003.00061-2017.

Asgharzadeh M, Ozma MA, Rashedi J, Mahdavi B, Agharzadeh V, Vegari A, et al. False-positive mycobacterium tuberculosis detection: ways to prevent cross-contamination. Tuberculosis and Respiratory Diseases 2020;83(3):211–7. Available from https://doi.org/10.4046/trd.2019.0087.

Allahyartorkaman Mohammadreza, Mirsaeidi Mehdi, Hamzehloo Gholamreza, Amini Sirus, Zakiloo Mona, Nasiri Mohammad Javad. Low diagnostic accuracy of Xpert MTB/RIF assay for extrapulmonary tuberculosis: A multicenter surveillance. Scientific Reports 2019;9. Available from https://doi.org/10.1038/s41598-019-55112-y.

Azyabi A, Khamaj A, Ali AM, Abushaega MM, Ghandourah E, Alam MM, et al. Predicting ergonomic risk among laboratory technicians using a cheetah optimizer-integrated deep convolutional neural network. Computers in Biology and Medicine 2024;183. Available from https://doi.org/10.1016/j.compbiomed.2024.109314.

Bashir S, Ali S, Yerlikaya S, Gaeddert M, Goscé L, Rangaka MX, M Denkinger C, et al. Cost-effectiveness of diagnostic technologies for mycobacterium tuberculosis infection in India and Brazil. PLOS Glob Public Health 2024;4(11):e0003638. Available from https://doi.org/10.1371/journal.pgph.0003638.

Berche Patrick. Laboratory-associated infections and biosafety. La Presse Médicale 2026;55(31142331RR–71513125678752414916174422921331231 0):104277. Available from 10.1016/j.lpm.2025.104277.

Boehme CC, Nicol MP, Nabeta P, Michael JS, Gotuzzo E, Tahirli R, et al. Feasibility, diagnostic accuracy, and effectiveness of decentralised use of the Xpert MTB/RIF test for diagnosis of tuberculosis and multidrug resistance: a multicentre implementation study. The Lancet 2011;377(9776):1495–505. Available from https://doi.org/10.1016/S0140-6736(11)60438-8.

Cao Y, Parmar H, Gaur RL, Lieu D, Raghunath S, Via N, Battaglia S, Cirillo DM, Denkinger C, Georghiou S, Kwiatkowski R, Persing D, Alland D, Chakravorty S, et al. Xpert MTB/XDR: a 10-Color Reflex Assay Suitable for Point-of-Care Settings To Detect Isoniazid, Fluoroquinolone, and Second-Line-Injectable-Drug Resistance Directly from Mycobacterium tuberculosis-Positive Sputum. J Clin Microbiol 2021;59(3). Available from https://doi.org/10.1128/JCM.02314-20.

Chakravorty S, Simmons AM, Rowneki M, Parmar H, Cao Y, Ryan J, et al. The new Xpert MTB/RIF ultra: improving detection of Mycobacterium tuberculosis and resistance to Rifampin in an assay suitable for point-of-care testing. mBio 2017;8(4). Available from https://doi.org/10.1128/mBio.00812-17.

Chaturvedi M, Patel M, Tiwari A, Dwivedi N, Mondal DP, Srivastava AK, et al. An insight to the recent advancements in detection of Mycobacterium tuberculosis using biosensors: a systematic review. Progress in Biophysics and Molecular Biology 2024;186:14–27. Available from https://doi.org/10.1016/j.pbiomolbio.2023.10.003.

Chew N, Yun S, See KC. Diagnostic ccuracy of breath tests to detect pulmonary tuberculosis: a systematic review. Lung 2025;203(1). Available from https://doi.org/10.1007/s00408-024-00779-y.

Chopra KK, Singh S. Tuberculosis: newer diagnostic tests: applications and limitations. Indian Journal of Tuberculosis 2020;67(4):S86. Available from https://doi.org/10.1016/j.ijtb.2020.09.025.

Coker R, Thomas M, Lock K, Martin R. Detention and the evolving threat of tuberculosis: evidence, ethics, and law. Journal of Law, Medicine & Ethics 2007;35(4):609–15. Available from https://doi.org/10.1111/j.1748-720x.2007.00184.x.

Coleman CH, Jaramillo E, Reis A, Selgelid M, World Health Organization.. Guidance on ethics of tuberculosis prevention, care and control. World Health Organization; 2010. Available from https://iris.who.int/bitstream/handle/10665/44452/9789241500531_eng.pdf?sequence=1.

Cruciani M, Scarparo C, Malena M, Bosco O, Serpelloni G, Mengoli C. Meta-analysis of BACTEC MGIT 960 and BACTEC 460 TB, with or without solid media, for detection of Mycobacteria. Journal of Clinical Microbiology 2004;42(5):2321–5. Available from https://doi.org/10.1128/jcm.42.5.2321-2325.2004.

da Silva EH, Lima E, dos Santos TR, Padoveze MC. Prevalence and incidence of tuberculosis in health workers: a systematic review of the literature. American Journal of Infection Control 2022;50(7):820–7. Available from https://doi.org/10.1016/j.ajic.2022.01.021.

Daftary A. HIV and tuberculosis: the construction and management of double stigma. Social Science & Medicine 2012;74(10):1512–9. Available from https://doi.org/10.1016/j.socscimed.2012.01.027.

Daniels N. Just Health: Meeting Health Needs Fairly. United States: Cambridge University Press; 2007. p. 1–397. Available from https://doi.org/10.1017/CBO9780511809514.

Das M, Sumana G, Nagarajan R, Malhotra BD. Zirconia based nucleic acid sensor for Mycobacterium tuberculosis detection. Applied Physics Letters 2010;96(13). Available from https://doi.org/10.1063/1.3293447.

Das M, Dhand C, Sumana G, Srivastava, Nagarajan, Nain L, et al. Electrophoretic fabrication of chitosan-zirconium-oxide nanobiocomposite platform for nucleic acid detection. Biomacromolecules 2011;12(3):540–7. Available from https://doi.org/10.1021/bm1013074.

Davoudian K, Nemtsov L, Thompson M. Challenges in the application of biosensor technology in the clinical biochemistry laboratory. Frontiers in Sensors 2025;6. Available from https://doi.org/10.3389/fsens.2025.1641221.

Denholm JT, McBryde ES, Brown GV. Ethical evaluation of immigration screening policy for latent tuberculosis infection. Journal of Public Health 2012;36(4):325–8. Available from https://doi.org/10.1111/j.1753-6405.2012.00895.x.

Denkinger CM, Kik SV, Cirillo DM, Casenghi M, Shinnick T, Weyer K, Gilpin C, Boehme CC, Schito M, Kimerling M, Pai M, et al. Defining the needs for next generation assays for tuberculosis. J Infect Dis 2015;211 Suppl 2(Suppl 2):S29–38. Available from 10.1093/infdis/jiu821.

Desalegn DM, Kitila KT, Balcha HM, Gebeyehu CS, Kidan YW, Amare K, et al. Misdiagnosis of pulmonary tuberculosis and associated factors in peripheral laboratories: a retrospective study, Addis Ababa, Ethiopia. BMC Research Notes 2018;11(1). Available from https://doi.org/10.1186/s13104-018-3414-6.

Drain PK, Hyle EP, Noubary F, Freedberg KA, Wilson D, Bishai WR, Rodriguez W, Bassett IV, et al. Diagnostic point-of-care tests in resource-limited settings. Lancet Infect Dis 2014;14(3):239–49. Available from https://doi.org/10.1016/S1473-3099(13)70250-0.

Edginton ME, Ornstein T, Denholm J, El Sony A, Kim SJ, Narain A, et al. Research ethics in the union: an 8-year review of the ethics advisory group. Public Health Action 2013;3(4):346–50. Available from https://doi.org/10.5588/pha.13.0061.

Eivazzadeh-Keihan R, Saadatidizaji Z, Mahdavi M, Maleki A, Irani M, Zare I. Recent advances in gold nanoparticles-based biosensors for tuberculosis determination. Talanta 2024;275:126099. Available from https://doi.org/10.1016/j.talanta.2024.126099.

Elbehiry A, Marzouk E, Edrees HM, AlShaqi R, Ellethy AT, Alzaben F, et al. Advancing the fight against tuberculosis: integrating innovation and public health in diagnosis, treatment, vaccine development, and implementation science. Frontiers in Medicine 2025;12. Available from https://doi.org/10.3389/fmed.2025.1596579 Available from: journal.frontiersin.org/journal/medicine.

Enemark C. Drug-resistant tuberculosis: security, ethics and global health. Global Society 2013;27(2):159–77. Available from https://doi.org/10.1080/13600826.2012.762345.

Engel N, Wolffs PFG. Aligning diagnostics to the point-of-care: lessons for innovators, evaluators and decision-makers from tuberculosis and HIV. BMJ Global Health 2020;5(11). Available from https://doi.org/10.1136/bmjgh-2020-003457 Available from: gh.bmj.com/.

Getahun H, Matteelli A, Abubakar I, Aziz MA, Baddeley A, Barreira D, et al. Management of latent Mycobacterium tuberculosis infection: WHO guidelines for low tuberculosis burden countries. European Respiratory Journal 2015;46(6):1563–76. Available from https://doi.org/10.1183/13993003.01245-2015.

Hasan MR, Anzar N, Sharma P, Singh S, Hassan H, Rawat C, et al. Mycobacterium tuberculosis diagnosis from conventional to biosensor-a systematic review. International Journal of Environmental Analytical Chemistry 2024;104(18):6519–34. Available from https://doi.org/10.1080/03067319.2022.2147427.

ISO 13485. (n.d.). In ISO 13485.

Jackson C, Gardy JL, Shadiloo HC, Silva DS. Trust and the ethical challenges in the use of whole genome sequencing for tuberculosis surveillance: a qualitative study of stakeholder perspectives. BMC Medical Ethics 2019;20(1). Available from https://doi.org/10.1186/s12910-019-0380-z.

Joshi Hemant, Kandari Divya, Maitra Subhrangsu Sundar, Bhatnagar Rakesh. Biosensors for the detection of*Mycobacterium tuberculosis*: a comprehensive overview. Critical Reviews in Microbiology 2022;48:784–812. Available from https://doi.org/10.1080/1040841x.2022.2035314.

Kanabalan RD, Lee LJ, Lee TY, Chong PP, Hassan L, Ismail R, et al. Human tuberculosis and Mycobacterium tuberculosis complex: a review on genetic diversity, pathogenesis and omics approaches in host biomarkers discovery. Microbiological Research 2021;246. Available from https://doi.org/10.1016/j.micres.2020.126674.

Khan FA, Pande T, Tessema B, Song R, Benedetti A, Pai M, et al. Computer-aided reading of tuberculosis chest radiography: moving the research agenda forward to inform policy. European Respiratory Journal 2017;50(1). Available from https://doi.org/10.1183/13993003.00953-2017.

Kruse RH, Puckett WH, Richardson JH. Biological safety cabinetry. Clinical Microbiology Reviews 1991;4(2):207–41. Available from https://doi.org/10.1128/cmr.4.2.207.

Layre E, Al-Mubarak R, Belisle J, Moody DB. Mycobacterial lipidomics. Microbiology Spectrum 2022;2(3):1–19. Available from https://doi.org/10.1128/microbiolspec.mgm2-0033-2013.

Lee JS, Kang YA, Kim HW,, et al. Time to treatment initiation and outcomes in patients with tuberculosis: systematic review and meta-analysis. BMC Infectious Diseases 2022;22(1):873. Available from https://doi.org/10.1186/s12879-022-07855-9.

Lee M-R, Chung K-P, Chen W-T, Huang Y-T, Lee L-N, Yu C-J, et al. Epidemiologic surveillance to detect false-positive Mycobacterium tuberculosis cultures. Diagnostic Microbiology and Infectious Disease 2012;73(4):343–9. Available from https://doi.org/10.1016/j.diagmicrobio.2012.04.014.

Lim SH, Martino R, Anikst V, Xu Z, Mix S, Benjamin R, et al. Rapid diagnosis of tuberculosis from analysis of urine volatile organic compounds. ACS Sensors 2016;1(7):852–6. Available from https://doi.org/10.1021/acssensors.6b00309.

Luukinen B, Aittoniemi J, Miikkulainen-Lahti T, Mentula S, Pätäri-Sampo A. Evaluation of the STANDARD M10 MDR-TB and MTB/NTM assays for the detection of Mycobacterium tuberculosis, rifampicin and isoniazid resistance, and nontuberculous mycobacteria in a low-incidence setting. Journal of Clinical Microbiology 2024;62(10). Available from https://doi.org/10.1128/jcm.00402-24.

Matias AO, de Oliveira GE, de Oliveira Farias EA, Wang J, Eiras C, Mendes AN. Advances in electrochemical biosensors for the diagnosis of tuberculosis: perspectives for control and eradication of the disease. Microchemical Journal 2025;213. Available from https://doi.org/10.1016/j.microc.2025.113622.

McLamore ES, Datta SPA. A connected world: system-level support through biosensors. Annual Review of Analytical Chemistry 2023;16:285–309. Available from https://doi.org/10.1146/annurev-anchem-100322-040914.

McNerney R, Daley P. Towards a point-of-care test for active tuberculosis: obstacles and opportunities. Nature Reviews. Microbiology 2011;9(3):204–13. Available from https://doi.org/10.1038/nrmicro2521.

Mehrotra P. Biosensors and their applications – a review. Journal of Oral Biology and Craniofacial Research 2016;6(2):153–9. Available from https://doi.org/10.1016/j.jobcr.2015.12.002.

Mengistu DA, Tolera ST, Demmu YM. Worldwide prevalence of occupational exposure to needle stick Injury among health care workers: a systematic review and meta-analysis. Journal of Infectious Diseases and Medical Microbiology 2021;2021. Available from https://doi.org/10.1155/2021/9019534.

Mennella C, Maniscalco U, De Pietro G, Esposito M. Ethical and regulatory challenges of AI technologies in healthcare: a narrative review. Heliyon 2024;10(4):e26297. Available from https://doi.org/10.1016/j.heliyon.2024.e26297.

Miller JM, Astles R, Baszler T, Chapin K, Carey R, Garcia L, et al. Guidelines for safe work practices in human and animal medical diagnostic laboratories. In Morbidity and Mortality Weekly Report 2012;Vol. 61.

Ministry of Health and Family Welfare, Government of India. National Guidelines for Management of Drug Resistant TB. National TB Elimination Programme Ministry of Health and Family Welfare Government of India, Central TB Division; 2024. Available from https://tbcindia.mohfw.gov.in/wp-content/uploads/2025/01/National-Guidelines-for-Management-of-DR-TB_Final.pdf.

Misera J, Melchert J, Bork-Hüffer T. Biosensing and biosensors—terminologies, technologies, theories and ethics. Geography Compass 2024;18(11). Available from https://doi.org/10.1111/gec3.70007.

Mishra R, Pal R, Mandal RK. Emerging technologies in tuberculosis diagnosis: a comprehensive review. International Journal of Pharmaceutical Research and Development 2025;7(1):265–77. Available from https://doi.org/10.33545/26646862.2025.v7.i1d.119.

Mousavi-Sagharchi SMA, Afrazeh E, Seyyedian-Nikjeh SF, Meskini M, Doroud D, Siadat SD. New insight in molecular detection of Mycobacterium tuberculosis. AMB Express 2024;14(1). Available from https://doi.org/10.1186/s13568-024-01730-3.

Nahid P, Dorman SE, Alipanah N, Barry PM, Brozek JL, Cattamanchi A, et al. Official American thoracic society/centers for disease control and prevention/infectious diseases society of america clinical practice guidelines: treatment of drug-susceptible tuberculosis. Clinical Infectious Diseases 2016;63(7):e147. Available from https://doi.org/10.1093/cid/ciw376.

Najjari M, Karimazar M, Ebrahimipour M, M. Budke C. A Review of ethical issues related to laboratory-based diagnostics. Medical Laboratory Journal 2022;16(2):1–6. Available from https://doi.org/10.61186/mlj.16.2.1.

Ogwu MC, Izah SC. Biosensors and wearable technologies for early detection and monitoring of tropical diseases. In Health Information Science 2025:57–81. Available from https://doi.org/10.1007/978-3-031-82622-1_3.

Pai NP, Vadnais C, Denkinger C, Engel N, Pai M. Point-of-care testing for infectious diseases: diversity, complexity, and barriers in low- and middle-income countries. PLoS Medicine 2012;9(9):e1001306. Available from https://doi.org/10.1371/journal.pmed.1001306.

Pang Y, Su B, Zheng H, Zhang Z, Ma A, Wang Y, et al. Factors associated with missed detection of Mycobacterium tuberculosis by automated BACTEC MGIT 960 system. BioMed Research International 2016;2016:1–4. Available from https://doi.org/10.1155/2016/5972021.

Papadopoli R, Nobile CGA, Trovato A, Pileggi C, Pavia M, Pavia M. Chemical risk and safety awareness, perception, and practices among research laboratories workers in Italy. Journal of Occupational Medicine and Toxicology 2020;15(1). Available from https://doi.org/10.1186/s12995-020-00268-x.

Parrish NM, Carroll KC. Role of the clinical mycobacteriology laboratory in diagnosis and management of tuberculosis in low-prevalence settings. Journal of Clinical Microbiology 2011;49(3):772–6. Available from https://doi.org/10.1128/JCM.02451-10.

Pathak A, Swargiary K, Kongsawang N, Jitpratak P, Ajchareeyasoontorn N, Udomkittivorakul J, et al. Recent advances in sensing materials targeting clinical volatile organic compound (VOC) biomarkers: a review. Biosensors 2023;13(1):114. Available from https://doi.org/10.3390/bios13010114.

Pawnikar V, Patel M. Biosensors in wearable medical devices: regulatory framework and compliance across US, EU, and Indian markets. Annales Pharmaceutiques Françaises 2025;83(4):637–48. Available from https://doi.org/10.1016/j.pharma.2025.02.007.

Peeling RW, Mcnerney R. Emerging technologies in point-of-care molecular diagnostics for resource-limited settings. Expert Review of Molecular Diagnostics 2014;14(5):525–34. Available from https://doi.org/10.1586/14737159.2014.915748.

Peeling RW, Holmes KK, Mabey D, Ronald A. Rapid tests for sexually transmitted infections (STIs): the way forward. Sexually Transmitted Infections 2006;82(suppl_5):v1. Available from https://doi.org/10.1136/sti.2006.024265.

Peters DH, Tran NT, Adam T. Implementation Research in Health: A Practical guide. Alliance for Health Policy and Systems Research, World Health Organization; 2013. Available from https://iris.who.int/bitstream/handle/10665/91758/9789241506212_eng.pdf?sequence=1.

Pillay S, Steingart KR, Davies GR, Chaplin M, De Vos M, Schumacher SG, et al. Xpert MTB/XDR for detection of pulmonary tuberculosis and resistance to isoniazid, fluoroquinolones, ethionamide, and amikacin. Cochrane Database of Systematic Reviews 2022;2022(5). Available from https://doi.org/10.1002/14651858.CD014841.pub2.

Price WN, Cohen IG. Privacy in the age of medical big data. Nature Medicine 2019;25(1):37–43. Available from https://doi.org/10.1038/s41591-018-0272-7.

Qin Q, Tseng C, Chen W, Tseng C-L, Kamrani E. Best practices for implementing biosafety inspections in a clinical laboratory: evidence from a multi-site experimental study. PLoS One 2023;18(10):e0292940. Available from https://doi.org/10.1371/journal.pone.0292940.

Ringshausen FC, Schlösser S, Nienhaus A, Schablon A, Schultze-Werninghaus G, Rohde G. In-hospital contact investigation among health care workers after exposure to smear-negative tuberculosis. Journal of Occupational Medicine and Toxicology 2009;4(1):11. Available from https://doi.org/10.1186/1745-6673-4-11.

Rubeis G, Schochow M, Steger F. Patient utonomy and quality of care in telehealthcare. Science and Engineering Ethics 2018;24(1):93–107. Available from https://doi.org/10.1007/s11948-017-9885-3.

Salvaraji L, Jeffree MS, Lukman KA, Saupin S, Avoi R. Electrical safety in a hospital setting: a narrative review. Annals of Medicine and Surgery 2022;78. Available from https://doi.org/10.1016/j.amsu.2022.103781.

Scandurra G, Degeling C, Douglas P, Dobler CC, Marais B. Tuberculosis in migrants – screening, surveillance and ethics. Pneumonia 2020;12(1). Available from https://doi.org/10.1186/s41479-020-00072-5.

Selgelid MJ. Ethics, Tuberculosis and Globalization. Public Health Ethics 2008;1(1):10–20. Available from https://doi.org/10.1093/phe/phn001.

Selgelid MJ, Reichman LB. Ethical issues in tuberculosis diagnosis and treatment. The International Journal of Tuberculosis and Lung Disease 2011;15(6):9–13. Available from https://doi.org/10.5588/ijtld.10.0434.

Sen S. Tuberculosis diagnosis in resource limited settings. Medical Journal Armed Forces India 2018;74(3):211–2. Available from https://doi.org/10.1016/j.mjafi.2018.06.003.

Sharma S, Zapatero-Rodríguez J, Estrela P, O'Kennedy R. Point-of-care diagnostics in low resource settings: present status and future role of microfluidics. Biosensors 2015;5(3):577–601. Available from https://doi.org/10.3390/bios5030577.

Sharp SE, Elder BL. Competency assessment in the clinical microbiology laboratory. Clinical Microbiology Reviews 2004;17(3):681–94. Available from https://doi.org/10.1128/CMR.17.3.681-694.2004.

Shickle D, Chadwick R. The ethics of screening: is 'screeningitis' an incurable disease? Journal of Medical Ethics 1994;20(1):12–8. Available from https://doi.org/10.1136/jme.20.1.12.

Shivakumar P, Shettigar KS. Tuberculosis diagnosis: updates and challenges. In Infectious diseases 2022. Available from https://doi.org/10.5772/intechopen.107168.

Singh JA, Bhan A, Upshur R. Diagnosis of drug-resistant TB and provision of second-line TB treatment in India: some ethical considerations. Indian Journal of Medical Ethics 2013;10(2):110–4. Available from https://doi.org/10.20529/ijme.2013.033.

Singh K. Laboratory-acquired infections. Clinical Infectious Diseases 2009;49(1):142–7. Available from https://doi.org/10.1086/599104.

Spiegel JM, Ehrlich R, Yassi A, Riera F, Wilkinson J, Lockhart K, et al. Using artificial intelligence for high-volume identification of silicosis and tuberculosis: a bio-ethics approach. Annals of Global Health 2021;87(1):58. Available from https://doi.org/10.5334/aogh.3206.

Srinivasan B, Tung S. Development and applications of portable biosensors. SLAS Technology 2015;20(4):365–89. Available from https://doi.org/10.1177/2211068215581349.

Steingart KR, Ng V, Henry M, Hopewell PC, Ramsay A, Cunningham J, et al. Sputum processing methods to improve the sensitivity of smear microscopy for tuberculosis: a systematic review. Lancet Infectious Diseases 2006;6(10):664–74. Available from https://doi.org/10.1016/S1473-3099(06)70602-8.

Swaminathan N, Perloff SR, Zuckerman JM. Prevention of Mycobacterium tuberculosis transmission in health care settings. Infectious Disease Clinics of North America 2021;35(4):1013–25. Available from https://doi.org/10.1016/j.idc.2021.07.003.

Thangarasu M, Nainar SNS, Piraman S, Sivasamy VV. Sensor and nanotechnology-based diagnostics in the field of mycobacteriology. Diagnosis of Mycobacterium. India: Springer Nature; 2023. p. 175–208.

Timimi H, Falzon D, Glaziou P, Sismanidis C, Floyd K. WHO guidance on electronic systems to manage data for tuberculosis care and control. Journal of the American Medical Informatics Association 2012;19(6):939–41. Available from https://doi.org/10.1136/amiajnl-2011-000755.

Varkey Basil. Principles of Clinical Ethics and Their Application to Practice. Medical Principles and Practice 2020;30:17–28. Available from https://doi.org/10.1159/000509119.

Verma G, Upshur REG, Rea E, Benatar SR. Critical reflections on evidence, ethics and effectiveness in the management of tuberculosis: public health and global perspectives. BMC Medical Ethics 2004;5:1–7. Available from https://doi.org/10.1186/1472-6939-5-1.

Wild V, Jaff D, Shah NS, Frick M. Tuberculosis, human rights and ethics considerations along the route of a highly vulnerable migrant from sub-Saharan Africa to Europe. The International Journal of Tuberculosis and Lung Disease 2017;21:1075–85. Available from https://doi.org/10.5588/ijtld.17.0324.

World Health Organization. Companion handbook to the WHO guidelines for the programmatic management of drug-resistant tuberculosis. Geneva: World Health Organization; 2014. Available from https://iris.who.int/bitstream/handle/10665/130918/9789241548809_eng.pdf.

de Vos M, David A, Duraisamy K, Nadarajan D, Noroc E, Penn-Nicholson A, et al. Accuracy of cobas MTB and MTB-RIF/INH for detection of Mycobacterium tuberculosis and drug resistance. The Journal of Molecular Diagnostics 2024;26(8):708–18. Available from https://doi.org/10.1016/j.jmoldx.2024.05.004.

World Health Organization. Ethics Guidance for the Implementation of the End TB strategy. Geneva: World Health Organization; 2020. Available from https://www.who.int/publications/i/item/9789241512114.

World Health Organization. Global Tuberculosis Report 2023. Geneva: World Health Organization; 2023. Available from https://iris.who.int/bitstream/handle/10665/373828/9789240083851-eng.pdf?sequence=1.

World Health Organization, Ismail NA, Nathanson C-M, Falzon D, Korobitsyn A, Kanchar A, et al. Coulter C, Hasan R, Kritski AL, Shanmugam SK, Lyadova I, Matteelli A, Muhwa JC, Omar SV, Shinnick T, editors. WHO operational handbook on tuberculosis: tests for tuberculosis infection. World Health Organization; 2022. [Handbook].

Yadav R, Sharma N, Khaneja R, et al. Cost-effectiveness analysis of molecular diagnostics for tuberculosis in India. Journal of Global Infectious Diseases 2022;14(4):147–54. Available from https://doi.org/10.4103/jgid.jgid_309_21.

CHAPTER 14

Challenges, limitations and future prospects in biosensor technology for *Mycobacterium tuberculosis* detection

Neeraj Kumari[1], Rajni Gautam[2], and Himani Shukla[1]
[1]Chemistry Department, School of Basic and Applied Sciences, K. R. Mangalam University, Gurugram, Haryana, India,
[2]Physics Department, School of Basic and Applied Sciences, K. R. Mangalam University, Gurugram, Haryana, India

14.1 Overview of biosensor technologies

Mycobacterium tuberculosis (MTB) infects millions of people annually, making tuberculosis (TB) one of the leading infectious disease killers in the world. Early detection is critical for infection control, public health surveillance, and breaking the chain of transmission in addition to individual patient care. However, the most popular diagnostic methods such as like culture methods, sputum smear microscopy, and nucleic acid amplification tests (such as GeneXpert) are constrained by things such as a long turnaround time, infrastructure needs, expense, and variable sensitivity, particularly in low-resource environments (World Health Organization, 2024).

In this regard, the potential of biosensor technologies to transform TB diagnostics is becoming more widely acknowledged. A biosensor is a small analytical instrument that detects the presence of a particular analyte by combining a physical or chemical transducer with a biological recognition element. Applications for biosensors can be found in many domains, such as biodefense, food safety, environmental monitoring, and medical diagnostics (Zhou et al., 2011).

14.2 Components of a biosensor

A biosensor typically comprises the following key components:

Biorecognition element:

Through interaction with the target analyte, this component guarantees specificity. It can consist of biological molecules such as aptamers, antibodies, enzymes, nucleic acids, or even entire cells that are responsive to biomarkers specific to MTB, such as volatile organic compounds, DNA sequences, or antigens (such as CFP-10 or ESAT-6) (Ghasemi-Varnamkhasti et al., 2018).

Advances in Biosensor Technology for Mycobacterium Tuberculosis Detection
DOI: https://doi.org/10.1016/B978-0-443-44589-7.00015-7

Transducer: This converts the biorecognition event into a quantifiable signal. Transducers can be of various types:

Electrochemical (measuring current, voltage, or impedance) (Shanbhag et al., 2023)
Optical (using light absorption, fluorescence, or surface plasmon resonance) (Jain et al., 2023).
Piezoelectric (detecting changes in mass or acoustic waves) (Ramaraj et al., 2025)
Thermal (measuring temperature changes during the biochemical reaction) (Jain et al., 2023).
Signal processor and display: This part of the system processes and amplifies the signal into a readable output, often via an integrated circuit or digital interface (Broesch, 2008).
Classification of biosensors

Biosensors can be classified based on either the transduction method or the type of biorecognition element:

By transduction method: *Electrochemical biosensors*: Most widely used for pathogen detection because of low cost, simplicity, and ease of miniaturization (Wang et al., 2021).*Optical biosensors*: Offer high sensitivity and are suitable for label-free detection (Peltomaa et al., 2018).*Piezoelectric biosensors*: Detect minute changes in mass, suitable for real-time monitoring (Zuber et al., 2019). *Thermal biosensors*: Less commonly used but beneficial in specific biochemical applications (Shanbhag et al., 2023).

By Biorecognition Type:

Immunosensors (use antibodies)
DNA biosensors (detect specific nucleic acid sequences)
Enzymatic biosensors
*Cell-based biosensors (*Warriner and Namvar, 2011*).*

14.3 Relevance to *Mycobacterium tuberculosis* detection (Campelo et al., 2021)

Biosensor platforms are particularly advantageous for TB diagnosis because of the following features:

Rapid detection: In contrast to the days or weeks needed for cultures, results can be obtained in a matter of minutes to hours.
High sensitivity: Able to identify latent TB and extrapulmonary TB by detecting low levels of MTB DNA, proteins, or antigens.
Portability: Since many biosensors are compact and light, they are perfect for point-of-care applications in isolated or underdeveloped locations.

Low sample volume requirement: Beneficial in situations involving children or extrapulmonary conditions where it is challenging to obtain large sample volumes.
Minimal infrastructure: Biosensors can function with very little equipment, in contrast to PCR-based techniques that need clean labs and heat cyclers.

Several biosensor platforms have been tested for MTB detection using different analyses, such as:

Mycobacterial DNA sequences (e.g., IS6110),
Lipoarabinomannan (LAM) antigen in urine,
Cell-free MTB-specific RNA or proteins, and
Host biomarkers such as cytokines.

14.4 Challenges and limitations of biosensing technology for *Mycobacterium tuberculosis* detection

MTB can be detected quickly, accurately, and affordably with Biosensors, especially in point-of-care (POC) settings. However, a number of practical and technical obstacles prevent biosensors from being widely used in clinical settings, even though research and development have advanced significantly. The main drawbacks of biosensing technology for MTB detection are examined in this section (Table 14.1.), with an emphasis on automation and sensitivity, stability and longevity, and the requirement for specialized sensor design and sample preparation (Akbari Nakhjavani et al., 2024).

14.4.1 Stability and longevity of sensors

The instability of biological recognition elements in a variety of environmental conditions is one of the main drawbacks of biosensors. DNA probes, aptamers, enzymes, and antibodies used to detect MTB are frequently sensitive to:

Table 14.1 Summary of key challenges and limitations of biosensing technology for *Mycobacterium tuberculosis* (MTB) detection (Wallis et al., 2013).

Category	Challenge
Stability	biological component deterioration and short shelf life
Sensitivity	Finding low-level MTB biomarkers can be challenging.
Automation	Samples must be processed manually; there are no integrated systems.
Design complexity	Specialized configurations are required for various sample types.
Usability	often requires skilled workers; lacks user-friendliness
Scalability and cost	Expensive production and upkeep; restricted availability

Temperature and humidity variations
pH fluctuations
Long-term storage and transport (Naresh and Lee, 2021)
In tropical or rural settings with inadequate cold-chain infrastructure, this becomes especially problematic. Field-use biosensors need to have a long shelf life without experiencing a noticeable decline in performance (Heidt et al., 2020).

Furthermore, a lot of biosensors are made for one-time use, which raises questions about environmental sustainability and raises operating costs. Long-term sensor performance and reusability are still unexplored, particularly for MTB diagnostics that might need regular screening or monitoring in high-risk populations (Williams et al., 2024).

14.4.2 Automation and sensitivity issues

It is consistently difficult to detect MTB with high sensitivity and selectivity, particularly when there is a low bacterial load (as in latent TB or extrapulmonary TB). There are multiple factors that contribute to this issue:

Low concentrations of biomarkers (e.g., DNA, proteins, antigens) in biological fluids such as blood, sputum, or urine.
Nonspecific binding of other biological materials presents in complex samples, leading to false positives or negatives.
Signal interference because of contaminants or matrix effects (Zaporojan et al., 2024).
Even though lab biosensors frequently exhibit high sensitivity under controlled settings, performance can drastically decline in the real world.

Furthermore, automated sample processing and analysis features are absent from many biosensor platforms. Manual handling raises the possibility of user error and contamination. Automation must be incorporated into every step of the TB diagnosis process, including sample preparation, biomarker recognition, signal processing, and result interpretation, for decentralized settings to be successful (Sin et al., 2014).

14.4.3 Specialized sensor design and sample preparation

The engineering challenges of designing biosensors for MTB detection are distinct. Since MTB is an intracellular pathogen that grows slowly and has a complex, lipid-rich cell wall, it is intrinsically challenging to detect directly (Takallu et al., 2024).

Some specific design and sample-preparation challenges include:

Need for Pretreatment: Prior to analysis, the majority of samples, particularly sputum need to be liquefied, decontaminated, and have their nucleic acids extracted. This restricts biosensors' ability to function as "sample-in, answer-out" devices.

Microfluidic Integration Gaps: Despite its ability to automate sample handling, microfluidics is still underutilized in MTB biosensors because of the expense and complexity of fabrication.
Customization for Diverse Samples: MTB may be found in different body parts. Biosensors that can analyze blood, cerebrospinal fluid (CSF), urine, or tissue samples are necessary for extrapulmonary tuberculosis (TB); each of these samples requires a different sensor configuration.
Biomarker Variability: The host immune response, coinfections such as HIV, and disease stage can all affect the expression of MTB-specific biomarkers. Because of this variability, biosensor platforms must be flexible and dynamic (Godary et al., 2025).

14.5 Prospects and innovations

Over the past decade, there has been notable progress in the development of biosensor technology for the detection of MTB. However, translating laboratory advancements into practical applications remains a key challenge. Future innovations must focus on addressing existing technical barriers, enhancing field-level reliability, expanding automation, and ensuring large-scale accessibility (Table 14.2). This section highlights three major areas that are expected to shape the next generation of MTB biosensing technologies: advanced automation for field applications, integration with intelligent robotic systems, and strategic innovations to overcome existing challenges (Mobed et al., 2024).

14.5.1 Integration with robotics and advanced digital technologies

The integration of robotics and digital platforms has already transformed conventional diagnostic techniques into portable efficient systems. Currently, mobile technology

Table 14.2 Summary of innovations and strategic directions (Ducrée, 2019).

Innovation area	Impact
Robotics and mobile technology	Enhanced speed, automation, and remote access
Microfluidics and lab-on-a-chip	Simplified workflow and integration in field settings
Disposable and automated cartridges	Ease of use, minimal human error, better compliance
Voice-guided interfaces and cloud support	Improved usability and centralized tracking
Modular and scalable sensor design	Adaptability, cost efficiency, long-term usability
Public-private collaboration	Financial sustainability and global implementation
Standardized validation protocols	Regulatory approval and clinical confidence

plays a crucial role in MTB biosensing, serving as an accessible and real-time interface for data visualization, analysis, and sharing. Moving forward, innovation will focus on enhancing these established integrations through AI-driven automation, remote robotic operations, and smart data networks. Conventional diagnostic techniques are being transformed into intelligent, effective, and portable systems through the combination of robotics and mobile technology. These developments are essential for making point-of-care MTB detection possible, especially in environments with limited resources.

14.5.1.1 Robotic integration

Robotic systems can play an instrumental role in automating key biosensor functions, such as:

Sample collection and handling: Sputum collection and transfer can be standardized by robotic arms to prevent contamination and lower human exposure.
Assay execution: Pipetting, reagent mixing, incubation, and signal detection can all be done with great accuracy and repeatability by automated systems.
Multi-sample processing: High-throughput testing, which is essential during TB outbreaks or mass screening, is made possible by robotic platforms, which process hundreds of samples at once.
By minimizing human intervention, robotics can improve the accuracy, speed, and safety of MTB diagnostics (Li, 2024).

14.5.1.2 Advanced digital and artificial intelligence-based systems

Artificial intelligence (AI) and machine learning (ML) are anticipated to refine data interpretation, enabling automated decision-making and improved diagnostic accuracy.

Integration with cloud and internet of medical things (IoMT) platforms will support real-time data sharing and epidemiological surveillance.

Predictive analytics can assist in mapping infection patterns and optimizing public health responses. Thus, while mobile-enabled biosensing is already operational, the future trajectory lies in *intelligent automation and AI integration* for improved precision and scalability.

14.5.2 Enhanced automation for field applications

Numerous shortcomings of existing biosensing platforms are addressed by automation, which lowers manual intervention, increases reproducibility, and facilitates smooth testing workflows, particularly in field settings (Raju et al., 2024).

14.5.2.1 Microfluidic automation

Miniaturized platforms known as lab-on-a-chip (LOC) systems combine several processes (sample preparation, mixing, amplification, and detection) onto a single chip.

These platforms typically run on solar or battery power, requiring no laboratory space and little infrastructure.

These systems provide user-friendly diagnostic solutions in areas with low literacy rates, reducing the need for trained staff (Lee and Kim, 2024).

14.5.2.2 Self-contained cartridge systems

Disposable cartridges preloaded with reagents and controls are anticipated to be the standard for future biosensors.

Users simply need to insert the sample and wait for an automated readout, which makes these devices perfect for community health workers to use in the field (Srinivasan and Tung, 2015).

14.5.2.3 Voice-guided interfaces and digital support

Biosensors embedded with voice prompts or touchscreen instructions can guide users through each step, reducing error and increasing compliance.

Integration with cloud-based dashboards allows central monitoring of diagnostic activity and trends.

Automation not only increases operational efficiency but also ensures consistency in test outcomes, which is critical for early diagnosis and treatment initiation (Baraneedharan et al., 2025).

14.5.3 Strategies for overcoming current limitations

To ensure sustainable implementation and global scalability of MTB biosensors, strategic actions must address existing scientific, infrastructural, and socioeconomic barriers.

14.5.3.1 Standardization and validation

Protocols for sensor performance evaluation, including sensitivity, specificity, and reproducibility, must be standardized across laboratories.

Clinical validation through multicentric field trials is essential to build trust in biosensor technology and secure regulatory approval (Yammouri and Ait Lahcen, 2024).

14.5.3.2 Modular and scalable design

Biosensors with modular architecture can be upgraded or customized for specific biomarkers, sample types, or detection modes.

Modular design also simplifies maintenance and repair, extending device life span and lowering costs (Sharma et al., 2021).

14.5.3.2.1 Policy and regulatory alignment

Collaboration with bodies such as WHO, ICMR, and national TB programs ensures alignment with global TB control strategies.

Simplified and fast-tracked approval mechanisms can encourage market entry and adoption (Singh et al., 2025).

14.5.3.3 Public-private partnerships

Partnerships between academic institutions, health care agencies, NGOs, and private industry can bridge the gap between R&D and deployment.

PPPs support funding, manufacturing, distribution, and capacity building, which are vital for large-scale impact (Chukwudalu et al., 2024).

14.5.3.4 Training and human resource development

Biosensor-based diagnosis requires training modules and certification programs to build a skilled workforce.

E-learning platforms and AI-based assistance tools can help remote or inexperienced users operate the systems effectively (Wasilewski et al., 2024).

14.5.3.5 Cost-efficiency and local manufacturing

Utilizing paper-based biosensors, locally sourced materials, and 3D-printing technologies can significantly lower production costs.

Supporting in-country fabrication ensures timely availability and reduces dependency on imports (Remaggi et al., 2022).

Biosensor technology holds immense promise for revolutionizing the detection of MTB, offering a much-needed alternative to conventional diagnostic tools that are often time-consuming, expensive, and infrastructure dependent. Advances in biosensor design have introduced possibilities for faster, more sensitive, and point-of-care detection, which is especially vital in high-burden, low-resource regions. These systems can reduce diagnostic turnaround time significantly and support early treatment initiation, helping to control the spread of the disease more effectively (Ma et al., 2021).

However, despite the promising advantages, biosensor technologies still face considerable challenges. Issues related to sensor stability, limited shelf life, and the influence of environmental conditions on sensor performance often hinder their long-term usability. Moreover, many biosensors require complex sample preparation or specialized detection environments, limiting their adaptability outside laboratory settings. Sensitivity and specificity, especially in distinguishing MTB from other mycobacterial species or contaminants, remain key concerns that need resolution. Additionally, the lack of standardization, difficulties in clinical validation, and insufficient automation present roadblocks to scaling up their use in real-world health care systems (Soltani et al., 2024).

Looking forward, technological innovation presents exciting opportunities to overcome these limitations. The convergence of biosensors with emerging technologies—such as robotics, mobile computing, and artificial intelligence—can enhance automation, simplify user interfaces, and allow real-time data analysis and remote monitoring. Microfluidic integration can enable fully automated lab-on-a-chip systems, while modular and scalable sensor platforms could provide more adaptable solutions suitable for various health care settings. Strategic collaborations among researchers, health care organizations, and industry can drive down costs, support local manufacturing, and enable regulatory streamlining to facilitate widespread deployment (Jin et al., 2020).

Furthermore, incorporating voice-guided instructions, smartphone interfaces, and cloud-based data systems can make biosensor platforms more user friendly and functional in the field. Training health care workers, especially in remote areas, to use such systems confidently and safely will be essential. Policy support, inclusive health care planning, and financial investment in research and development will be cornerstones of successful implementation.

In essence, while several scientific and logistical hurdles remain, biosensor-based detection of MTB is on the brink of becoming a practical, scalable, and impactful diagnostic solution. By aligning technological innovation with accessibility, regulatory preparedness, and public health needs, these biosensing systems have the potential to become integral components in global tuberculosis control and eradication strategies (Ross et al., 2022).

References

Akbari Nakhjavani S, Mirzajani H, Carrara S, Onbaşlı MC. Advances in biosensor technologies for infectious diseases detection. TrAC Trends in Analytical Chemistry 2024;180:117979. Available from https://doi.org/10.1016/j.trac.2024.117979.

Baraneedharan P, Kalaivani S, Vaishnavi S, Somasundaram K. Revolutionizing health care: a review on cutting-edge innovations in raspberry pi-powered health monitoring sensors. Computers in Biology and Medicine 2025;190:110109. Available from https://doi.org/10.1016/j.compbiomed.2025.110109.

Broesch JD. Chapter 3 - DSP system general model. In: Broesch JD, editor. Digital Signal Processing: Instant Access. Newnes; 2008. p. 27–36. Available from https://doi.org/10.1016/b978-0-7506-8976-2.00003-1.

Campelo TA, Cardoso de Sousa PR, Nogueira Lde L, Frota CC, Zuquim Antas PR. Revisiting the methods for detecting *Mycobacterium tuberculosis*: what has the new millennium brought thus far? Access Microbiology 2021;3(7):000245. Available from https://doi.org/10.1099/acmi.0.000245.

Chukwudalu C, Ebulue R, Ekesiobi S. Public-private partnerships in health sector innovation: lessons from around the world. International Medical Science Research Journal 2024;4(4):484–99. Available from https://doi.org/10.51594/imsrj.v4i4.1051.

Ducrée J. Efficient development of integrated lab-on-a-chip systems featuring operational robustness and manufacturability. Micromachines 2019;10(12):886. Available from https://doi.org/10.3390/mi10120886.

Ghasemi-Varnamkhasti M, Apetrei C, Lozano J, Anyogu A. Potential use of electronic noses, electronic tongues and biosensors as multisensor systems for spoilage examination in foods. Trends in Food Science & Technology 2018;80:71–92. Available from https://doi.org/10.1016/j.tifs.2018.07.018.

Godary T, Binkley B, Liu Z, Awoyemi O, Overby A, Yuliantoro H, et al. Acoustofluidics: Technology advances and applications from 2022 to 2024. Analytical Chemistry 2025;97(13):6847–70. Available from https://doi.org/10.1021/acs.analchem.4c06803.

Heidt B, Siqueira W, Eersels K, Diliën H, van Grinsven B, Fujiwara R, et al. Point of care diagnostics in resource-limited settings: a review of the present and future of PoC in its most needed environment. Biosensors 2020;10(10):133. Available from https://doi.org/10.3390/bios10100133.

Jain U, Chauhan N, Saxena K. Chapter 3 - Fundamentals of sensors and biosensors: an overview. In: Singh L, Mahapatra D, Kumar S, editors. Multifaceted Bio-sensing Technology. Academic Press; 2023. p. 31–44. Available from https://doi.org/10.1016/b978-0-323-90807-8.00005-1.

Jin X, Liu C, Xu T, Su L, Zhang X. Artificial intelligence biosensors: challenges and prospects. Biosensors and Bioelectronics 2020;165:112412. Available from https://doi.org/10.1016/j.bios.2020.112412.

Lee I, Kim H-Y. Lab-on-a-chip devices for nucleic acid analysis in food safety. Micromachines 2024;15(12):1524. Available from https://doi.org/10.3390/mi15121524.

Li Y. Advances in the application of artificial intelligence robots in critical care - a promising way in the future: scoping review. Journal of Medical Internet Research 2024;26:e54095. Available from https://doi.org/10.2196/54095.

Ma J, Du M, Wang C, Xie X, Wang H, Li T, et al. Rapid and sensitive detection of *mycobacterium tuberculosis* by an enhanced nanobiosensor. ACS Sensors 2021;6(9):3367–76. Available from https://doi.org/10.1021/acssensors.1c01227.

Mobed A, Darvishi M, Kohansal F, Dehfooli FM, Alipourfard I, Tahavvori A, et al. Biosensors; nanomaterial-based methods in diagnosing of *Mycobacterium tuberculosis*. Journal of Clinical Tuberculosis and Other Mycobacterial Diseases 2024;34:100412. Available from https://doi.org/10.1016/j.jctube.2023.100412.

Naresh V, Lee N. A Review on biosensors and recent development of nanostructured materials-enabled biosensors. Sensors (Basel, Switzerland) 2021;21(4):1109. Available from https://doi.org/10.3390/s21041109.

Peltomaa R, Glahn-Martínez B, Benito-Peña E, Moreno-Bondi M. Optical biosensors for label-free detection of small molecules. Sensors (Basel, Switzerland) 2018;18(12):4126. Available from https://doi.org/10.3390/s18124126.

Raju CM, Elpa DP, Urban PL. Automation and computerization of (bio)sensing systems. ACS Sensors 2024. Available from https://doi.org/10.1021/acssensors.3c01887.

Ramaraj SG, Alrebh A, Elamaran D, Zhou H, Huang K, Almansoori M, et al. Surface acoustic wave gas sensors: recent developments and their role in sensing technology. Materials Science and Engineering: B 2025;317:118157. Available from https://doi.org/10.1016/j.mseb.2025.118157.

Remaggi G, Zaccarelli A, Elviri L. 3D printing technologies in biosensors production: recent developments. Chemosensors 2022;10(2):65. Available from https://doi.org/10.3390/chemosensors10020065.

Ross GMS, Zhao Y, Bosman AJ, Geballa-Koukoula A, Zhou H, Elliott CT, et al. Data handling and ethics of emerging smartphone-based (bio)sensors – part 1: best practices and current implementation. TrAC Trends in Analytical Chemistry 2022:116863. Available from https://doi.org/10.1016/j.trac.2022.116863.

Shanbhag MM, Manasa G, Mascarenhas RJ, Mondal K, Shetti NP. Fundamentals of bio-electrochemical sensing. Chemical Engineering Journal Advances 2023;16:100516. Available from https://doi.org/10.1016/j.ceja.2023.100516.

Sharma A, Badea M, Tiwari S, Marty JL. Wearable biosensors: an alternative and practical approach in health care and disease monitoring. Molecules 2021;26(3):748. Available from https://doi.org/10.3390/molecules26030748.

Singh UB, Rade K, Rao R, Kumar N, Mattoo SK, Nair S, et al. Lessons and updates from India's national tuberculosis elimination program – bold decisions and innovative ways of fast-tracking progress toward ending tuberculosis. IJID Regions 2025;14:100599. Available from https://doi.org/10.1016/j.ijregi.2025.100599.

Sin ML, Mach KE, Wong PK, Liao JC. Advances and challenges in biosensor-based diagnosis of infectious diseases. Expert Review of Molecular Diagnostics 2014;14(2):225–44. Available from https://doi.org/10.1586/14737159.2014.888313.

Soltani SS, Rahimi M, Mehraveh Hosseinali S, Guilandokht Z, Raahemifar K. Recent advancements in microfluidic-based biosensors for detection of genes and proteins: applications and techniques. Biosensors and Bioelectronics X 2024;19. Available from https://doi.org/10.1016/j.biosx.2024.100489. 100489–100489.

Srinivasan B, Tung S. Development and applications of portable biosensors. Journal of Laboratory Automation 2015;20(4):365–89. Available from https://doi.org/10.1177/2211068215581349.

Takallu S, Aiyelabegan HT, Zomorodi AR, Alexandrovna KV, Aflakian F, Asvar Z, et al. Nanotechnology improves the detection of bacteria: recent advances and future perspectives. Heliyon 2024;10(11). Available from https://doi.org/10.1016/j.heliyon.2024.e32020. e32020–e32020.

Wallis RS, Kim P, Cole S, Hanna D, Andrade BB, Maeurer M, et al. Tuberculosis biomarkers discovery: developments, needs, and challenges. The Lancet Infectious Diseases 2013;13(4):362–72. Available from https://doi.org/10.1016/s1473-3099(13)70034-3.

Wang C, Liu M, Wang Z, Li S, Deng Y, He N. Point-of-care diagnostics for infectious diseases: from methods to devices. Nano Today 2021;37:101092. Available from https://doi.org/10.1016/j.nantod.2021.101092.

Warriner K, Namvar A. 4.54 - Biosensors for Foodborne Pathogen Detection. In: Moo-Young M, editor. Comprehensive Biotechnology (Second Edition), Vol. 4. Pergamon; 2011. p. 659–74. Available from https://doi.org/10.1016/b978-0-08-088504-9.00322-6.

Wasilewski T, Kamysz W, Gębicki J. AI-assisted detection of biomarkers by sensors and biosensors for early diagnosis and monitoring. Biosensors 2024;14(7):356. Available from https://doi.org/10.3390/bios14070356.

Williams A, Aguilar MR, Pattiya Arachchillage KGG, Chandra S, Rangan S, Gupta SG, et al. Biosensors for public health and environmental monitoring: the case for sustainable biosensing. ACS Sustainable Chemistry & Engineering 2024;12(28):10296–312. Available from https://doi.org/10.1021/acssuschemeng.3c06112.

World Health Organization. (2024). Tuberculosis. World Health Organization. https://www.who.int/health-topics/tuberculosis#tab=tab_1.

Yammouri G, Ait Lahcen A. AI-reinforced wearable sensors and intelligent point-of-care tests. Journal of Personalized Medicine 2024;14(11). Available from https://doi.org/10.3390/jpm14111088. 1088–1088.

Zaporojan N, Negrean RA, Hodişan R, Zaporojan C, Csep A, Zaha DC. Evolution of laboratory diagnosis of tuberculosis. Clinics and Practice 2024;14(2):388–416. Available from https://doi.org/10.3390/clinpract14020030.

Zhou L, He X, He D, Wang K, Qin D. Biosensing technologies for*mycobacterium tuberculosis*detection: status and new developments. Clinical and Developmental Immunology 2011;2011:1–8. Available from https://doi.org/10.1155/2011/193963.

Zuber AA, Klantsataya E, Bachhuka A. David L. Andrews, Robert H. Lipson and Thomas Nann. 3.06 - Biosensing. In: Andrews DL, Lipson RH, Nann T, editors. Comprehensive Nanoscience and Nanotechnology (second Edition), Vol. 3. Academic Press; 2019. p. 105–26. Available from https://doi.org/10.1016/b978-0-12-803581-8.10410-2.

Index

Note: Page numbers followed by "*f*" and "*t*" indicates figures and tables respectively.

Printed and bound by CPI Group (UK) Ltd, Croydon, CR0 4YY
14/09/2026
02216291-0003